Design and Application of Additive Manufacturing

Design and Application of Additive Manufacturing

Editor

Rubén Paz

MDPI • Basel • Beijing • Wuhan • Barcelona • Belgrade • Manchester • Tokyo • Cluj • Tianjin

Editor
Rubén Paz
University of Las Palmas de Gran Canaria
Spain

Editorial Office
MDPI
St. Alban-Anlage 66
4052 Basel, Switzerland

This is a reprint of articles from the Special Issue published online in the open access journal *Materials* (ISSN 1996-1944) (available at: https://www.mdpi.com/journal/materials/special_issues/design_application_AM).

For citation purposes, cite each article independently as indicated on the article page online and as indicated below:

LastName, A.A.; LastName, B.B.; LastName, C.C. Article Title. *Journal Name* **Year**, *Volume Number*, Page Range.

ISBN 978-3-0365-5077-0 (Hbk)
ISBN 978-3-0365-5078-7 (PDF)

Contents

About the Editor

Rubén Paz

Dr. Rubén Paz (1987) completed his Industrial Engineering degree at the University of Las Palmas de Gran Canaria (ULPGC) in 2011. Since then, he has been linked to the research group of Integrated and Advanced Manufacturing at ULPGC. In 2012, he finished his Master's Degree in Intelligent Systems and Numerical Applications in Engineering and presented his PhD in 2014 (ULPGC). Since 2015, he has been a lecturer at ULPGC in the department of Mechanical Engineering, in the field of Manufacturing Processes. His main research lines are related to manufacturing processes, Additive Manufacturing, optimization, polymer processing, numerical simulations, material characterization and natural fibers. He has taken part in 16 research projects and 5 educational innovation projects, 25 scientific papers, 7 book chapters, 27 participations in conferences, 1 national patent and 1 international patents and was the reviewer of 49 contributions to JCR journals.

Preface to "Design and Application of Additive Manufacturing"

Additive Manufacturing (AM) comprises a set of technologies with amazing capabilities to improve the manufacturing industry. Despite the continuous evolution of AM technologies and materials, the potential of these technologies is still wasted since there are many applications that could take advantage of AM's characteristics such as design freedom, reduction in material waste or low-cost prototyping. For this reason, this book focuses on the design and application of AM, since the proper combination of design, materials and technologies, together with their corresponding characterizations, may boost the uptake of AM as a replacement of conventional manufacturing techniques.

This book presents several research works carried out in this field and with a high variety of applications which may be of special interest for engineers and researchers working in the manufacturing sector, particularly in the areas of research, development and innovation.

I would like to thank all the authors and reviewers that have contributed in this Special Issue, as well as the MDPI staff and the editorial team of the *Materials* journal, especially the assistant editor of this Special Issue, Mr. Felix Guo.

Rubén Paz
Editor

 *materials*

Editorial

Special Issue "Design and Application of Additive Manufacturing"

Rubén Paz

Mechanical Engineering Department, Campus de Tafira Baja, Universidad de Las Palmas de Gran Canaria, 35017 Las Palmas, Spain; ruben.paz@ulpgc.es

Citation: Paz, R. Special Issue "Design and Application of Additive Manufacturing". *Materials* **2022**, *15*, 4554. https://doi.org/10.3390/ma15134554

Received: 20 June 2022
Accepted: 27 June 2022
Published: 28 June 2022

Publisher's Note: MDPI stays neutral with regard to jurisdictional claims in published maps and institutional affiliations.

Additive manufacturing (AM) is continuously improving and offering new opportunities in the manufacturing industry. The advantages of AM, such as design freedom, reduction in material waste or low-cost prototyping, can be exploited to improve or replace traditional manufacturing methods. To realize this, the combination of design, materials and technology must be thoroughly analyzed for every specific application. This Special Issue presents several studies related to the design and application of AM technologies, covering different technologies, methods and applications useful for the promotion and uptake of AM technologies.

Nam and Pei [1] analyzed the influence of some process parameters (print pattern and infill density) of material extrusion AM (MEX) in the recovery properties of polylactic acid (PLA) samples (4D printing). The 'Quarter-cubic' pattern with a 100% infill density showed the best recovery result, whereas the 'Cubic-subdivision' pattern with a 20% infill density demonstrated the shortest recovery time. On the other hand, a high recovery temperature and high infill density resulted in better recovery, and a low temperature and low infill density resulted in poor recovery.

In the research by Paraschiv et al. [2], laser defocusing in the selective laser melting (SLM) process (powder bed fusion, PBF) was investigated to assess the influence on the surface quality, melt pool shape, tensile properties, and densification of the produced samples (IN 625 material). It was observed that the melting height increased while the melting depth decreased from positive to negative defocusing. The use of negative defocusing distances yielded the maximum density and ultimate tensile strength.

In [3], a new method for the automatic detection of powder spreading defects in the SLM process was developed, with a high recognition rate and rapid detection speed, which could not only meet the SLM forming efficiency, but also improve the quality of the formed parts through feedback control.

Kusano et al. [4] proposed a novel calibration strategy for the heat source of a finite element thermal simulation model of the SLM process, thus improving the accuracy of the temperature field estimated during the process. The new calibration strategy combined with the finite element thermal simulation model may be useful to understand the complex thermal history applied in the layer of metal powder (which cyclically melts and solidifies) and that affects the microstructure, material properties, and performance.

Regarding the finite element analysis of the mechanical properties of porous structures (scaffolds made by AM for tissue engineering), Vega et al. [5] developed a new geometry-based modelling methodology to model the deposited part in MEX AM (starting from the manufacturing file), which enabled the mechanical simulation of the deposited geometry by finite element analysis. They compared this methodology with a voxel-based modelling technique in terms of the accuracy of the simulations (with respect to experimental mechanical results) and computational efficiency. Geometry-based modelling performed better for simple or larger parts, whereas voxel-based modelling was more advantageous for small and complex geometries. Both methodologies led to certain inaccuracy in the prediction of the mechanical properties, but enabled the comparison of different deposited geometries, which is useful for the design optimization of porous structures.

Adiaconitei et al. [6] manufactured a closed impeller for high-pressure mechanically pumped fluid loop (MPFL) systems by SLM, fulfilling the quality requirements and avoiding the technical challenges related to the conventional manufacturing process, thus demonstrating the feasibility of AM for this space application.

Zhong et al. [7] investigated the deformation mechanism and energy absorption behavior of 316 L triply periodic minimal surface (TPMS) structures with uniform and graded wall thicknesses fabricated by SLM. They observed that the stress level and length of each plateau in the deformation process could be adjusted by changing the wall thickness and position of the barrier layer between different segments. The gradient design of TPMS structures may find applications where the energy absorption requires a double-level feature or a warning function.

In [8], a flexible pressure sensor (widely used in human motion, robot monitoring, and medical treatment) was developed with a flat top plate and a microstructured bottom plate, both made of polydimethylsiloxane (PDMS) molded from a 3D-printed template. The contact surfaces of the top and bottom plates were coated with a mixture of poly (3,4-ethylenedioxythiophene) poly (styrene sulfonate) (PEDOT:PSS) and polyurethane dispersion (PUD) as stretchable film electrodes with carbon nanotubes on the electrode surface. Using digital light processing (DLP), the fabrication of the sensor was rapid and low-cost.

In [9], the authors optimized a hip implant that reduced the strain shielding effect, using the solid isotropic material with penalization (SIMP) topology optimization method. Finite element analyses and experimental measurements (hip manufactured with 316L-0407 stainless steel in PBF technology) were conducted to measure stem stiffness and predict the reduction in stress shielding. The developed topology optimization process enabled compliant hip implant design for more natural load transfer, reduced strain shielding, and improved implant survivorship.

Lam et al. [10] examined SLM-fabricated 15-5 PH steel with the 8% transient–austenite phase towards the fully reversed strain-controlled low-cycle fatigue (LCF) test. The cyclic deformation response and microstructural evolution were investigated.

In [11], ABS, PETG, and PLA polymers, which are common in MEX, were joined to grit-blasted aluminum substrates. They demonstrated the suitability of a polymer and a thermal processing condition to form a polymer–aluminum joint through MEX.

Finally, in [12], the authors applied topological optimization and a technological co-design to combine the advantages of 3D model printing (PMMA material by Binder Jetting technology) and conventional investment casting production methods. The PMMA material was more accurate than standard injected wax models, but the surface quality of the printed model was significantly lower.

Funding: This research received no external funding.

Institutional Review Board Statement: Not applicable.

Informed Consent Statement: Not applicable.

Data Availability Statement: Data sharing not applicable.

Conflicts of Interest: The author declares no conflict of interest.

References

1. Nam, S.; Pei, E. The influence of shape changing behaviors from 4D printing through material extrusion print patterns and infill densities. *Materials* **2020**, *13*, 3754. [CrossRef] [PubMed]
2. Paraschiv, A.; Matache, G.; Condruz, M.R.; Frigioescu, T.F.; Ionică, I. The influence of laser defocusing in selective laser melted in 625. *Materials* **2021**, *14*, 3447. [CrossRef] [PubMed]
3. Lin, Z.; Lai, Y.; Pan, T.; Zhang, W.; Zheng, J.; Ge, X.; Liu, Y. A new method for automatic detection of defects in selective laser melting based on machine vision. *Materials* **2021**, *14*, 4175. [CrossRef] [PubMed]

4. Kusano, M.; Kitano, H.; Watanabe, M. Novel calibration strategy for validation of finite element thermal analysis of selective laser melting process using Bayesian optimization. *Materials* **2021**, *14*, 4948. [CrossRef] [PubMed]

5. Vega, G.; Paz, R.; Gleadall, A.; Monzón, M.; Alemán-Domínguez, M.E. Comparison of cad and voxel-based modelling methodologies for the mechanical simulation of extrusion-based 3d printed scaffolds. *Materials* **2021**, *14*, 5670. [CrossRef] [PubMed]

6. Adiaconitei, A.; Vintila, I.S.; Mihalache, R.; Paraschiv, A.; Frigioescu, T.F.; Popa, I.F.; Pambaguian, L. Manufacturing of closed impeller for mechanically pump fluid loop systems using selective laser melting additive manufacturing technology. *Materials* **2021**, *14*, 5908. [CrossRef] [PubMed]

7. Zhong, M.; Zhou, W.; Xi, H.; Liang, Y.; Wu, Z. Double-level energy absorption of 3d printed tpms cellular structures via wall thickness gradient design. *Materials* **2021**, *14*, 6262. [CrossRef] [PubMed]

8. Shao, Y.; Zhang, Q.; Zhao, Y.; Pang, X.; Liu, M.; Zhang, D.; Liang, X. Flexible pressure sensor with micro-structure arrays based on pdms and pedot:Pss/pud&cnts composite film with 3d printing. *Materials* **2021**, *14*, 27–32.

9. Tan, N.; van Arkel, R.J. Topology optimisation for compliant hip implant design and reduced strain shielding. *Materials* **2021**, *14*, 7184. [CrossRef] [PubMed]

10. Lam, T.N.; Wu, Y.H.; Liu, C.J.; Chae, H.; Lee, S.Y.; Jain, J.; An, K.; Huang, E.W. Transient Phase-Driven Cyclic Deformation in Additively Manufactured 15-5 PH Steel. *Materials* **2022**, *15*, 777. [CrossRef] [PubMed]

11. Examining, J.; Strength, S.; Melt, P.; Bechtel, S.; Schweitzer, R.; Frey, M.; Busch, R.; Herrmann, H. Material Extrusion of Structural Polymer–Aluminum. *Materials* **2022**, *15*, 3120.

12. Castings, H.T.; Novosad, P. Requirements for Hybrid Technology Enabling the Production. *Materials* **2022**, *15*, 3805. [CrossRef] [PubMed]

Article

The Influence of Shape Changing Behaviors from 4D Printing through Material Extrusion Print Patterns and Infill Densities

Seokwoo Nam * and Eujin Pei

Department of Design, Brunel University, London UB8 3PH, UK; Eujin.Pei@brunel.ac.uk
* Correspondence: Seokwoo.Nam@brunel.ac.uk

Received: 14 July 2020; Accepted: 21 August 2020; Published: 25 August 2020

Abstract: Four-dimensional printing (4DP) is an approach of using Shape Memory Materials (SMMs) with additive manufacturing (AM) processes to produce printed parts that can deform over a determined amount of time. This research examines how Polylactic Acid (PLA), as a Shape Memory Polymer (SMP), can be programmed by manipulating the build parameters of material extrusion. In this research, a water bath experiment was used to show the results of the shape-recovery of bending and shape-recovery speed of the printed parts, according to the influence of the print pattern, infill density and recovery temperature (Tr). In terms of the influence of the print pattern, the 'Quarter-cubic' pattern with a 100% infill density showed the best recovery result; and the 'Line' pattern with a 20% infill density showed the worst recovery result. The 'Cubic-subdivision' pattern with a 20% infill density demonstrated the shortest recovery time; and the 'Concentric' pattern with a 100% infill density demonstrated the longest recovery time. The results also showed that a high temperature and high infill density provided better recovery, and a low temperature and low infill density resulted in poor recovery.

Keywords: 4D printing; material extrusion; shape changing behavior; shape memory polymers; print pattern; infill density; polylactic acid

---

1. Introduction

According to Pei et al [1], the development of four-dimensional printing (4DP) has been encouraged by the rapid development of Additive Manufacturing (AM) and Shape Memory Materials (SMMs). Notably, 4DP parts can morph or change shape, depending on the passage of time and the given environment conditions [2]. They have a potential to enable complex actuation devices and moving components to be built, in which sensors, mechanical parts and batteries can be eliminated. Smart materials, including Shape Memory Alloys (SMAs) or Shape Memory Polymers (SMPs) and other organic materials such as paper and wood, that are stimuli-responsive bio-composites, can be adapted for 4DP [3]. These SMMs can be restored to their original shape by environmental stimulation without external force after the programming step [4]. For shape recovery and deformation in 4DP, a stimuli is required to facilitate the shape change. Examples of stimuli include water, heat, pH, UV light, magnets, or combined sources of stimuli. Combining AM with smart materials through 4D printing, offers opportunities to design and build smart and active structures [5–8]. Statistics indicate that the market size for SMPs is expected to grow from US$1bn in 2021 to US$3.4bn in 2025 [9]. With the development of widely available SMPs, new 4DP applications within the product design industry are expected to grow. This paper focuses on SMPs, as polymeric materials are lighter, cheaper and easier to manufacture than SMAs, and they have a better recovery strain when compared to SMAs [10,11]. Felton et al. [12], demonstrated the use of AM with SMPs to produce self-assembly and self-folding parts. However, current research has not extensively examined how material extrusion build parameters may influence the shape memory effect of 4DP parts.

The aim of this study is to present a framework that will guide users to effectively predict the change in shape memory effect of bending, according to the influence of the build parameters using Polylactic acid (PLA). PLA has been chosen for this work due to its commercial availability and cost effectiveness. The outcome of this research will provide a better understanding of how shape changing behaviors can be effectively controlled through material extrusion print patterns and infill densities. Raviv et al. [13], found that intentionally specified build parameters can have a positive effect on the quality of the printed components and their mechanical properties. Therefore, research into the determination of optimal parameters is an important factor when designing 4DP parts [14]. A review of the literature has found that very little guidance is available for the selection of SMPs and whether build settings could influence the shape changing behavior.

2. Experimental Work

The first experiment consists of a material selection protocol to compare 16 different types of commercially available PLA materials, and to shortlist suitable PLA filaments. The second experiment is to produce printed samples based on the selection, to identify the best results. The third experiment is to investigate the use of 12 print patterns with different infill densities to ascertain the results of the shape memory effect (SME). The final experiment is to validate the results of the print patterns using a combination of different infill densities. The experiments were conducted with a consistent approach, by using the same AM hardware, software, and the same water bath set up.

2.1. Process for Material Selection and Shape Recovery Grades

Overall, 16 different types of PLA samples were purchased from different material suppliers as shown in Figure 1. Notably, 1.75 mm filaments were used, and they had a melting point for printing between 180 °C to 210 °C. Each reel of material was cut into strips 80 mm long. To record and grade the recovery effect, a grading system of 1 represented excellent recovery, returning to its full original shape; and a grade of 9 refers to no recovery. This is an important step to quantify the SME. A flat shape with a grade of 1 therefore indicates the best achievable recovery quality.

No.	16 Filaments	Manufacturers	Printing Temperature	Image
1	Material A	Filament PM	200–220 °C	
2	Material B	3D JAKE	195–215 °C	
3	Material C	ColorFabb	195–220 °C	
4	Material D	Bq	200–220 °C	
5	Material E	Kyoraku	195–220 °C	
6	Material F	Dremel	180–230 °C	
7	Material G	FlashForge	195–220 °C	
8	Material H	AMZ3D	180–210 °C	
9	Material I	Pxmalion	190–210 °C	
10	Material J	Prusa research	190–210 °C	
11	Material K	FilaPrint	180–210 °C	
12	Material L	Utsaline 3D	190–210 °C	
13	Material M	Fiberlogy	200–210 °C	
14	Material N	Fillamentum	190–210 °C	
15	Material O	3Dom	195–210 °C	
16	Material P	Qidi tech	180–220 °C	

Figure 1. Sixteen different types of polylactic acid (PLA) samples.

The SME process comprises three steps, encompassing the original shape, a temporary shape and the recovered shape [15]. A visual summary of the SME process is shown in Figure 2, which explains the systematic steps of programming, heating, cooling and recovery. The SME of the material is also determined by many factors, such as the material composition, the applied shape changing behavior, deformation rate, deformation temperature and recovery temperature [16].

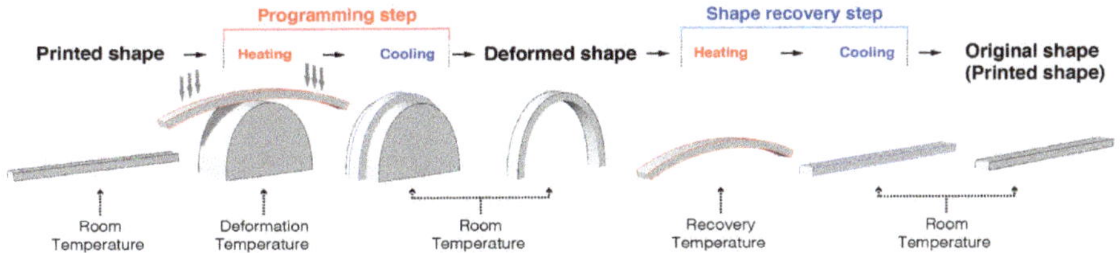

Figure 2. The processes involved in the shape change effect.

For this experiment, the procedures described in [17,18] were referenced. For the water bath experiment, hot water was used to trigger the filaments and the AM parts. Programming of the raw PLA filaments first begins with a change in temperature of the water. An immersion circulator was used to heat the water in the water-bath to the deformation temperature. The 16 types of PLA filaments (Figure 2), each 80 mm long, were placed in 80 °C of water in a water-bath for 60 s, and they were manually deformed to the bending shape by using a deformation device. The fixation of the deformed PLA filaments was achieved by allowing them to naturally cool for 60 s at room temperature (25 °C). Next, the water-bath temperature was changed and set to the recovery temperature. The recovery temperature for the filament test was performed with four samples, with four different levels of recovery temperatures of 60 °C, 65 °C, 70 °C and 75 °C. This meant that there were four recovery temperature samples for each material, and a total of 16 types of PLA materials were applied. The deformed PLA filaments were placed into the recovery temperature of the water-bath, and the shape recovery result was recorded using a high resolution video camera mounted on a tripod. Effort was also made to ensure that the tests conducted were consistent for all samples being used.

2.2. Results of Shape Recovery with Shortlisted Materials

The results of the first stage described in Section 2.1 were systematically documented and placing the results of the PLA filaments onto a grading card. The grading card shows a grade of 1 that indicates the best achievable recovery quality, and a grade of 9 that indicates the worst performing recovery result. From the 16 shortlisted materials, Material D and Material N showed the best results, since a grading system of 1 represents excellent recovery to its original shape (Figure 3). In addition, the experiment enabled us to record the time it took to recover to its original shape. Figure 3 shows the recovery of four samples from Material D, and the recovery of four samples from Material N.

Next, the results of the recovery quality and the recovery time were compiled and quantitatively compared in a form of a histogram. To determine the best overall recovery result based on the 4 recorded temperatures, the total grade for the 4 samples of each material was calculated. For example, the calculation of Material N whereby 3 + 1 + 1 + 1 gave a total score of 6, which was the lowest score and the best recovery. It also showed that a better recovery result was achieved between 65 °C to 75 °C. A line graph was also plotted in Figure 4 to determine the recovery time, whereby the sum from the 16 filaments of each different temperatures was calculated. It was also found that generally, a higher temperature in the water bath produced a much faster reaction. Material O achieved the fastest recovery at 36 s. Material L and E achieved the second fastest recovery level at 41 s, and Material N achieved the third fastest recovery level at 44 s to return to its original shape. A decision was made to shortlist Material D and Material N, based on the performance achieved, and the availability and cost of the PLA filaments.

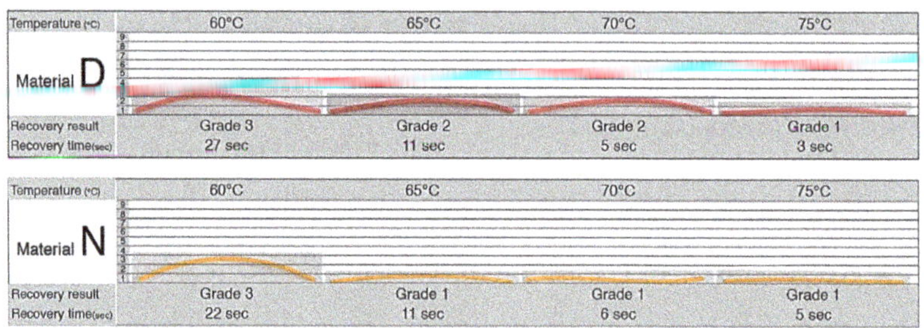

Figure 3. Material D and Material N showed the best recovery to their original shapes.

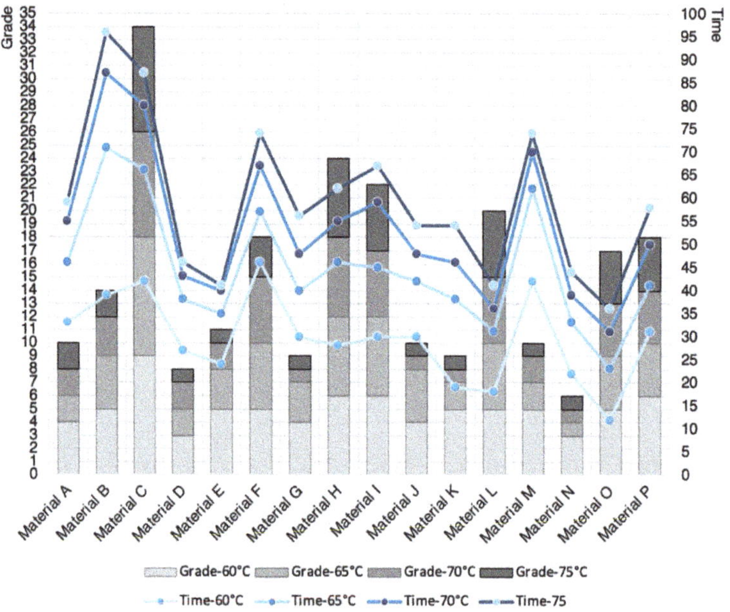

Figure 4. Recovery results of Materials A to P.

2.3. Additively Manufactured Samples

For the second step, standard rectangular samples measuring 80 mm × 6 mm × 3 mm were produced using filaments D and N selected as suitable PLA materials from the first step. The Qidi X-Pro Material Extrusion Printer was used, in which a consistent build setting of print temperatures of 210 °C, print speed at 50 mm/s, retraction time at 30 mm/min, retraction distance at 1.5 mm and infill density of 100% were used. The PLA samples were programmed for 60 s using deformation temperature of 80 °C and bent using the same deformation device. The fixation of the deformed PLA samples was allowed to cool for 60 s at room temperature (25 °C) after fixing to the deformation device. After PLA samples D and N were completely cooled down and programmed, the water-bath temperature was changed to the recovery temperature. To obtain more accurate results, the recovery temperature for the filament test was performed five times, by setting five different levels of recovery temperatures (55 °C, 60 °C, 65 °C, 70 °C and 75 °C). The deformed PLA samples were placed in the water-bath, and the progress of the shape recovery was recorded using the same video camera. The results in Figure 5 showed that almost the same results were obtained for both materials, for temperatures between 60 °C to 75 °C. When compared in detail, Material N showed better recovery quality. In terms of time, Materials D and N showed a slight difference, and both materials showed faster recovery time as the temperature increased. Overall, Material N showed a better recovery quality and a faster recovery than Material D. From this experiment, we can conclude that Material N would be used for the next step, focusing on aspects of print pattern and infill density.

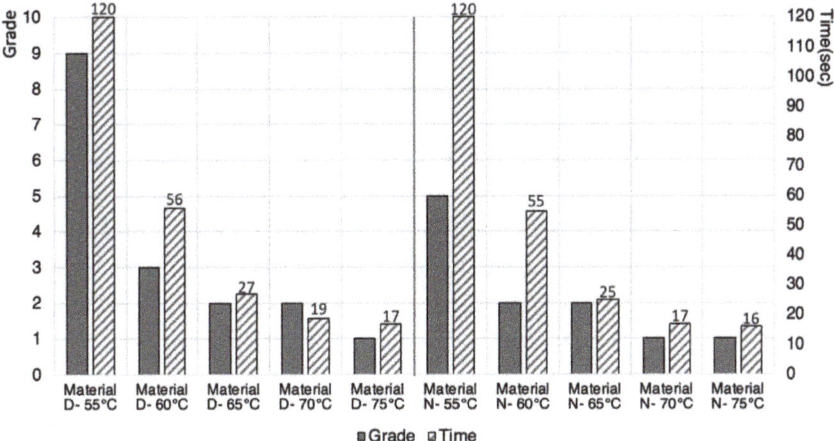

Figure 5. Histogram comparing the recovery grade and recovery time.

2.4. Print Patterns and Infill Densities

The Qidi X-Pro Material Extrusion Printer was supplied with a Qidi Print slicer software, although other third-party programs such as Cura and Simplify3D are also compatible. There were total of 12 pre-defined patterns, including concentric, cross, cubic, cubic-subdivision, grid, gyroid, line, octet, quarter-cubic, triangle, tri-hexagon and zigzag patterns, as shown in Figure 6. The use of these patterns can affect the strength and flexibility of the printed parts. According to Zolfagharian et al [19], the bending angle of the 4DP actuators can be significantly affected by the type of pattern and number of layers. Therefore, the AM build parameters such as print patterns and the layer height could potentially influence the SME.

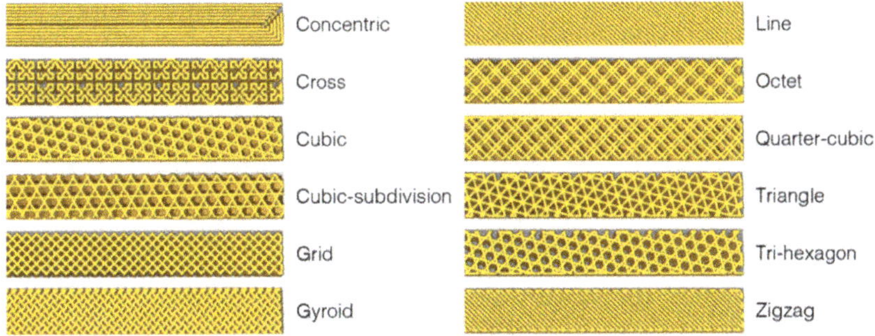

Figure 6. The 12 different types of print patterns available with Qidi Print.

In terms of infill density, this refers to the filling up of the space within the printed object or the density. The infill density is measured by means of a percentage. An object with a 100% infill volume represents a 100% object density. Objects with a higher percentage of the infill process consume more material, produce harder objects, and consequently take longer to produce. In contrast, objects with a lower percentage of the infill pattern consume less material, are lighter, and can be produced much faster [20]. A 0% infill density generally means that the object is empty, and 100% infill density means that the inside of the printed object is completely filled. Infill density can also affect the print strength, flexibility and the amount of material used. Figure 7 provides a visual example of the percentages of infill densities.

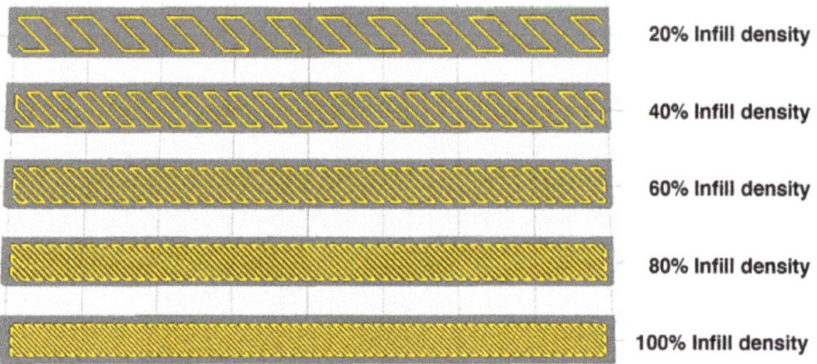

Figure 7. Example of infill densities from 20% to 100%.

For this experiment, 12 patterns and 5 infill densities would be used for Material N, encompassing 20%, 40%, 60%, 80% and 100%. The same procedure performed in the experiments described in 2.1 and 2.3 was carried out. The printed sample measuring 80 mm × 6 mm × 3 mm would be produced. The experimental process through programming, cooling and recovery would be used, although it was decided to utilize only recovery temperatures of 65 °C, 70 °C and 75 °C to accelerate the study. This is because temperatures of 55 °C and 60 °C were found to be insufficient to produce a thermo-mechanical response from 4DP parts.

3. Results and Development of a Tool

Firstly, the experiments showed that a 20% infill density produced the best recovery result of the tri-hexagon pattern, with a grade of 2 at each 65 °C, 70 °C and 75 °C; and the worst result was the line pattern, with a grade of 9 at each degree. In terms of time, the cubic-subdivision pattern took the shortest duration of time of 6 s at 75 °C; and the concentric pattern took the longest duration of 21 s at 65 °C. Secondly, the experiments showed that a 40% infill density produced the best recovery result of the quarter-cubic, tri-hexagon and zigzag pattern, with a grade of 2 at 75 °C; and worst result from the line pattern, with a grade of 8 at 65 °C. In terms of time, the cross, gyroid and line pattern took the shortest duration of 10 s at 75 °C; and the concentric pattern took the longest duration of 25 s at 65 °C. Thirdly, the experiments showed that a 60% infill density produced the best recovery result of the cubic and quarter-cubic pattern with a grade of 2 at 75 °C; and worst result was from the line pattern, with a grade of 7 at 65 °C. In terms of time, the cross, cubic-subdivision, grid and gyroid pattern took the shortest duration of 12 s at 75 °C; and the concentric pattern took the longest duration of 28 s at 65 °C. Next, the experiments showed that a 80% infill density produced the best recovery result of the grid and quarter-cubic pattern with a grade of 1 at 75 °C; and worst result of the cubic-subdivision and line pattern, with a grade of 7 at 65 °C. In terms of time, the cross pattern took the shortest duration of 12 s at 75 °C; and the concentric pattern took the longest duration of 29 s at 65 °C. Lastly, the experiments showed that a 100% infill density produced the best recovery result of the octet and quarter-cubic pattern, with a grade of 1 at 70 °C and 75 °C; and worst result was the cubic-subdivision pattern, with a grade of 8 at 65 °C. In terms of time, the cross and cubic-subdivision pattern took the shortest duration of 13 s at 75 °C; and the concentric pattern took the longest duration of 29 s at 65 °C.

The experiments showed that the best recovery result came from an infill density of 100% for the octet and quarter-cubic pattern with a grade of 1; and the worst recovery result came from an infill density of 20% for the line pattern, with a grade of 9. The shortest period of recovery time was achieved from an infill density of 20% for the cubic-subdivision pattern with a time of 6 s; and longest period of recovery time was achieved from an infill density of 100% for the concentric pattern, with a time of 29 s. The findings are in line with Yang et al. [21], who investigated the relationship between the part density and extrusion temperature, concluding that dense SMP structures caused greater recovery stress during recovery.

To disseminate the findings, and to make the results usable by the research community, the collected information was compiled in the form of a matrix, for designers, engineers and manufacturers to be able to appreciate and implement the possibilities of applying shape changing behaviors when developing 4DP parts. This is achieved by the selection of the material, the print pattern and the infill density, as well as the activation temperature. In addition, the information gathered would help users to determine whether the accuracy of the shape recovery or the time it took to return to its original form was more important. This toolkit has two main advantages. Firstly, 4DP parts can be designed and programmed through the control and selection of patterns and infill densities. Secondly, it can potentially facilitate the diagnosis of errors in the print output, and to analyze the interaction between the programming elements and the AM machine. The initial information from the experimental results described in Section 2.4 was quantified in the table shown in Figure 8.

20% infill density recovery quality and time result

Pattern (°C)	Concentric	Cross	Cubic	Cubic-subdivision	Grid	Gyroid	Line	Octet	Quarter-cubic	Triangles	Tri-hexagon	Zigzag
65°C	21 sec / 7	14 sec / 7	15 sec / 5	11 sec / 3	12 sec / 5	14 sec / 4	16 sec / 9	15 sec / 5	13 sec / 5	16 sec / 4	17 sec / 2	17 sec / 3
70°C	16 sec / 4	11 sec / 4	09 sec / 3	08 sec / 2	10 sec / 4	11 sec / 3	13 sec / 9	12 sec / 5	12 sec / 3	13 sec / 3	15 sec / 2	14 sec / 2
75°C	16 sec / 4	09 sec / 4	08 sec / 3	06 sec / 2	09 sec / 3	07 sec / 3	09 sec / 9	10 sec / 5	11 sec / 3	11 sec / 3	08 sec / 2	11 sec / 2

40% infill density recovery quality and time result

Pattern (°C)	Concentric	Cross	Cubic	Cubic-subdivision	Grid	Gyroid	Line	Octet	Quarter-cubic	Triangles	Tri-hexagon	Zigzag
65°C	25 sec / 7	16 sec / 6	17 sec / 5	14 sec / 4	14 sec / 5	17 sec / 4	18 sec / 8	16 sec / 5	15 sec / 5	19 sec / 4	20 sec / 3	19 sec / 3
70°C	22 sec / 4	13 sec / 4	14 sec / 3	13 sec / 3	12 sec / 4	15 sec / 3	15 sec / 6	12 sec / 4	14 sec / 3	15 sec / 3	18 sec / 2	17 sec / 2
75°C	17 sec / 3	10 sec / 4	13 sec / 3	11 sec / 3	11 sec / 3	10 sec / 3	10 sec / 6	11 sec / 4	12 sec / 2	12 sec / 3	12 sec / 2	12 sec / 2

60% infill density recovery quality and time result

Pattern (°C)	Concentric	Cross	Cubic	Cubic-subdivision	Grid	Gyroid	Line	Octet	Quarter-cubic	Triangles	Tri-hexagon	Zigzag
65°C	28 sec / 6	17 sec / 6	23 sec / 5	17 sec / 5	19 sec / 5	21 sec / 4	20 sec / 7	18 sec / 4	19 sec / 4	20 sec / 5	22 sec / 4	21 sec / 4
70°C	23 sec / 3	14 sec / 4	16 sec / 3	15 sec / 4	15 sec / 3	17 sec / 3	17 sec / 6	15 sec / 3	15 sec / 3	17 sec / 4	20 sec / 3	20 sec / 3
75°C	18 sec / 3	12 sec / 3	14 sec / 2	12 sec / 3	12 sec / 3	12 sec / 3	15 sec / 6	13 sec / 3	14 sec / 2	14 sec / 3	14 sec / 3	14 sec / 3

80% infill density recovery quality and time result

Pattern (°C)	Concentric	Cross	Cubic	Cubic-subdivision	Grid	Gyroid	Line	Octet	Quarter-cubic	Triangles	Tri-hexagon	Zigzag
65°C	29 sec / 6	19 sec / 5	23 sec / 4	19 sec / 7	20 sec / 5	22 sec / 3	22 sec / 7	21 sec / 3	21 sec / 3	24 sec / 6	23 sec / 6	24 sec / 5
70°C	24 sec / 3	15 sec / 4	18 sec / 3	16 sec / 4	16 sec / 3	19 sec / 3	19 sec / 6	18 sec / 2	17 sec / 2	18 sec / 4	21 sec / 3	22 sec / 3
75°C	22 sec / 2	12 sec / 3	16 sec / 2	13 sec / 3	14 sec / 1	13 sec / 3	17 sec / 5	14 sec / 2	14 sec / 1	14 sec / 3	15 sec / 3	16 sec / 3

100% infill density recovery quality and time result

Pattern (°C)	Concentric	Cross	Cubic	Cubic-subdivision	Grid	Gyroid	Line	Octet	Quarter-cubic	Triangles	Tri-hexagon	Zigzag
65°C	29 sec / 4	22 sec / 5	24 sec / 4	19 sec / 8	21 sec / 3	22 sec / 3	27 sec / 7	21 sec / 2	25 sec / 2	26 sec / 7	27 sec / 6	28 sec / 5
70°C	26 sec / 2	18 sec / 4	19 sec / 2	17 sec / 4	16 sec / 2	20 sec / 3	22 sec / 6	18 sec / 1	17 sec / 1	18 sec / 5	24 sec / 3	24 sec / 4
75°C	24 sec / 1	13 sec / 3	17 sec / 3	13 sec / 4	14 sec / 1	15 sec / 2	19 sec / 5	15 sec / 1	15 sec / 1	15 sec / 4	17 sec / 3	20 sec / 3

Figure 8. Results from 5 infill density showing recovery quality and time.

It was then represented as a graph in Figure 9 as a single visual reference guide. As shown in the bottom of Figure 9, the distribution diagram from the results was divided according to the infill density, temperature and pattern. The 12 patterns were expressed in the form of a numerical value from 1 to 12. Patterns according to numbers are arranged alphabetically. No. 1—Concentric pattern, No. 2—Cross Line pattern, No. 3—Cubic pattern, No. 4—Cubic-subdivision pattern, No. 5—Grid pattern, No. 6—Gyroid pattern, No. 7—Line pattern, No. 8—Octet pattern, No. 9—Quarter-cubic pattern, No. 10—Triangle pattern, No. 11—Tri-hexagon pattern and No. 12—Zigzag pattern. The *X*-axis represents the time it takes for the part to recover to its original printed shape. The *Y*-axis represents the grade of the shape recovery result, and the lower the grade, the better the shape recovery. The graph shows the distribution of a total of 180 shape recovery, and the time results with 5 print densities, 12 print patterns and 3 shape recovery temperatures of 65 °C, 70 °C and 75 °C. The results with excellent shape recovery and fast recovery time are shown in the upper right of the matrix. Conversely, the results of poor shape recovery and slow recovery time are shown at the bottom left of the matrix. The blue arrows indicate the sequential direction of the printing parameters. As a result of the printing parameters, the shape recovery result and time are generally evenly distributed in the graph, as shown within the red circle. This means that except for a specific pattern, the print parameters greatly affect the shape recovery quality result and time.

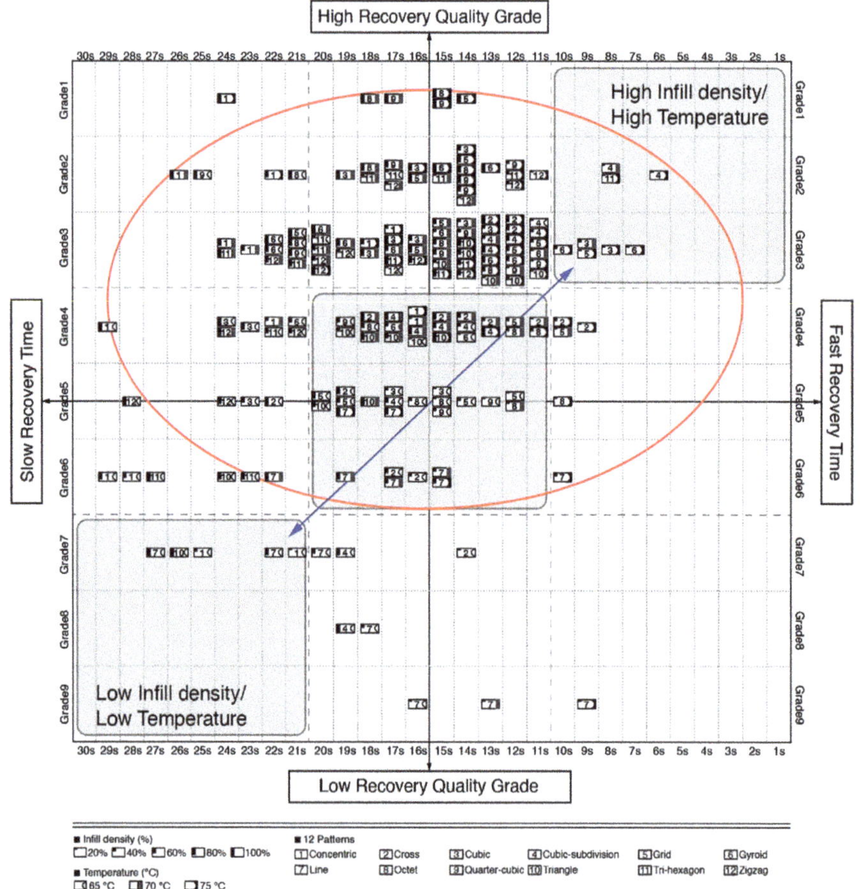

Figure 9. Shape recovery result and time distribution according to printing parameters.

The information from Figure 9 was further developed into a graphical layout in Figure 10, to improve access to the information. The single digits in the horizontal rows colored in red describe the grade of the shape recovery result, where a lower grade represents better shape recovery. The horizontal rows colored in blue describe the time taken for the parts to recover to its original printed shape. Elements of the 12 print patterns, as well as the 5 print densities, are also illustrated in the image, in addition to 3 different shape recovery temperatures, of 65 °C, 70 °C and 75 °C. Figure 10 shows that the lowest recovery grade pattern in the 20% infill density is the line pattern (since it has a numerical value of 9), with poor results at 65 °C, 70 °C and 75 °C. Similarly, the longest recovery time was recorded by observing the concentric pattern with a 100% infill density, taking over 65 s and at 65 °C. Using this graphical method, it is possible to check all the recorded results from the 5 infill densities, 12 patterns, and 3 different temperatures.

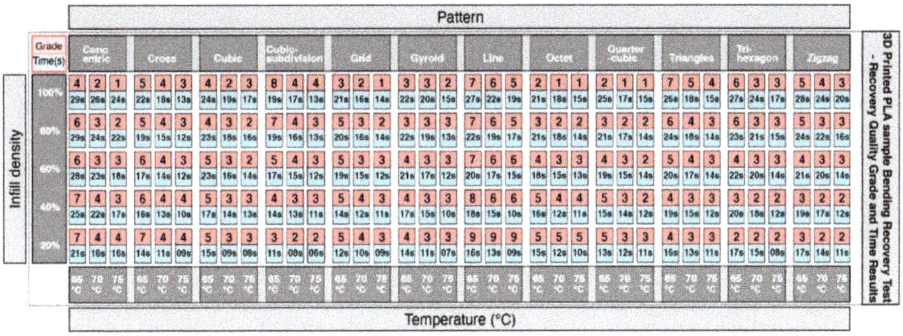

Figure 10. Graphical table representing the results of 12 patterns and 5 infill densities.

From the table in Figure 10, the quarter-cubic and octet print patterns showed the best recovery quality, and the line pattern showed the worst recovery quality. Taking it a step further, we also wanted to analyze which print patterns could provide a good recovery of 4DP parts. The information showing the quality of recovery was divided into 3 groups. Grades from 1–3 (Figure 11) showed the highest quality of recovery and reflected with the represented print patterns. Grades from 4–6 (Figure 12) showed a medium quality of recovery; and grades from 7–9 (Figure 13) showed print patterns that represent the worst quality of recovery.

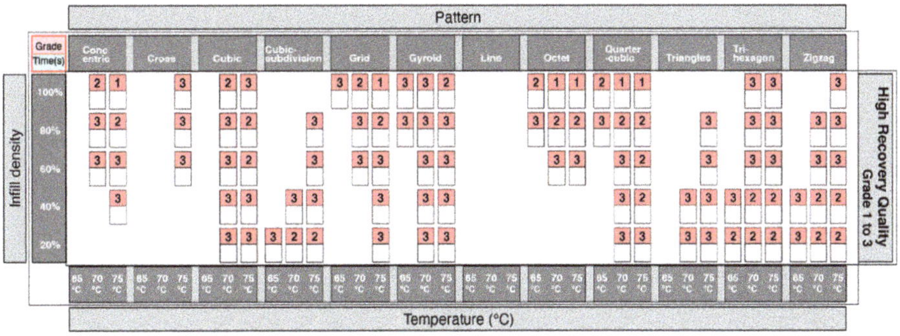

Figure 11. Graphical table representing best results of shape recovery.

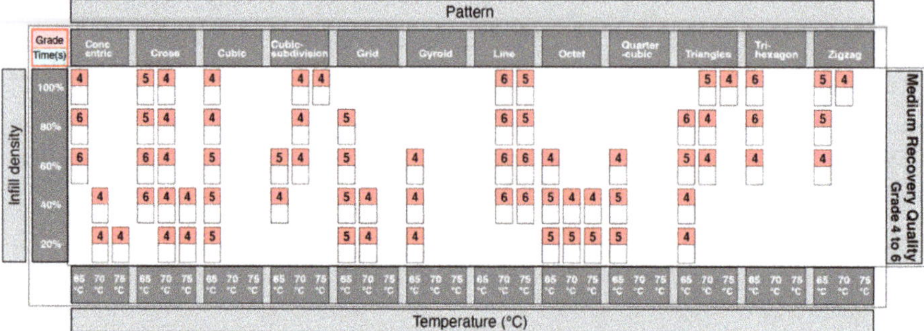

Figure 12. Graphical table representing moderate results of shape recovery.

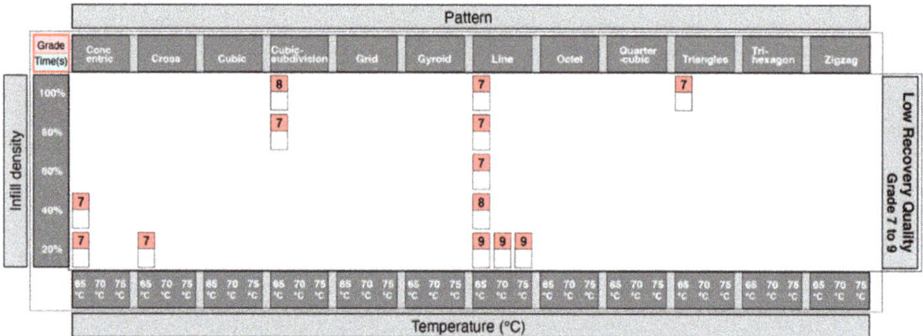

Figure 13. Graphical table representing poor results of shape recovery.

Adopting the same approach using the data from Figure 10, we were able to identify which print patterns influenced the recovery time of the 4DP parts. The information showing the time taken for recovery was divided into 3 groups. Figure 14 showed the shortest time for recovery, Figure 15 with a moderate time for recovery, and Figure 16 with the longest time for recovery. Moreover, 1–10 s was the fastest recovery time, 11–20 s reflects a moderate recovery time, and 21–30 s reflects the longest recovery time. The shortest recovery time is mainly due to a low percentage of infill density of 20% and 40%; and using high temperatures with patterns such as cross, cubic-subdivision, grid, gyroid, line, octet and tri-hexagon being used. The moderate recovery time took up the largest proportion. Slow recovery times were mainly due to having a high percentage of infill density of 80% and 100% and with a low temperature. Therefore, a low percentage infill density with a high temperature results in a much shorter recovery time. Conversely, a high percentage infill density with a low temperature would result in a slow recovery time. In conclusion, the results show that the recovery time is largely influenced by the pattern, infill density and the temperature.

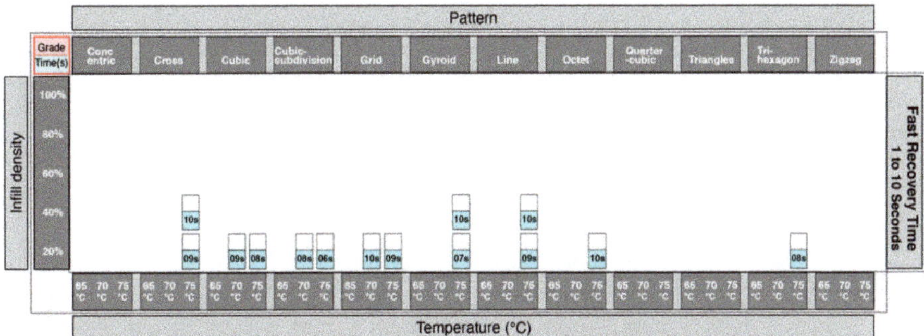

Figure 14. Graphical table representing the shortest recovery time.

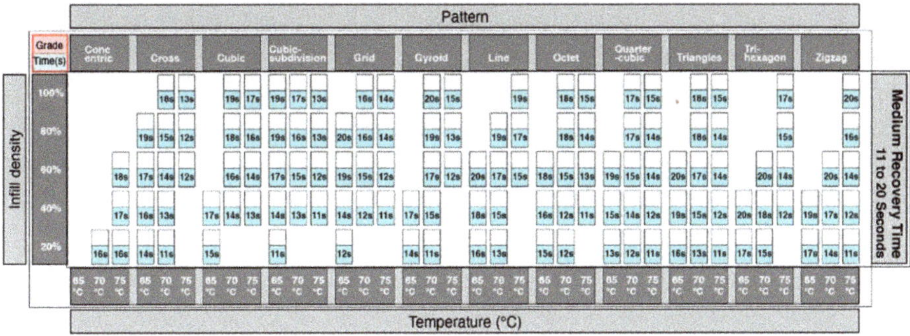

Figure 15. Graphical table representing moderate recovery time.

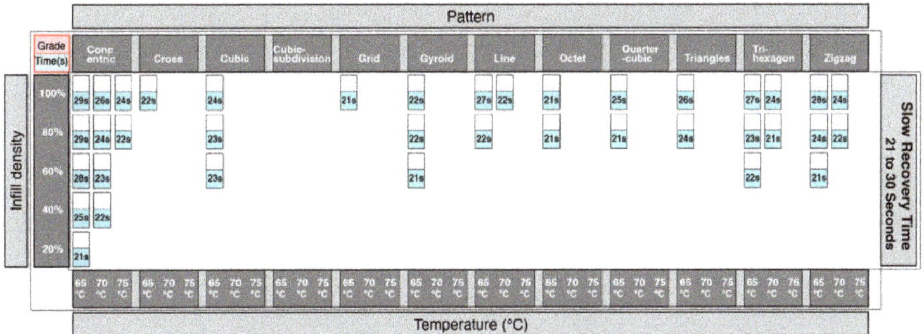

Figure 16. Graphical table representing the longest recovery time.

Finally, a physical version of the toolkit in Figure 17 was produced, with each of the information sheets from Figures 10–16 printed onto transparent acetate film, so that users could intuitively filter information or identify overlapping areas. This physical version was tested with potential users for feedback, which received a very positive response for the accuracy of information and ease of use. However, future work would aim to undertake more thorough testing of the kit, involving more participants, and with a view of making this more widely accessible and available in a digital format.

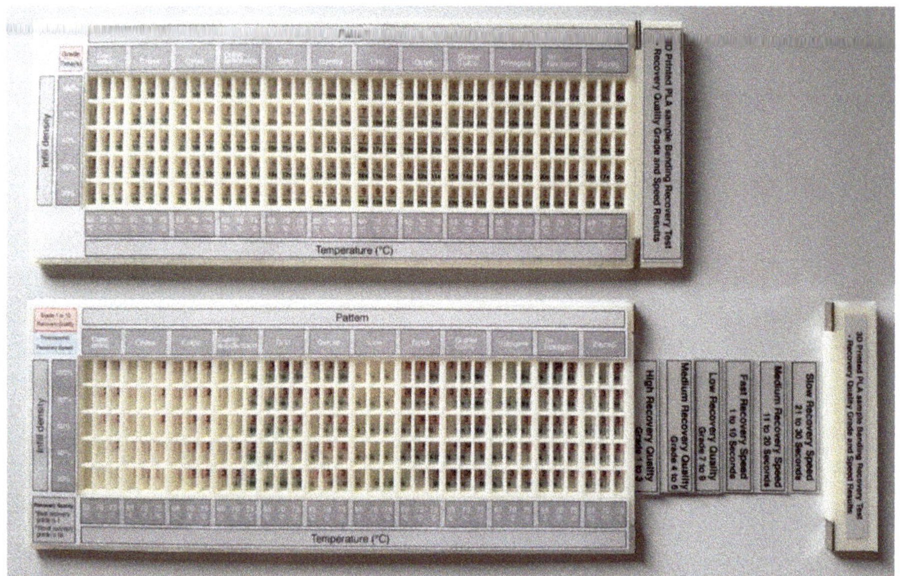

Figure 17. Physical version of the 4D Printing toolkit.

4. Discussion

In conclusion, the print pattern and infill density of PLA samples were tested using the same water-bath process to ensure consistency and rigor. The purpose of the test was to confirm the findings from Section 3, and to also validate the use of the toolkit. Due to time and resource constraints, it was not possible to evaluate every single element. An objective comparison method was used, in which the pattern and infill density of the highest and lowest recovery results, and the pattern and infill density of the fastest and slowest recovery time, would be tested. The 'Quarter-cubic' pattern with a 100% infill density (with an assumed best recovery result); and the 'Line' pattern with a 20% infill density (with an assumed worst recovery result) were selected. In addition, the 'Cubic-subdivision' pattern with a 20% infill density (with an assumed shortest recovery time); and the 'Concentric' pattern with a 100% infill density (with an assumed longest recovery time) were selected (Figure 18). The PLA samples were fabricated using the same Qidi X-Pro printer using the identical print parameters. To further extend our understanding of the subject matter, and to validate the results, each sample would contain two different print patterns with two different infill densities using a rectangular sample of 160 mm × 6 mm × 3 mm (each pattern consists of 80 mm × 6 mm × 3 mm). Having two different patterns and infill densities in a single sample should show a clear difference when they are immersed in the water bath.

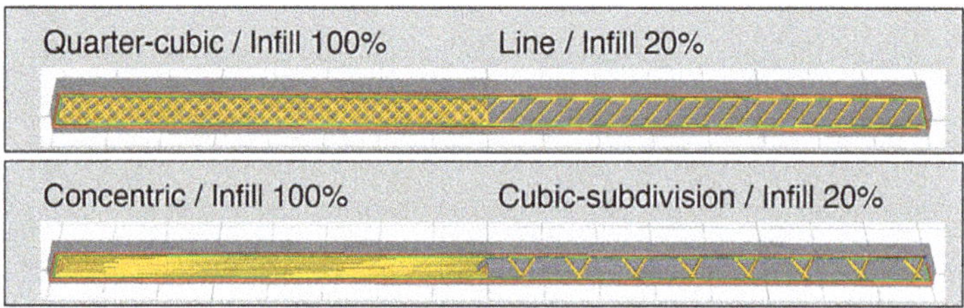

Figure 18. A computer generated file showing a combination of an assumed best and worst recovery setting (**top**), as well as fastest and slowest recovery time (**bottom**).

After the physical samples were produced, the results were similar from the previous experiments. In terms of recovery quality, the 'Line' pattern achieved a grade of 9 at 65 °C, a grade of 8 at 70 °C, and a grade of 7 at 75 °C, showing a slight difference of grades 1 and 2 from the previous experiment. The 'Quarter-cubic' pattern recovery quality showed the same results, of a grade of 3 at 65 °C, a grade of 1 at 70 °C, and a grade of 1 at 75 °C. In terms of recovery time, there was only a difference of between one to two seconds when compared to earlier experiments. The comparison of the 'Concentric' pattern and the 'Cubic-subdivision' pattern also showed similar results. The 'Concentric' pattern achieved a grade of 4 at 65 °C, a grade of 2 at 70 °C, and a grade of 1 at 75 °C. The 'Cubic-subdivision' pattern achieved a grade of 3 at 65 °C, a grade of 2 at 70 °C, and of grade 2 at 75 °C. Both patterns showed the same recovery quality, except 70 °C, as shown in Figure 19. In terms of recovery time, there was only a difference of one to two seconds when compared to earlier experiments. The samples on the left in Figure 19 show the use of patterns that display the best and worst recovery results; and the samples on the right display the fastest and slowest recovery time. The results are in line with Wu et al [16], who conducted an experiment on the influence of process parameters of the SME. The results revealed that the SME of AM parts using PLA were mostly influenced by the recovery temperature, followed by the deformation temperature and print parameters (layer thickness and raster angle). However, results from this research showed that parameters such as pattern and infill density also largely affect the SME. Further experiments are still required to verify the results with greater accuracy and repeatability.

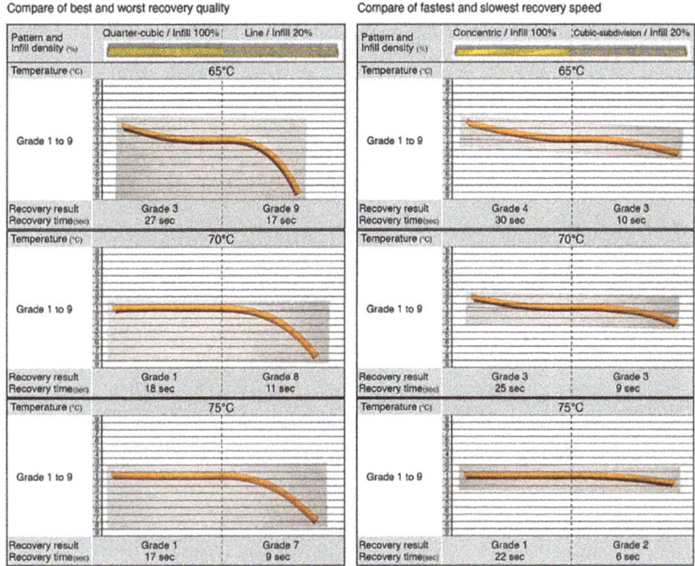

Figure 19. Physical samples showing the combination of mixed pattern and different infill densities.

5. Conclusions

This research has enabled a statistical approach of categorizing the SME of 16 different filaments, followed by tailoring the SME through the print pattern and infill density. The findings provide designers, engineers and manufacturers with a fundamental understanding that print parameters of pattern and infill density significantly influence the shape recovery quality and the recovery time of 4DP parts. This experiment shows that the use of different infill patterns has a different resultant SME. The temperature affects both recovery quality and recovery time. A low percentage infill density with a high temperature results in a much shorter recovery time. Conversely, a high percentage infill density with the low temperature would result in a slow recovery time. In general, a high temperature and high infill density provide better recovery quality, and a low temperature and low infill density result in low recovery quality. In conclusion, Figure 20 presents a flowchart that documents the entire process of this research. At the input stage, four experiments were conducted sequentially, and each result was performed using the same water bath. Although these experiments have provided a fundamental basis that the print parameters of pattern and infill density influence the shape recovery quality and the recovery time, there are still limitations of this study. For example, even more trials involving different PLA materials, the size and shape of the sample, and other types of shape changing behaviors and stimuli still need to be investigated. This study has provided an understanding of the SME in the use of 4DP parts, to enable better control of the shape change behavior through intelligent design and the modeling of CAD.

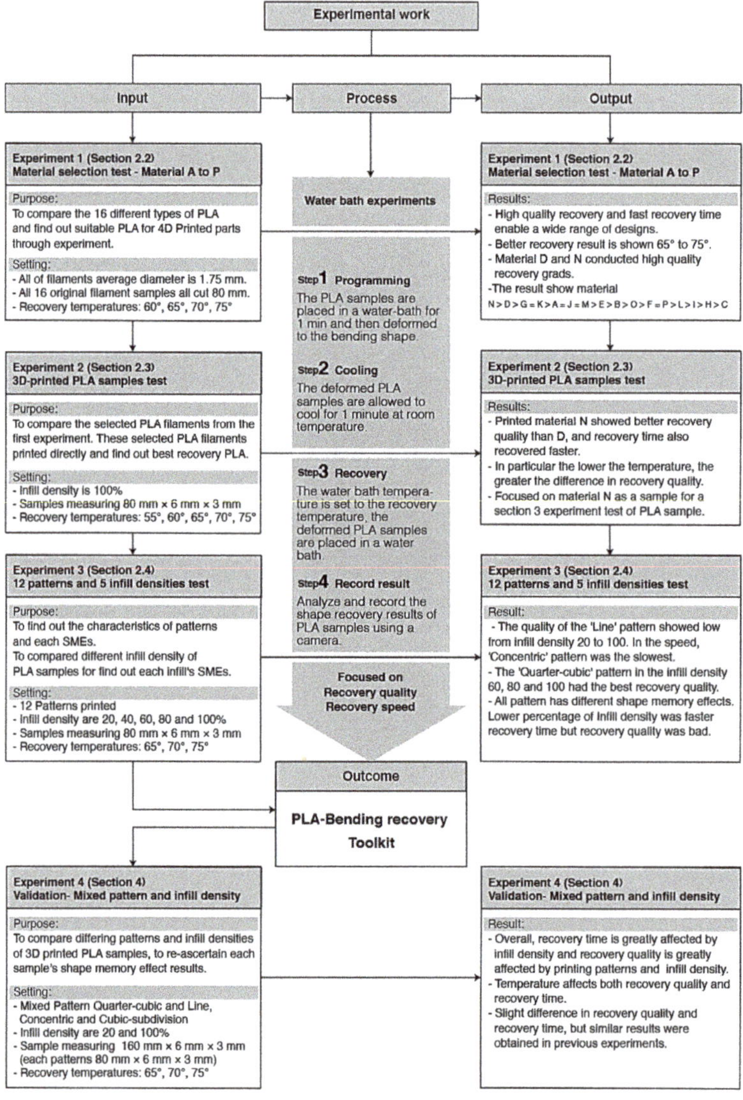

Figure 20. The flowchart showing the inputs, processes, outputs and outcomes.

Author Contributions: Conceptualization, S.N., E.P.; Methodology, S.N., E.P.; Validation, S.N., E.P.; Formal analysis, S.N., E.P.; Investigation, S.N., E.P.; Resources, S.N., E.P.; Data curation, S.N., E.P.; Writing—Original draft preparation, S.N., E.P.; Writing—Review and editing, S.N., E.P.; Visualization, S.N. Supervision, E.P. Project administration, S.N. All authors have read and agreed to the published version of the manuscript.

Funding: This research received no external funding.

Conflicts of Interest: The authors declare no conflict of interest.

References

1. Pei, E.; Loh, G.H.; Nam, S. Concepts and Terminologies in 4D Printing. *Appl. Sci.* **2020**, *10*, 4443. [CrossRef]
2. Teoh, J.E.M.; Chua, C.K.; Liu, Y.; An, J. 4D printing of customised smart sunshade: A conceptual study. In *Challenges for Technology Innovation: An Agenda for the Future*; Da Silva, F.M., Bártolo, H., Bártolo, P., Almendra, R., Roseta, F., Almeida, H.A., Eds.; CRC Press: London, UK, 2017; pp. 105–108.
3. Nkomo, N. A Review of 4D Printing Technology and Future Trends. In Proceedings of the 11th South African Conference on Computational and Applied Mechanics, Vanderbijlpark, South Africa, 17–19 September 2018.
4. Lei, M.; Chen, Z.; Lu, H.; Yu, K. Recent progress in shape memory polymercomposites: Methods, properties, applicationsand prospects. *Nanotechnol. Rev.* **2019**, *8*, 327–351. [CrossRef]
5. Balk, M.; Behl, M.; Wischke, C.; Zotzmann, J.; Lendlein, A. Recent advances in de-gradable lactide-based shape-memory polymers. *Adv Drug Deliv Rev.* **2016**, *107*, 136–152. [CrossRef] [PubMed]
6. Lee, A.Y.; An, J.; Chua, C.K. Two-Way 4D Printing: A Review on the Reversibility of 3D-Printed Shape Memory Materials. *Engineering* **2017**, *3*, 663–674. [CrossRef]
7. Pei, E.; Shen, J.; Watling, J. Direct 3D Printing of Polymers onto Textiles: Experimental Studies and Applications. *Rapid Prototyp. J.* **2015**, *21*, 556–571. [CrossRef]
8. Wu, J.; Yuan, C.; Ding, Z.; Isakov, M.; Mao, Y.; Wang, T.; Dunn, M.; Qi, H. Multi-shape active composites by 3D printing of digital shape memory polymers. *Sci. Rep.* **2016**, *6*, 24224. [CrossRef]
9. N-tech Research. Future Opportunities for Metamaterials in Aerospace and Defense Markets 2016. Available online: http://2o5ql235mlob1jjn7g4147f5-wpengine.netdna-ssl.com/wp-content/uploads/2017/12/metamaterials.pdf (accessed on 13 June 2020).
10. Hornat, C.C.; Yang, Y.; Urban, M.W. Quantitative predictions of shape-memory effects in polymers. *Adv. Mater.* **2017**, *29*, 1603334. [CrossRef]
11. Lan, X.; Liu, Y.; Lv, H.; Wang, X.; Leng, J.; Du, S. Fiber reinforced shape-memory polymer composite and its application in a deployable hinge. *Smart Mater. Struct.* **2009**, *18*, 024002.1–024002.6. [CrossRef]
12. Felton, S.M.; Tolley, M.T.; Shin, B.; Onal, C.D.; Demaine, E.D.; Rus, D.; Wood, R.J. Self-folding with shape memory composites. *Soft Matter.* **2013**, *9*, 7688–7694. [CrossRef]
13. Raviv, D.; Zhao, W.; McKnelly, C.; Papadopoulou, A.; Kadambi, A.; Shi, B.; Hirsch, S.; Dikovsky, D.; Zyracki, M.; Olguin, C.; et al. Active printed materials for complex self-evolving deformations. *Sci. Rep.* **2014**, *4*, 7422. [CrossRef] [PubMed]
14. Türk, D.A.; Brenni, F.; Zogg, M.; Meboldt, M. Mechanical characterization of 3D printed polymers for fiber reinforced polymers processing. *Mater. Des.* **2017**, *18*, 256–265. [CrossRef]
15. Sobota, M.; Jurczyk, S.; Kwiecień, M.; Smola-Dmochowska, A.; Musioł, M.; Domański, M.; Janeczek, H.; Kawalec, M.; Kurcok, P. Crystallinity as a tunable switch of poly(L-lactide) shape memory effects. *J. Mech. Behav. Biomed. Mater.* **2016**, *66*, 144–151. [CrossRef] [PubMed]
16. Wu, W.; Ye, W.; Wu, Z.; Geng, P.; Wang, Y.; Zhao, J. Influence of Layer Thickness, Raster Angle, Deformation Temperature and Recovery Temperature on the Shape Memory Effect of 3D-Printed Polylactic Acid Samples. *Materials* **2017**, *8*, 970. [CrossRef] [PubMed]
17. Wang, G.; Yang, H.; Yan, Z.; Ulu, N.G.; Tao, Y.; Gu, J.; Kara, L.B.; Yao, L. 4DMesh: 4D Printing Morphing Non-Developable Mesh Surfaces. *ACM* **2018**, 623–635. [CrossRef]
18. Wang, G.; Tao, Y.; Capunaman, O.; Yang, H.; Yao, L. *A-line: 4D Printing Morphing Linear Composite Structures*; ACM: New York, NY, USA, 2019; ISBN 978-1-4503-5970-2/19/05. [CrossRef]
19. Zolfagharian, A.; Kaynak, A.; Khoo, S.Y.; Kouzani, A. Pattern-driven 4D printing. *Sens. Actuat. A Physical.* **2018**, *274*, 231–243. [CrossRef]

20. Yarwindran, M.; Sa'aban, N.A.; Ibrahim, M.; Periyasamy, R. Thermoplastic elastomer infill pattern impact on mechanical properties 3D printed customized orthotic insole. *ARPN* **2016**, *11*, 10.
21. Yang, Y.; Chen, Y.; Wei, Y.; Li, Y. 3D printing of shape memory polymer for functional part fabrication. *Int. J. Adv. Manuf. Technol.* **2016**, *84*, 2079–2095. [CrossRef]

materials

Article

The Influence of Laser Defocusing in Selective Laser Melted IN 625

Alexandru Paraschiv [1,*], Gheorghe Matache [1], Mihaela Raluca Condruz [1], Tiberius Florian Frigioescu [1] and Ion Ionică [2]

[1] Special Components for Gas Turbines Department, Romanian Research and Development Institute for Gas Turbines COMOTI, 220D Iuliu Maniu, 061126 Bucharest, Romania; gheorghe.matache@comoti.ro (G.M.); raluca.condruz@comoti.ro (M.R.C.); tiberius.frigioescu@comoti.ro (T.F.F.)

[2] SC Plasma Jet SRL, 401E Atomiștilor, 077125 Magurele, Romania; officeplasmajet@gmail.com

[*] Correspondence: alexandru.paraschiv@comoti.ro; Tel.: +40-726-791-478

Abstract: Laser defocusing was investigated to assess the influence on the surface quality, melt pool shape, tensile properties, and densification of selective laser melted (SLMed) IN 625. Negative (-0.5 mm, -0.3 mm), positive ($+0.3$ mm, $+0.5$ mm), and 0 mm defocusing distances were used to produce specimens, while the other process parameters remained unchanged. The scanning electron microscopy (SEM) images of the melt pools generated by different defocusing amounts were used to assess the influence on the morphology and melt pool size. The mechanical properties were evaluated by tensile testing, and the bulk density of the parts was measured by Archimedes' method. It was observed that the melt pool morphology and melting mode are directly related to the defocusing distances. The melting height increases while the melting depth decreases from positive to negative defocusing. The use of negative defocusing distances generates the conduction melting mode of the SLMed IN 625, and the alloy (as-built) has the maximum density and ultimate tensile strength. Conversely, the use of positive distances generates keyhole mode melting accompanied by a decrease of density and mechanical strength due to the increase in porosity and is therefore not suitable for the SLM process.

Keywords: SLM; defocusing; IN 625; melt pool; tensile testing; density

Citation: Paraschiv, A.; Matache, G.; Condruz, M.R.; Frigioescu, T.F.; Ionică, I. The Influence of Laser Defocusing in Selective Laser Melted IN 625. *Materials* **2021**, *14*, 3447. https://doi.org/10.3390/ma14133447

Academic Editor: Rubén Paz

Received: 22 May 2021
Accepted: 15 June 2021
Published: 22 June 2021

Publisher's Note: MDPI stays neutral with regard to jurisdictional claims in published maps and institutional affiliations.

1. Introduction

Selective laser melting (SLM) is one of the most promising laser powder bed fusion (LPBF) techniques that has gained increasing attention in the last decade because of its ability to produce customized and functional parts with a complex geometry that would be difficult or even impossible to produce with standard subtractive manufacturing technologies [1–5].

Nevertheless, the lack of a full understanding of the impact of all process parameters on the quality of SLMed parts is still a limitation of this technology. Over 50 process parameters are involved in the SLM process that must be optimized to obtain high-density parts with tailored microstructures and high strength [6–8]. The process parameters are considered globally critical in the SLM process in terms of melt pool characteristics (shape and size), mechanical properties, and density, and many studies have focused on optimizing them to maximize the potential of the SLM technology [4,8,9].

As in any pulsed laser technology (laser additive manufacturing, laser welding, laser-cladding process), the laser beam has a high impact on the final microstructure of the melt pool [4,10], weld seams [11,12], or coating layer, [13] and on the mechanical performance of SLMed parts [14], joints [12,15] or clad coatings [16].

The laser's power in the SLM process is concentrated on a spot, allowing for complete melting of the powder layer and partial melting of the previously molten layer. The SLM machines use a small laser beam focus diameter of less than 100 μm when the laser has a power of up to 600 W and a larger laser beam focus diameter (100–500 μm) when the laser has a power of up to 1 kW. In both cases, there are several advantages and disadvantages. Using a small spot of laser beam generates a small melt pool size and is beneficial in terms of microstructure and mechanical properties but reduces the manufacturing productivity [1] and increases the temperature gradients and cooling rates of the material, which may lead to the development of high residual stresses [5,17], keyhole mode melting [11], distortion, balling, evaporation [17], and microstructural defects [1,5]. However, when a higher spot is used, the manufacturing productivity can be slightly increased by reducing the number of scan lines [9], but the part's dimensional accuracy and surface roughness can be compromised [1].

The spot of the laser beam and, consequently, the energy density influence the laser-melting modes, known in the domain of laser welding as the "conduction mode" and "keyhole mode" [1,4,8,9,14,18–20]. In the conduction mode, the depth of the melt pool is controlled by thermal conduction [10] and results in a semicircular melt pool shape [4,20], while in the keyhole mode, the formation of the melt pool is controlled by the evaporation of material and its depth is higher than in the conduction mode [10].

Due to these particularities, the melting mode is identified based on the ratio between the depth and width of the melt pool [4]. Additionally, a transition threshold between the melting modes is considered to be between 10^5 and 10^6 W/cm^2 [18].

Many studies have reported that the keyhole mode in the LPBF process is unsuitable [1,4,10,18,19]. Metelkova et al. [4] showed that the melt pool is stable during the conduction mode, but becomes deep with spherical voids during the keyhole mode. King et al. [10] found the conditions in terms of laser parameters (power, speed, and spot size of the laser) required to avoid the transition from conduction mode to keyhole mode for a SLMed 316L stainless steel.

The transition from keyhole mode to conductive mode, or vice versa, can be influenced using a negative or positive defocusing distance, as was shown in several studies [4,14,19]. The defocusing distance represents the depth of penetration of the laser beam into the powder layer up to a certain distance from the surface of the powder bed that can affect the solidification path, melt pool morphology, density, and mechanical properties of the SLMed alloy [4,8]. McLouth et al. [19] changed the focal heights of the laser at three values (−3, 0, and 3 mm) and found that the focal shift alters the power density and microstructure of SLMed IN 718. Leo et al. [14] found that the defocusing distance (−1, 0, and 1 mm) affects the sizes of the melt pool and grain of SLMed 17-4PH but does not affect the number of defects. However, the specimens built with 0- and 1-mm defocusing distance had the best properties in terms of tensile strength and hardness.

Besides optimising other process parameters in the SLM process, finding an appropriate defocusing distance for the manufacturing of a part with specific requirements in terms of quality surface, density, microstructure, and mechanical properties can be challenging.

As shown by McLouth et al. [19], despite the similarities between different additive manufacturing processes, the influence of the defocus on porosity and microstructure cannot be generalized among processes, not even for the same type of alloy.

In recent years, extensive studies have been done mainly on the influence of the process parameters, such as laser power, laser speed, layer thickness, hatch spacing, and scanning strategy, on the microstructure and properties of the alloys, but the laser defocus has not been investigated as thoroughly.

Recent studies were done on the laser defocus influence on the microstructure of the additive manufactured alloys IN 718 [19], 17-4PH, 316L steels [14], AlSi10Mg [8], and Ti-6Al-4V [21] but less on IN 625 produced by selective laser melting.

In order to advance the knowledge on the influence of process parameters on material properties, the present study aimed to assess the influence of negative and positive defocusing distances on the melt pool size and morphology, densification, and tensile properties of selective laser melted IN 625.

2. Materials and Methods

For this study, prismatic specimens for the microstructural analysis and densification and round tensile specimens were manufactured using a Lasertec 30 SLM machine (DMG MORI, Bielefeld, Germany) equipped with a 600W Yb: YAG fiber laser and IN 625 metal powder manufactured by LPW Technology Ltd. (Runcorn, UK) with the chemical composition presented in Table 1. The actual chemical composition of the metal powder was supplied by the manufacturer. The metal powder particle size distribution measured by the authors in a previous work was D10 = 22 μm, D50 = 34 μm, and D90 = 42 μm [22].

Table 1. Chemical composition of IN 625.

Elements (%wt.)	Al	C	Co	Cr	Fe	Mn	Mo	Nb	Si	Ti	Ni
Specification	<0.4	<0.1	<1.0	20–23	3–5	<0.5	8–10	3.15–4.15	<0.5	<0.4	Bal.
Actual composition	0.06	0.02	0.1	20.7	4.1	0.01	8.9	3.77	0.01	0.07	62.26

The specimens were built with no contour in a vertical position, using five defocusing distances (−0.5 mm, −0.3 mm, 0 mm, 0.3 mm, and 0.5 mm), while the following process parameters were kept constant: 250 W laser power, 750 mm/s laser speed, 40 μm layer thickness, 0.11 mm hatch distance, and scan pattern of 90° between two successive layers. In the case of the Lasertec 30 SLM machine, when a negative defocusing distance is set, the focal plane of the laser beam is located at a certain distance above the powder bed, while for a positive defocusing, the focal plane is moved into the depth of the powder bed. The size of the laser beam with different defocusing distances was measured using an Ophir-Optronics beam analyzer equipped with a charge-coupled device (CCD) camera and a commercial beam profiling software (BeamGage Professional, version 6.12, Ophir Optronics Solutions Ltd., Jerusalem, Israel).

The specimens were manufactured on a preheated building plate (80 °C), under argon flow to maintain 0.2% oxygen in the building chamber, using a cross-type support structure with the geometry, dimensions, and process parameters presented elsewhere [23].

The roughness of the top surface of the specimens manufactured with different defocusing distances was evaluated using a mobile roughness measuring instrument (MarSurf-PS-10, Mahr Inc., Providence, RI, USA).

The effects of defocusing distances on the melt pool depth, width, and height were analyzed by scanning electron microscopy (SEM) on sets of two specimens of 30 mm × 10 mm × 5 mm in size with a series of 50 single tracks manufactured on the last layer at a distance of 500 μm from each other (Figure 1a). Before the microstructural investigation, the single-track specimens were mounted in a resin (hot mounting) and fixed in the stainless steel clamping devices and T-slotted table of the abrasive disc-cutting machine. The specimens were cross sectioned, and the surface to be investigated was metallograpically prepared with sandpaper of varying grits (up to 1200) and polishing wheels with varying diamond suspension (3 and 1 microns) and etched with Aqua Regia reagent for 20 s. For each defocusing distance, 100 single tracks were analyzed in cross-section, transverse to the melting direction under the scanning electron microscope at 800× magnification. The melt pool dimensions were measured with the operating program of the microscope (XT microscope control 4.1.4.2010, FEI Company, Brno, Czech Republic).

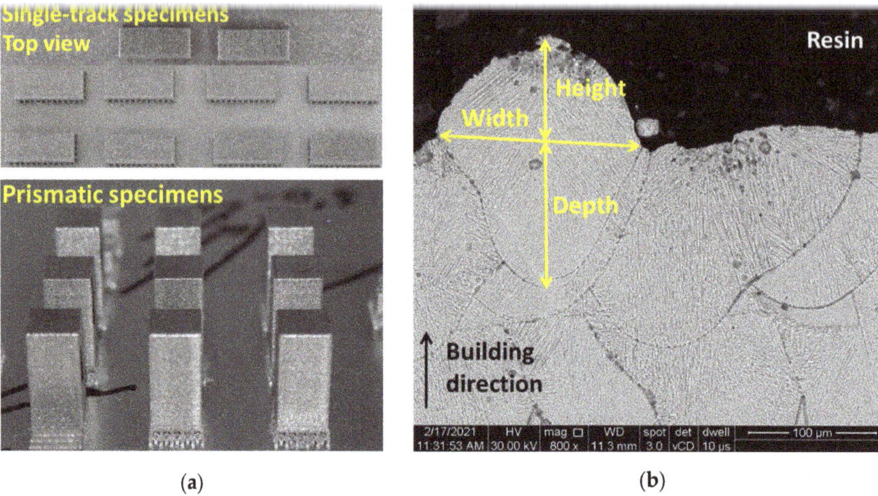

(a)   (b)

Figure 1. Prismatic specimens design used for the density measurements and melt pool analysis: (**a**) single-track and density specimens and (**b**) representation of the measured melt pool width, height, and depth of penetration on the cross section of single-track specimens.

The bulk density of sets of two prismatic specimens of 10 mm × 10 mm × 20 mm in size (Figure 1a) was measured via the Archimedes' method using an analytical balance equipped with a density measurements kit with an accuracy of 0.0001 g. (Pioneer PX224, Ohaus Europe GmbH, Nänikon, Switzerland). The specimens were immersed in an auxiliary fluid (99.3% purity ethanol) with a known density variation with temperature according to the International Organization for Standardization (ISO) 3369:2006 [24] after the support material was removed by grinding with sandpaper. A theoretical density of 8.49 g/cm^3 was calculated based on the actual chemical composition presented in Table 1 and was used as a reference to calculate the relative density of SLMed IN 625.

Sets of three cylindrical rods of 11 mm in diameter and 80 mm long (Figure 2a) were manufactured with respect to the building direction (Z) in the same conditions and using the same process parameters as the prismatic specimens to assess the influence of defocusing distances on the tensile properties of IN 625.

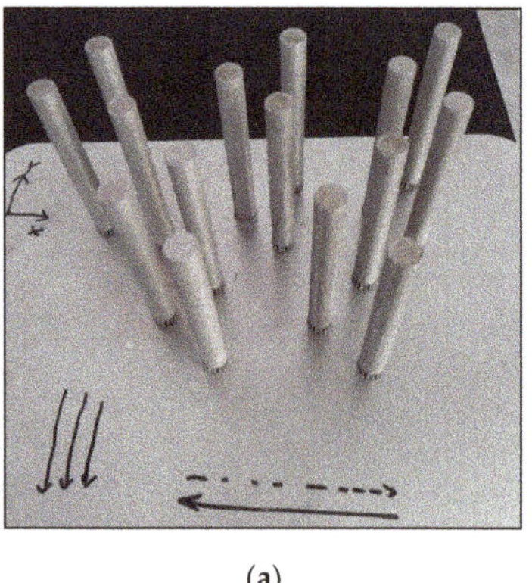

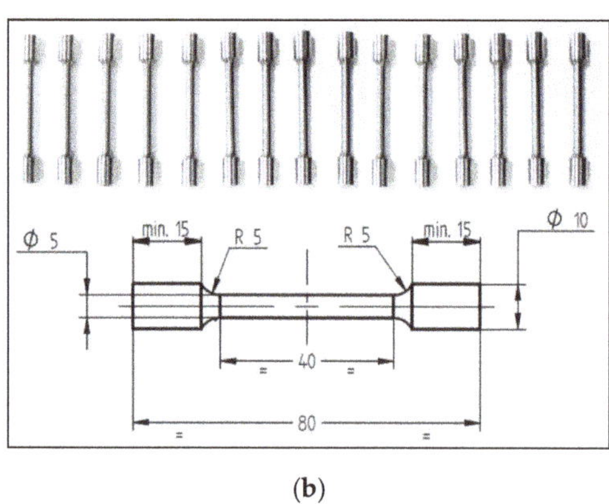

(a)   (b)

Figure 2. Tensile specimens: (**a**) cylindrical rods manufactured and (**b**) tensile specimen dimensions after machining (units in millimeters).

The cylindrical rods were subjected (prior machining) to heat treatment in air using an electric air furnace (Nabertherm LH 30/14 GmbH, Lilienthal/Bremen, Germany) that consists of stress relief heat treatment (heating with 10 °C/min up to 870 °C, held for 1 h, followed by air cooling) and annealing heat treatment (heating with 10 °C/min up to 1000 °C, held for 1 h, followed by fast cooling and oil quenching). The heat-treated rods were machined to obtain standard specimens (Figure 2b) and tested by using the Instron 3369 mechanical testing system (Instron, Norwood, MA, USA) according to ISO 6892-1:2009 [25]. The same heat treatment, geometry, and dimensions of the tensile specimens were used by the authors in other studies [23,26] to investigate the mechanical properties of SLMed IN 625.

3. Results

3.1. Macrostructural Analysis

The spot size of the laser beam changes in the x- or y-axis with respect to negative or positive defocusing according to the 2D beam displays presented in Figure 3a–e, where the macroscopic top-view images of the manufactured specimens and the top-surface roughness value (Ra) are also presented.

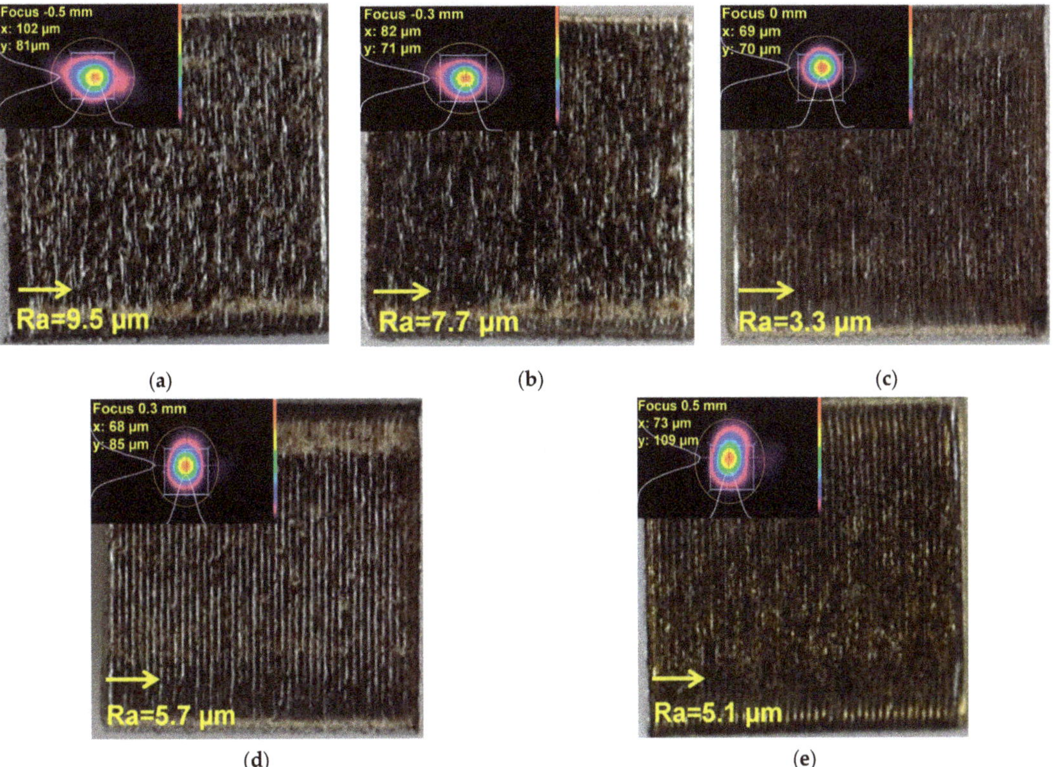

Figure 3. Macroscopic top-view images of the manufactured specimens built with different defocusing distances: (**a**) −0.5 mm, (**b**) −0.3 mm, (**c**) 0 mm, (**d**) 0.3 mm, and (**e**) 0.5 mm.

As shown in Figure 3a–e, the change of defocusing distance from positive to negative had a high impact on the surface roughness and quality. The surface roughness was measured perpendicular to the scanning rotation (90°) for all specimens, and the lowest value (Ra = 3.3 µm) was obtained when the defocusing distance was set to 0 mm. Generally, increasing the laser beam diameter [4] or decreasing the offset focus [21] increases the surface roughness of parts manufactured by the LPBF process due to the flow instability induced by high-intensity laser irradiation that affects the melt pool behavior during the melting process.

3.2. Melt Pool Behavior

The SEM images of the cross-section of the single-track specimens built with defocusing distances of 0.5 mm, 0.3 mm, 0 mm, −0.3 mm, and −0.5 mm are presented in Figure 4a–e and were acquired to evaluate the influence of defocusing amounts on the dimensions and morphology of the melt pool.

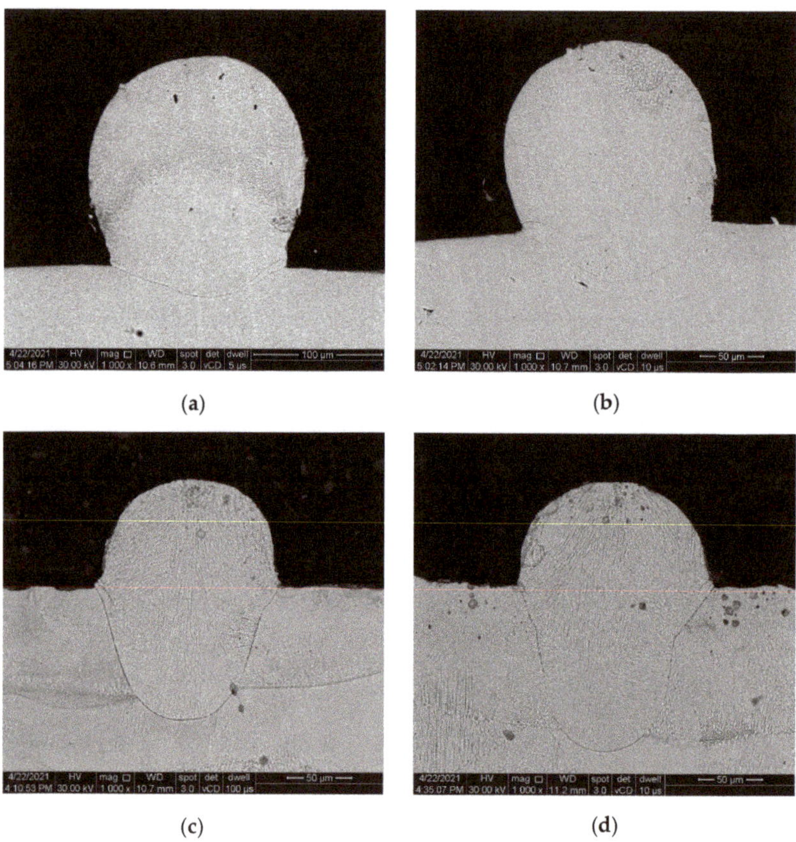

(a)

(b)

(c)

(d)

Figure 4. *Cont.*

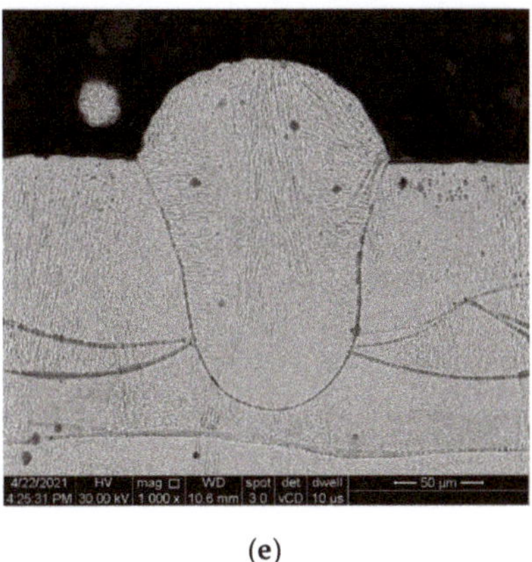

(e)

Figure 4. Cross-section SEM images of the single-track specimens built with different defocusing distance: (**a**) −0.5 mm, (**b**) −0.3 mm, (**c**) 0 mm, (**d**) 0.3 mm, and (**e**) 0.5 mm.

As shown in the SEM images presented in Figure 4a–e, the defocusing distance substantially impacts the melt pool dimensions and morphology. As a general observation, the melting height increases while the melting depth decreases from positive to negative defocusing. The quantitative measurements of the melt pool depth, width, and height are presented in Table 2.

Table 2. Melt pools size for different defocusing amounts.

Defocusing, mm	−0.5	−0.3	0	0.3	0.5
Width, μm	151 ± 32	146 ± 26	150 ± 17	137 ± 12	143 ± 10
Depth, μm	30 ± 11	42 ± 16	89 ± 23	114 ± 21	115 ± 22
Height, μm	125 ± 40	107 ± 33	86 ± 33	80 ± 17	65 ± 16

As shown in Figure 5, the height/width ratio decreases and the depth/height ratio increases with increasing the defocusing distance.

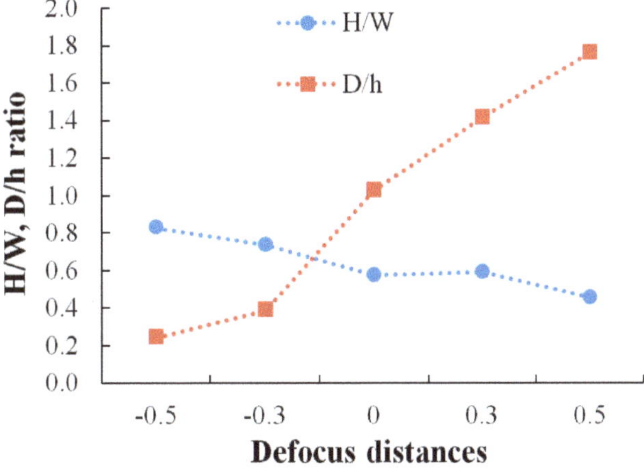

Figure 5. Influence of defocusing on melt pools height/width and depth/height ratio.

Based on the microstructural analyses of the melt pool behavior, it was found that even when using the same energy density, the melt pool became wider and less deep at negative defocusing distances (conduction mode) and narrower and deeper at positive defocusing distances (keyhole mode).

3.3. Density Measurements

The relative density of IN 625 specimens built with different defocusing distances was expressed as the ratio between the average density of specimens determined by Archimedes' method and the theoretical density of IN 625 (8.49 g/cm^3). The density of specimens was measured using an analytical balance equipped with a density measurement kit with an accuracy of 0.0001 g. The average values of density for each case (defocusing distance) were obtained based on 12 measurements with a slight standard deviation, as presented in Table 3.

Table 3. The relative density of SLMed IN 625 as a function of defocusing amounts.

Defocusing, mm	−0.5	−0.3	0	0.3	0.5
Relative density, %	99.52	99.52	99.37	99.27	99.28
Standard deviation, %	0.02	0.03	0.02	0.09	0.06

Based on the density measurements presented in Table 3, it can be stated that the relative density of SLMed IN 625 with 0 mm and positive defocusing distances is maintained close to 99.3%, while at negative defocusing distances, the density slightly increases up to 99.5%. An explanation for this density evolution is the porosity that increases at positive values of the defocusing distances. Figure 6b highlights a higher porosity of the specimen built with a +0.5 mm defocusing distance than the specimen built with a −0.5 mm defocusing distance (Figure 6a). The micrographs shown in the windows present the details at higher magnification (8000×), highlighting porosities generated inherently by the melting and solidification processes. The differences between the pore size by defocusing shifting from negative to positive are visible.

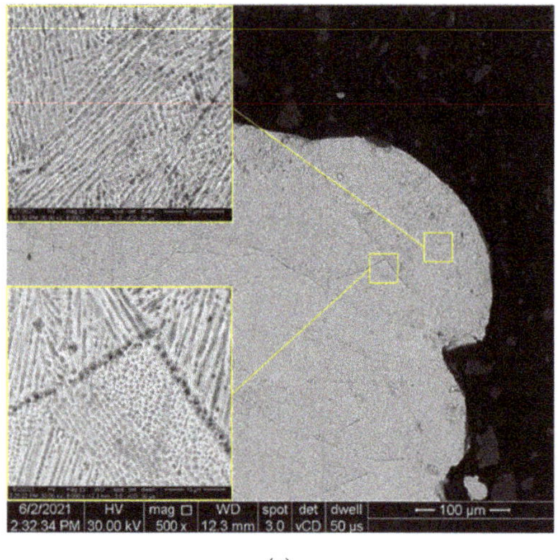

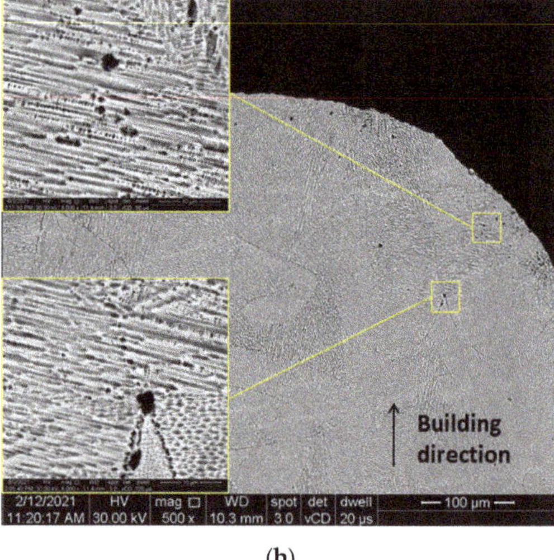

(a) (b)

Figure 6. Porosity in SLMed IN 625: (**a**) with negative (conduction mode) and (**b**) positive (keyhole mode) defocusing distances.

3.4. Tensile Testing

Tensile tests on standard tensile test pieces machined from cylindrical rods built with different defocusing distances (−0.5 mm, −0.3 mm, 0 mm, +0.3 mm, and +0.5 mm) were performed at room temperature. The specimens were evaluated with respect to the ultimate tensile strength (UTS), 0.2% yield strength (YS), reduction of area (RA), and elongation (El).

The average tensile test results and standard deviations of each set of three specimens are presented in Table 4.

Table 4. Tensile properties of specimens built with different defocusing distances.

Tensile Properties	−0.5	−0.3	0	0.3	0.5
Ultimate tensile strength, MPa	831 ± 3.3	826 ± 5.7	834 ± 6.4	821 ± 3.9	816 ± 4.1
0.2% Yield strength, MPa	536 ± 3.3	539 ± 3.3	542 ± 5.4	534 ± 5	532 ± 4
Reduction of area, %	57 ± 1	50 ± 2.6	54 ± 1.9	54 ± 1.6	55 ± 0.5
Elongation, %	48 ± 0.8	46 ± 1	49 ± 1	47 ± 0.3	49 ± 0.3

The SLMed alloys are very sensitive in terms of tensile properties due to the porosity, even in the case of up to 1% porosity, as was presented by Plessis et al. [27]. They conclude that the strength and ductility of LPBF materials are reduced with increasing porosity, and the failure is determined by the largest pores. The inherent presence of porosities and microcracks, because of the manufacturing method, act as stress concentrators, and therefore, fracture crack initiation and propagation under tensile load are influenced by their spatial distribution [28]. It was previously shown that cracks initiate on voids near the surface or in the subsurface and propagate radially [29].

Figure 7 presents SEM images of the tensile fracture surfaces of test pieces built with negative (−0.5 mm), 0, and positive (+0.5 mm) defocusing distances. Micrographs in the windows present the details of the final fracture surfaces showing pre-existing porosities generated by the manufacturing process. Notably, it can be seen that positive defocusing is associated with a higher porosity level, which was also highlighted by relative density measurements.

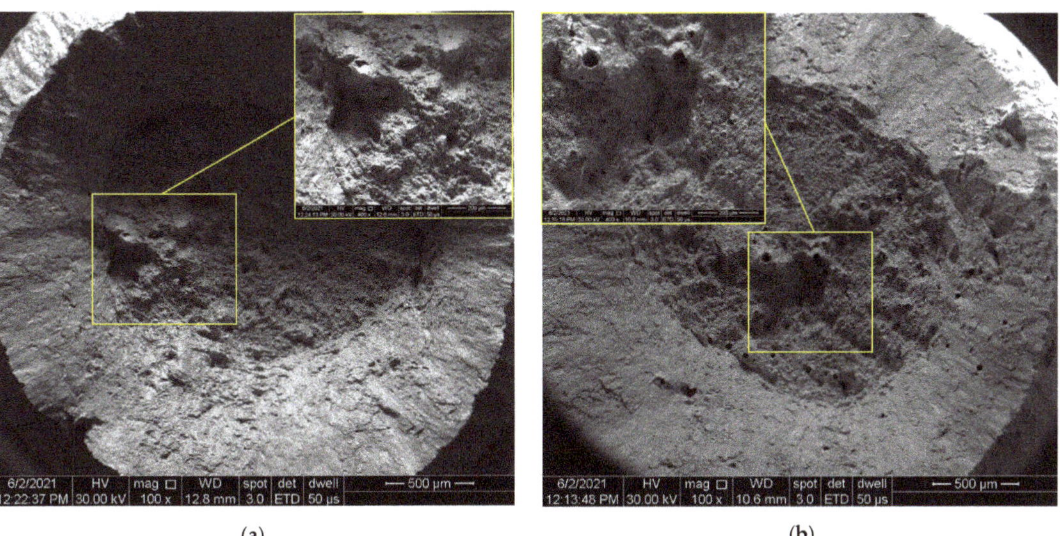

(a)   (b)

Figure 7. *Cont.*

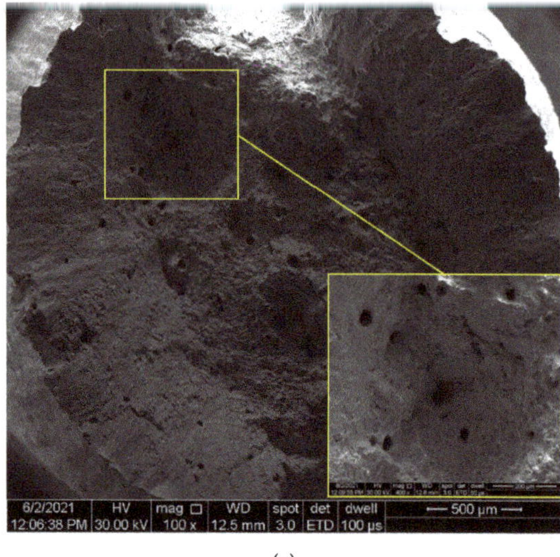

(c)

Figure 7. SEM images of the fracture surface of tensile test pieces built with different defocusing distances: (**a**) −0.5 mm, (**b**) 0 mm, and (**c**) +0.5 mm.

The anisotropic behavior of SLMed IN 625 can also be related to the grain growth orientation over multiple layers (in the build direction) [30] and the sizes and morphology of the melt pool, which is strongly influenced by the defocusing distances. As was shown in Figure 4a–e, the melt pool became wider and thinner at negative defocusing. Therefore, the columnar grains grew on shorter distances and had lower mechanical strength than in the case of the 0 mm defocusing distances. However, at positive defocusing, the melt pool was deeper, and a higher mechanical strength would have been expected than in the other defocusing distances, but the higher porosity generated by the keyhole mode was a determining factor that weakened the ultimate tensile strength.

4. Discussion

The present study investigated the influence of laser defocusing on the surface quality, melt pool shape, densification, and tensile properties of selective laser melted IN 625.

When the defocusing distance was changed to positive or negative values, the melt pool was affected by the fluctuations from the thermocapillary convection and generated an increase of the surface roughness, especially in the case of negative distances, where the highest values were obtained (Ra = 7.7–9.5 μm).

During the solidification of the melt pool, the columnar dendrites become elongated in the growth direction normal to the solidification front [3], and the melt pool width and depth increase linearly with the increase of laser energy density [1]. These interactions between the laser beam and powder bed can generate a conduction mode, where the melt pool has a small depth and spherical shape, or a keyhole mode, where the depth of the melt pool is higher than half of its width [4]. The microstructural analysis revealed that the melting height and width decreases while the melting depth increases as the negative defocusing amount decreases. Another study [31] found the opposite effect of using negative defocusing distances, which can be explained by the beam divergence that can be reversed from one SLM system to another. The melting mode, and consequently the melt pool morphology and size, can also be influenced by a high laser power, low beam size, and low scanning speed [4,10]. However, the semicircular melt pool is obtained in the conduction mode (negative distances) and becomes narrower and deeper when the heat flow has a keyhole mode (positive distances) [4,10,14].

When varying the defocusing distances from negative distances to positive distances, the relative density of IN 625 prismatic specimens decreased from 99.52% (conduction mode) to 99.28% because of the increasing of the porosity, which usually occurs in the keyhole mode melting due to the vaporization of the overheated material [8,10,32]. Another effect of this variation is the weakening of the tensile properties. Both ultimate tensile strength (UTS) and yield strength 0.2% (YS) tend to slightly decrease with the variation of defocusing distance from negative to positive values, but no more than 2%.

Comparative experimental data obtained on SLMed IN 625 have not been identified in the literature. A similar variation was observed in another study by Zhou et al. [8] on a SLMed AlSi10Mg alloy. They found that the defocusing distances affected both the tensile mechanical properties and density of the SLMed alloy, and the highest relative densities and tensile properties were achieved in the conduction mode melting. However, in the study done by Leo et al. [14], the highest tensile strength of SLMed 17-4PH was obtained for the positive, with 0 and 1 mm defocusing distances.

The anisotropy of SLMed materials is well known, and many studies have demonstrated the influence of grain morphology, texture, building orientation, and scanning strategy on the anisotropic microstructure and tensile properties of SLMed Ni-based superalloys [5,26,33].

However, the tensile properties of IN 625 specimens built along the Z-axis with different defocusing distances exceed the minimum specification requirements for both conventional and additively manufactured IN 625 alloys according to the American Society for Testing and Materials, ASTM B 443 [34] and ASTM F3056-14e1 [35], respectively, as is presented in Figure 8. The tensile results presented in Table 4 are also presented in a graphic form in Figure 8 to compare them with the minimum requirements of the ultimate tensile strength and yield strength of additively manufactured IN 625 and hot-rolled IN 625.

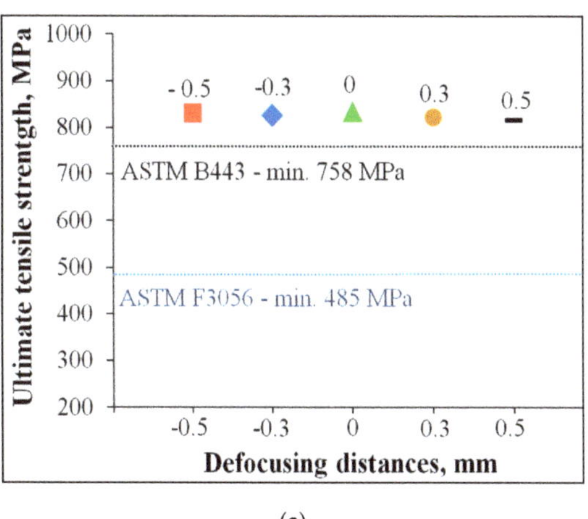

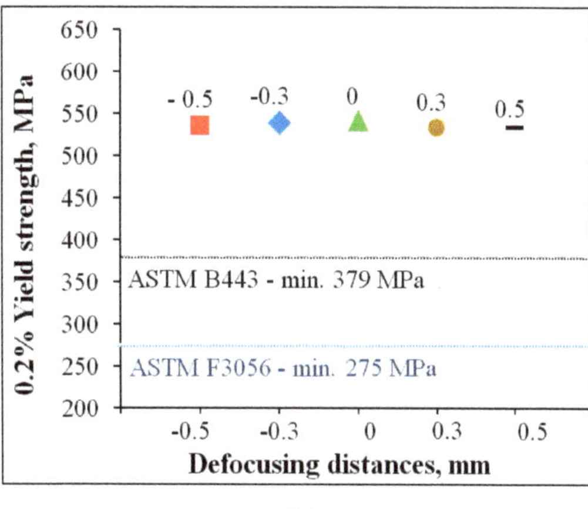

(a)

(b)

Figure 8. *Cont.*

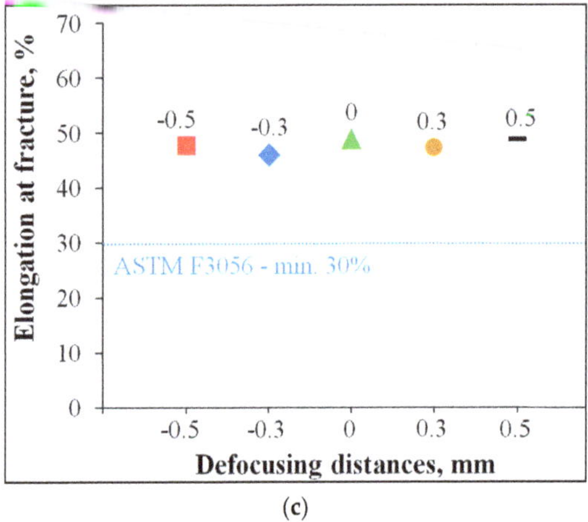
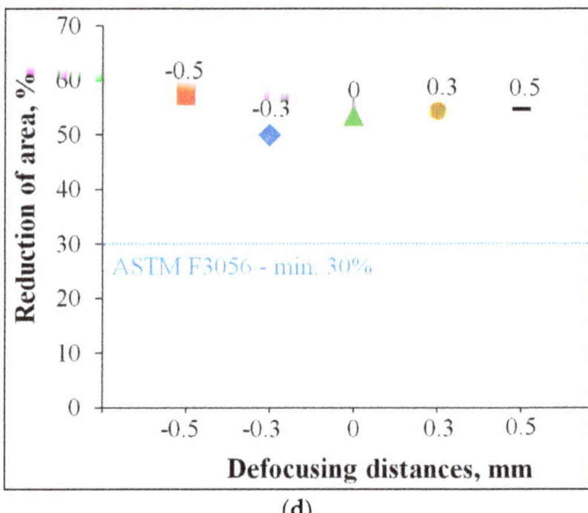

Figure 8. Tensile properties of specimens built with different defocusing distances: (**a**) UTS; (**b**) YS; (**c**) El; (**d**) RA.

Figure 8a–d presents the tensile test results of specimens built with different defocusing distances, where the dashed lines represent the minimum values for additively manufactured IN 625 according to ASTM F3056-14e1 [35]. Additional dashed lines were also plotted for the hot-rolled IN 625 plates according to ASTM B 443 [34]. In addition to the higher strength values of SLMed specimens relative to the minimum values, the elongation and reduction of area after fracture are higher than the minimum values of both ASTM, which shows a higher ductility due to the orientation of the columnar grains relative to the loading direction.

5. Conclusions

Laser defocusing was investigated to assess the influence of negative (−0.5 mm, −0.3 mm), 0 mm, and positive (−0.3 mm, 0.5 mm) defocusing distances on the melt pool width, height, and depth of penetration, surface roughness, densification, and tensile properties of selective laser melted IN 625.

The defocusing distances had a high impact on the quality of the surface. The lowest surface roughness (Ra = 3.3 μm) was generated when the defocusing distance was set to 0 mm but increased from positive to negative defocusing distances up to Ra = 9.5 μm (for −0.5 mm defocusing distance).

The use of negative defocusing distances generates a conduction melting mode of the SLMed IN 625, where the melt pool has a small depth and spherical shape. Conversely, the use of positive distances generates a keyhole mode melting, where the depth of the melt pool is higher than half of its width. As a general trend, the melt pools tended to become wider and thinner at negative defocusing (conduction mode) and narrower and deeper at positive defocusing distances (keyhole mode).

When negative defocusing amounts (−0.3 and −0.5 mm) were used, a relative density higher than 99.5% was obtained, while in the cases of 0 mm and positive defocusing distances (0.3 and 0.5 mm), the relative density was slightly reduced close to 99.3%.

The tensile test results showed that the defocusing distances slightly influenced the tensile properties of IN 625. Both ultimate tensile strength (UTS) and yield strength (YS) tend to decrease very slightly with the variation of defocusing distance from negative to positive values due to the loading direction, the orientation of the columnar grains grown over multiple layers, and the anisotropy of the SLMed IN 625.

Author Contributions: Conceptualization, A.P., G.M. and M.R.C.; methodology, A.P., G.M. and M.R.C.; software, T.F.F. and I.I.; investigation, A.P., T.F.F., M.R.C. and I.I; writing—original draft preparation, A.P.; writing—review and editing, G.M., T.F.F. and I.I. All authors have read and agreed to the published version of the manuscript.

Funding: This research was carried out within POC-A1-A1.2.3-G-2015, ID/SMIS code: P_40_422/105884, "TRANSCUMAT" Project, Grant no. 114/09.09.2016 (Subsidiary Contract no. 5/D.1.3/114/18.12.2017). The project is supported by the Romanian Minister of Research and Innovation.

Institutional Review Board Statement: Not applicable.

Informed Consent Statement: Not applicable.

Data Availability Statement: Data sharing not applicable.

Acknowledgments: This paper was presented at the 8th International Conference "Materials Science and Technologies RoMat 2020", organized by the Faculty of Materials Science and Engineering from University "POLITEHNICA" of Bucharest, Romania.

Conflicts of Interest: The authors declare no conflict of interest.

References

1. Kumar, P.; Farah, J.; Akram, J.; Teng, C.; Ginn, J.; Misra, M. Influence of laser processing parameters on porosity in Inconel 718 during additive manufacturing. *Int. J. Adv. Manuf. Technol.* **2019**, *103*, 1497–1507. [CrossRef]
2. Bean, G.E.; Witkin, D.B.; McLouth, T.D.; Patel, D.N.; Zaldivar, R.J. Effect of laser focus shift on surface quality and density of Inconel 718 parts produced via selective laser melting. *Addit. Manuf.* **2018**, *22*, 207–215. [CrossRef]
3. Shi, R.; Khairallah, S.A.; Roehling, T.T.; Heo, T.W.; McKeown, J.T.; Matthews, M.J. Microstructural control in metal laser powder b e d fusion additive manufacturing using laser beam shaping strategy. *Acta Mater.* **2020**, *184*, 284–305. [CrossRef]
4. Metelkova, J.; Kinds, Y.; Kempen, K.; de Formanoir, C.; Witvrouw, A.; Hooreweder, B.V. On the influence of laser defocusing in Selective Laser Melting of 316L. *Addit. Manuf.* **2018**, *23*, 161–169. [CrossRef]
5. Tian, Z.; Zhang, C.; Wang, D.; Liu, W.; Fang, X.; Wellmann, D.; Zhao, Y.; Tian, Y. A Review on Laser Powder Bed Fusion of Inconel 625 Nickel-Based Alloy. *Appl. Sci.* **2020**, *10*, 81. [CrossRef]
6. Popovich, A.A.; Sufiiarov, V.S.; Borisov, E.V.; Polozov, I.A.; Masaylo, D.V. Design and manufacturing of tailored microstructure with selective laser melting. *Mater. Phys. Mech.* **2018**, *38*, 1–10.
7. Spears, T.G.; Gold, S.A. In-process sensing in selective laser melting (SLM) additive manufacturing. *Integr. Mater. Manuf. Innov.* **2016**, *5*, 16–40. [CrossRef]
8. Zhou, S.; Su, Y.; Gu, R.; Wang, Z.; Zhou, Y.; Ma, Q.; Yan, M. Impacts of Defocusing Amount and Molten Pool Boundaries on Mechanical Properties and Microstructure of Selective Laser Melted AlSi10Mg. *Materials* **2019**, *12*, 73. [CrossRef]
9. Sow, M.C.; de Terris, T.; Castelnau, O.; Hadjem-Hamouche, Z.; Coste, F.; Fabbro, R.; Peyre, P. Influence of beam diameter on Laser Powder Bed Fusion (L-PBF) process. *Addit. Manuf.* **2020**, *36*, 1–11. [CrossRef]
10. King, W.E.; Barth, H.D.; Castillo, V.M.; Gallegos, G.F.; Gibbs, J.W.; Hahn, D.E.; Kamath, C.; Rubenchik, A.M. Observation of keyhole-mode laser melting in laser powder-bed fusion additive manufacturing. *J. Mater. Process. Techol.* **2014**, *214*, 2915–2925. [CrossRef]
11. Chiang, M.-F.; Lo, T.-Y.; Chien, P.-H.; Chi, C.-H.; Chang, K.-C.; Yeh, A.-C.; Shiue, R.-K. The Dilution Effect in High-Power Disk Laser Welding the Steel Plate Using a Nickel-Based Filler Wire. *Metals* **2021**, *11*, 874. [CrossRef]
12. Mohammed, G.R.; Ishak, M.; Ahmad, S.N.A.S.; Abdulhadi, H.A. Fiber Laser Welding of Dissimilar 2205/304 Stainless Steel Plates. *Metals* **2017**, *7*, 546. [CrossRef]
13. Susanto, I.; Tsai, C.-Y.; Ho, Y.-T.; Tsai, P.-Y.; Yu, I.-S. Temperature Effect of van derWaals Epitaxial GaN Films on Pulse-Laser-Deposited 2D MoS2 Layer. *Nanomaterials* **2021**, *11*, 1406. [CrossRef]
14. Leo, P.; Cabibbo, M.; Del Prete, A.; Giganto, S.; Martínez-Pellitero, S.; Barreiro, J. Laser Defocusing Effect on the Microstructure and Defects of 17-4PH Parts Additively Manufactured by SLM at a Low Energy Input. *Metals* **2021**, *11*, 588. [CrossRef]
15. Chludzinski, M.; dos Santos, R.E.; Churiaque, C.; Ortega-Iguña, M.; Sánchez-Amaya, J.M. Pulsed LaserWelding Applied to Metallic Materials—A Material Approach. *Metals* **2021**, *11*, 640. [CrossRef]
16. Ma, Q.; Dong, Z.; Ren, N.; Hong, S.; Chen, J.; Hu, L.; Meng, W. Microstructure and Mechanical Properties of Multiple in-situ-Phases-Reinforced Nickel Composite Coatings Deposited by Wide-Band Laser. *Coatings* **2021**, *11*, 36. [CrossRef]
17. Heeling, T.; Wegener, K. The effect of multi-beam strategies on selective laser melting of stainless steel 316L. *Addit. Manuf.* **2018**, *22*, 334–342. [CrossRef]
18. Tenbrock, C.; Fischer, F.G.; Wissenbach, K.; Schleifenbaum, J.H.; Wagenblast, P.; Meiners, W.; Wagner, J. Influence of keyhole and conduction mode melting for top-hat shaped beam profiles in laser powder bed fusion. *J. Mater. Process. Technol.* **2020**, *278*, 1–10. [CrossRef]

19. McLouth, T.D.; Bean, G.E.; Witkin, D.B.; Sitzman, S.D.; Adams, P.M.; Patel, D.N.; Park, W.; Yang, J.M.; Zaldivar, R.J. The effect of laser focus shift on microstructural variation of Inconel 718 produced by selective laser melting. *Mater. Des.* **2018**, *149*, 205–213. [CrossRef]

20. Soylemez, E. High deposition rate approach of selective laser melting through defocused single bead experiments and thermal finite element analysis for Ti-6Al-4V. *Addit. Manuf.* **2020**, *31*, 1–14. [CrossRef]

21. Safdar, A.; He, H.; Wei, L.; Snis, A.; de Paz, L.C. Effect of process parameters settings and thickness on surface roughness of EBM produced Ti-6Al-4V. *Rapid Prototyp. J.* **2012**, *18*, 401–408. [CrossRef]

22. Condruz, M.R.; Matache, G.; Paraschiv, A. Characterization of IN 625 recycled metal powder used for selective laser melting. *Manuf. Rev.* **2020**, *7*, 1–8. [CrossRef]

23. Matache, G.; Paraschiv, A.; Condruz, M.R. Tensile Notch Sensitivity of Additively Manufactured IN 625 Superalloy. *Materials* **2020**, *13*, 4859. [CrossRef]

24. *Impermeable Sintered Metal Materials and Hard Metals—Determination of Density*; ISO 3369:2006; International Organization for Standardization: Geneva, Switzerland, 2006.

25. *Metallic Materials—Tensile Testing—Part 1: Method of Test at Room Temperature*; ISO 6892-1:2009; International Organization for Standardization: Geneva, Switzerland, 2009.

26. Condruz, M.R.; Matache, G.; Paraschiv, A.; Frigioescu, T.F.; Badea, T. Microstructural and Tensile Properties Anisotropy of Selective Laser Melting Manufactured IN 625. *Materials* **2020**, *13*, 4829. [CrossRef]

27. Plessis, A.d.; Igor, I.; Yadroitsev, Y. Effects of defects on mechanical properties in metal additive manufacturing: A review focusing on X-ray tomography insights. *Mater. Des.* **2020**, *187*, 1–19. [CrossRef]

28. Stef, J.; Poulon-Quintin, A.; Redjaimia, A.; Ghanbaja, J.; Ferry, O.; De Sousa, M.; Goune, M. Mechanism of porosity formation and influence on mechanical properties in selective laser melting of Ti-6Al-4V parts. *Mater. Des.* **2018**, *156*, 480–493. [CrossRef]

29. Gong, H.; Rafi, K.; Gu, H.; Ram, G.J.; Starr, T.; Stucker, B. Influence of defects on mechanical properties of Ti–6Al–4 V components produced by selective laser melting and electron beam melting. *Mater. Des.* **2015**, *86*, 545–554. [CrossRef]

30. Chen, Z.; Chen, S.; Wei, Z.; Zhang, L.; Wei, P.; Lu, B.; Zhang, S.; Xiang, Y. Anisotropy of nickel-based superalloy K418 fabricated by selective laser melting. *Pro. Nat. Sci. Mater.* **2018**, *28*, 496–504. [CrossRef]

31. Gao, P.; Wang, Z.; Zeng, X. Effect of process parameters on morphology, sectional characteristics and crack sensitivity of Ti-40Al-9V-0.5Y alloy single tracks produced. *Int. J. Lightweight Mater. Manuf.* **2019**, *2*, 355–361. [CrossRef]

32. Koutiri, I.; Pessard, E.; Peyre, P.; Amlou, O.; de Terris, T. Influence of SLM process parameters on the surface finish, porosity rate and fatigue behavior of as-built Inconel 625 parts. *J. Mater. Process Technol.* **2018**, *255*, 536–546. [CrossRef]

33. Ni, M.; Chen, C.; Wang, X.; Wang, P.; Li, R.; Zhang, X.-Y.; Zhou, K. Anisotropic tensile behavior of in situ precipitation strengthened Inconel 718 fabricated by additive manufacturing. *Mater. Sci. Eng. A* **2017**, *701*, 344–351. [CrossRef]

34. *Standard Specification for Nickel-Chromium-Molybdenum-Columbium Alloy and Nickel-Chromium-Molybdenum- Silicon Alloy Plate, Sheet, and Strip*; ASTM B 443; ASTM International: West Conshohocken, PA, USA, 2019.

35. *Standard Specification for Additive Manufacturing Nickel Alloy (UNS N06625) with Powder Bed Fusion*; ASTM F3056-14e1; ASTM International: West Conshohocken, PA, USA, 2014.

Article

A New Method for Automatic Detection of Defects in Selective Laser Melting Based on Machine Vision

Zhenqiang Lin [1], Yiwen Lai [2], Taotao Pan [1], Wang Zhang [1], Jun Zheng [1,*], Xiaohong Ge [1,*] and Yuangang Liu [3,*]

1   College of Materials Science and Engineering, Xiamen University of Technology, Xiamen 361024, China; zhenqianglin@3dmetalwerks.com (Z.L.); taotaopan@3dmetalwerks.com (T.P.); wangzhang@3dmetalwerks.com (W.Z.)
2   College of Electrical Engineering and Automation, Xiamen University of Technology, Xiamen 361024, China; yiwenlai@3dmetalwerks.com
3   College of Chemical Engineering, Huaqiao University, Xiamen 362021, China
*   Correspondence: jason@3dmetalwerks.com (J.Z.); xhge@xmut.edu.cn (X.G.); ygliu@hqu.edu.cn (Y.L.)

Abstract: Selective laser melting (SLM) is a forming technology in the field of metal additive manufacturing. In order to improve the quality of formed parts, it is necessary to monitor the selective laser melting forming process. At present, most of the research on the monitoring of the selective laser melting forming process focuses on the monitoring of the melting pool, but the quality of forming parts cannot be controlled in real-time. As an indispensable link in the SLM forming process, the quality of powder spreading directly affects the quality of the formed parts. Therefore, this paper proposes a detection method for SLM powder spreading defects, mainly using industrial cameras to collect SLM powder spreading surfaces, designing corresponding image processing algorithms to extract three common powder spreading defects, and establishing appropriate classifiers to distinguish different types of powder spreading defects. It is determined that the multilayer perceptron (MLP) is the most accurate classifier. This detection method has high recognition rate and fast detection speed, which cannot only meet the SLM forming efficiency, but also improve the quality of the formed parts through feedback control.

Keywords: selective laser melting; powder spreading defect; machine vision; classifier

Citation: Lin, Z.; Lai, Y.; Pan, T.; Zhang, W.; Zheng, J.; Ge, X.; Liu, Y. A New Method for Automatic Detection of Defects in Selective Laser Melting Based on Machine Vision. *Materials* **2021**, *14*, 4175. https://doi.org/10.3390/ma14154175

Academic Editor: Rubén Paz

Received: 10 June 2021
Accepted: 19 July 2021
Published: 27 July 2021

Publisher's Note: MDPI stays neutral with regard to jurisdictional claims in published maps and institutional affiliations.

1. Introduction

As one of the most promising technologies in metal additive manufacturing, selective laser melting (SLM) mainly uses the idea of layer-by-layer accumulation to construct metal parts [1]. SLM technology can produce complex parts that cannot be formed by traditional manufacturing technology, so it has a broad application prospect in aerospace, medical, molds, and other fields [2–4].

It is undoubtedly the ultimate goal of SLM technology to directly produce metal parts with excellent performance and quality, but the SLM forming process is a complex physical and chemical process. The quality of SLM forming parts is influenced by many factors, including laser parameters (laser power, spot size, and laser wavelength), scanning parameters (scanning strategy, scan spacing, and scanning speed), environmental conditions (oxygen content and humidity), and material conditions (thick, spread powder, and metal powder particle size) [5–8].

Therefore, it is very necessary to monitor the formation process of SLM to ensure the quality of finished products. According to the selected laser melting process steps, the selected laser melting process detection can be divided into the powder spreading process detection, powder bed detection, melting process detection, and molten layer detection.

For example, Reinarz et al. [9] detected the stability of the powder spreading device by the principle that the powder spreading device would produce a speed change when passing through the raised molten layer. Tom Craeghs et al. [10] used the column-wise cross-sectional view of the average value of the gray pixels in each row of the powder bed image to detect whether the scraper was damaged. Clijsters et al. [11] used information such as molten pool heat, molten pool area, molten pool length, molten pool width, etc. with good performance of formed parts as reasonable data, compared and evaluated the collected molten pool information with a reasonable data interval, and then identified the melting abnormal pool and assess the forming quality of parts. Foster et al. [12] collected molten layer images through industrial cameras, used edge detection algorithms to extract the contours of the molten layer, analyzed and processed the original images of each layer, superimposed them into a three-dimensional model, and detected defects by evaluating three-dimensional data.

To monitor the SLM forming process effectively, machine vision inspection is a safe, reliable, and highly automated non-contact inspection technology that can mimic human eye functions through industrial cameras and extract effective information from various environments [13]. The application of this technology in the industrial field is relatively large, mainly involving the detection of surface defects in metals, fabrics, glass, electronic devices, etc. [14–17].

Among the SLM process steps, the quality of the powder spreading has a huge impact on the final quality of SLM molded parts. Packing quality refers to the smoothness, uniformity, and whether there are impurities in the surface of the powder during each laying operation by SLM equipment. The most concern in the SLM powder coating process is the flatness and layer thickness error of the powder coating surface. Poor flatness can easily cause incomplete melting or over-melting when the laser is applied to the convex or concave surface of the powder coating surface. If the thickness is too thick, the layers may not be fully fused together. If the thickness is too thin, the thermal stress may be too large. Therefore, when the surface of the powder coating is defective, the surface quality of the single-layer processing will deteriorate. The accumulation of the single-layer surface with poor quality will cause pores, cracks, and surface spheroidization in the final formed part, which will seriously affect the performance of the part [18–20]. In order to further improve the performance and quality of SLM forming parts, it is necessary to ensure the quality of each layer of the powder coating surface during the SLM forming process.

In this paper, to solve the problem of poor performance of forming part caused by powder spreading defect, the machine vision inspection technology is used to monitor the powder spreading condition in the material condition factors, the construction of SLM powder spreading defect collection system, and the detection algorithm of SLM powder spreading defects are mainly studied.

2. Experiment Setup and Powder Spreading Defect

2.1. Experiment Setup

As the laser galvanometer system (Scancube 3, Scanlab, Siemensstr. 2a82178 Puchheim Germany) is already installed directly above the forming bin of the selective laser melting equipment (WXL-120, 3D Metalwerks Co. Ltd., Xiamen, China), the image of the powdered surface can only be collected by the side axis. The specific installation location is shown in Figure 1. In order to improve the resolution capability of the system, because the size of the forming bin of the SLM equipment used in this system is $120 \times 120 \times 100$ (mm), and to enable the detection system to resolve at least 0.5 mm of details, each detail needs to be composed of 4–8 pixels. This system takes six pixels corresponding to a 0.5 mm defect, so the number of pixels in the length and width direction of the forming chamber needs to be greater than 1440, and due to the angle of the camerac, the powder bed of the forming cylinder will appear on the image in a diamond shape, so the final pixels in all directions of the camera number needs to be greater than that $1440 \times \sqrt{2} \approx 2037$, so we used a Hikvision MV-CA050-10GM camera with sensor type IMX264, pixel size 3.45 μm × 3.45 μm, sensor area 2/3", resolution 2448 × 2448, frame rate 23.5 fps, and camera interface gigabit ethernet.

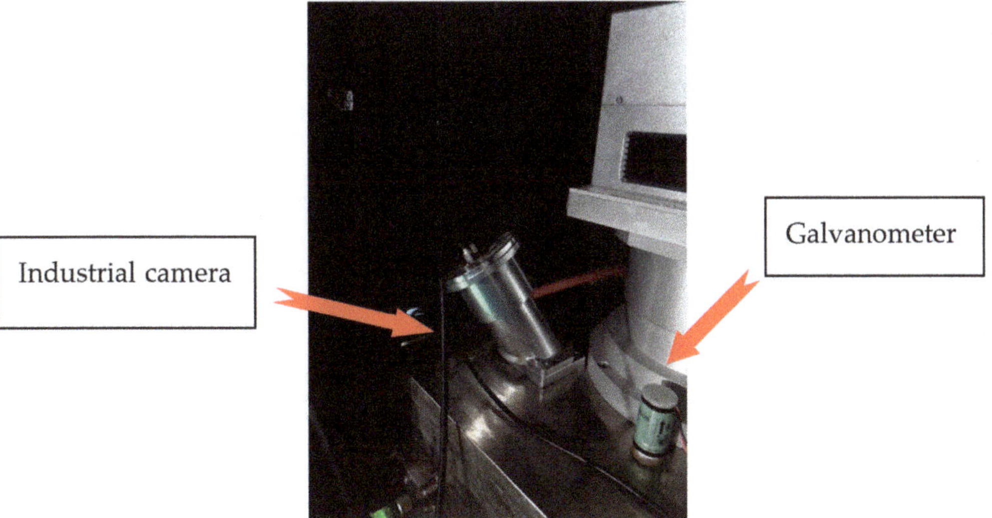

Figure 1. Industrial camera installation location.

After the industrial camera is selected, high-quality images cannot be acquired without a matching lens. The function of the camera lens is similar to that of the crystalline lens. It mainly focuses the light reflected by the object on the sensor. The distance from the camera to the powder spreading surface of this system is 215 mm, thus the magnification is $\beta = 6.6/120\sqrt{2} = 0.0389$, and the focal length of the objective lens is $215/(1 + 1/\beta) = 8.05$ mm. Therefore, we chose the FL-CC0814A-2M Ricoh 8 mm lens with format size 2/3″, focal length 8 mm, iris range 1.4–16, and minimum object distance 0.1 mm.

In this work, the strip Light-Emitting Diode (LED) light is used as the light source of the detection system. The installation position of the strip LED light source is shown in Figure 2.

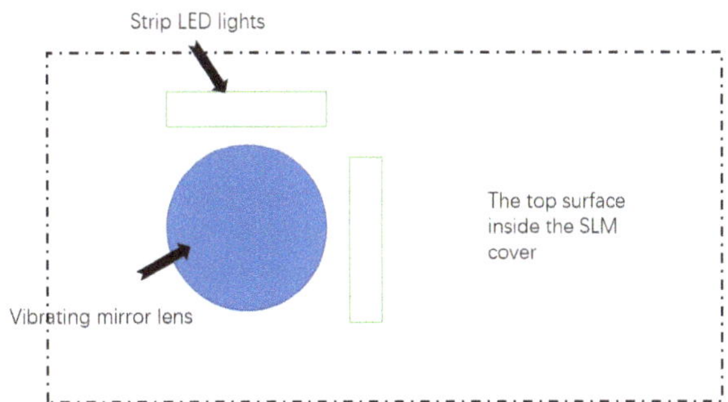

Figure 2. Light-Emitting Diode (LED) light installation position.

2.2. SLM Powder Spreading Defect Type

In the SLM powder spreading process, defects are easily formed on the powder coating surface due to improper selection of process parameters or wear of the scraper strips. Several common powder spreading defects mainly include the following.

(1) Cladding layer increased. Due to the phenomenon of spheroidization, warpage, and incomplete powder melting during the molding process of the selected laser melting equipment, it is difficult to control the molding process. The spheroidization phenomenon

occurs when the surface of the metal solution turns into a spherical surface under the action of interfacial tension between the molten metal and the surrounding powder particles. Therefore, spheroidization often causes unevenness on the surface of the part, which leads to the phenomenon of local high. In the process of powder spreading, it is difficult for the metal powder to cover the local high area, as shown in Figure 3 (1).

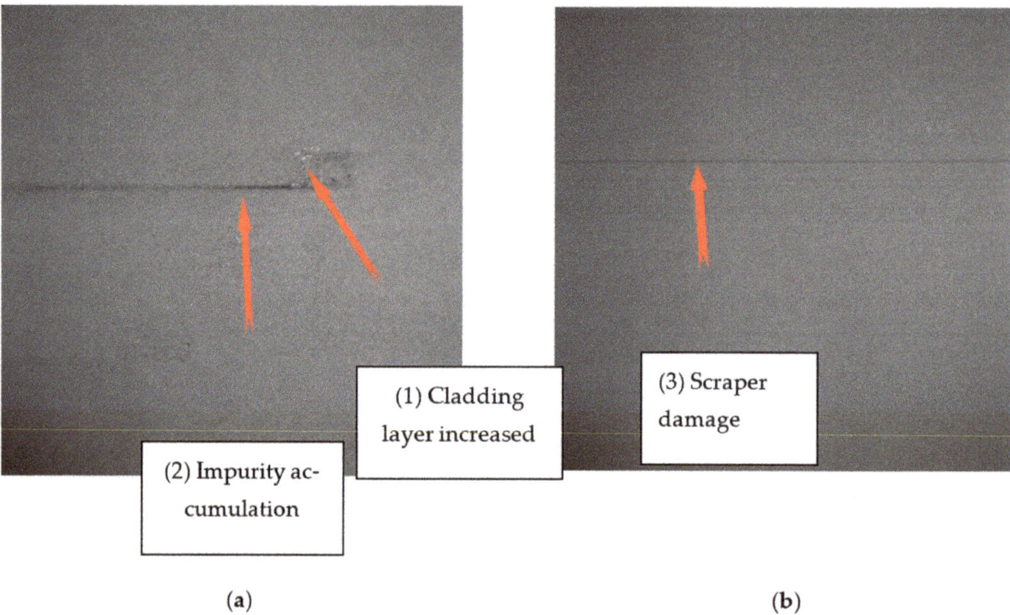

Figure 3. SLM powder spreading defect: (**a**) Defects of cladding (1) and impurity accumulation (2). (**b**) Scraper damage defects (3).

(2) Impurity accumulation. Due to the excessive energy of the molten pool, the liquid metal evaporates to produce black smoke impurities. During the laser scanning process, due to the influence of the laser scanning direction and the laser scanning speed, part of the black smoke impurities falling on the metal surface cannot be brought into the airflow generated by the fan into the gas circulation system, and thus black smoke impurities accumulate. When the powder spreading device is activated, the scraper will drive the impurities and form this kind of defect, as shown in Figure 3 (2).

(3) Scraper damage. Due to the friction of the scraper with the higher part of the cladding layer for a long time, the surface of the scraper is partially missing. When the powder spreading device is activated, part of the metal powder driven by the scraper will leak out from the position where the scraper strip is missing, forming a scraper damage defect, as shown in Figure 3 (3).

As impurity accumulation and scraper damage appear as stripes on the image, these two defects are collectively referred to as strip defects (3).

2.3. Defect Overlap

For the overlap of defects, there is only the probability that the scratch damage and impurities will overlap in the above defects, and the impurity defects will be disposed of in the first powder laying, while the scratch defects will not. The cladding layer is raised, because the detection threshold is different, it is not necessary to classify the calculation together, so defect overlap has no effect on the overall detection.

3. Image Processing Method

3.1. Perspective Transformation

In this paper, the industrial camera mainly adopts the side-axis method, so the collected powder spreading images have some linear distortion. Linear distortion can easily cause distortion of the powder spreading image and increase the difficulty of the defect detection, so corresponding measures should be taken to correct it.

The process of correction is mainly to project each pixel of the image to a new plane at one time. The specific equation is as follows:

$$\begin{bmatrix} u_{corr} \\ v_{corr} \\ z_{corr} \end{bmatrix} = \begin{bmatrix} h_{11} & h_{12} & h_{13} \\ h_{21} & h_{22} & h_{23} \\ h_{31} & h_{32} & h_{33} \end{bmatrix} \cdot \begin{bmatrix} u \\ v \\ 1 \end{bmatrix} \tag{1}$$

In the equation, (u, v) is the original coordinates of P, $(u_{corr}, v_{corr}, z_{corr})$ is the coordinates of P_{corr}, $(h_{11}, h_{12}; h_{21}, h_{22})$ is the linear transformation matrix, $(h_{13}; h_{23})$ is the translation matrix, (h_{31}, h_{32}) is the perspective matrix, because the final corrected image is located in the two-dimensional plane, so the coordinates of P_{corr} are divided by z_{corr}, and the transformation equation is as follows:

$$\begin{cases} u'_{corr} = \frac{h_{11}u + h_{12}v + h_{13}}{h_{31}u + h_{32}v + 1} \\ v'_{corr} = \frac{h_{21}u + h_{22}v + h_{23}}{h_{31}u + h_{32}v + 1} \\ z'_{corr} = 1 \end{cases} \tag{2}$$

In the equation, $(u'_{corr}, v'_{corr}, 1)$ is the final corrected coordinate.

It is not difficult to see from Equation (2) that only four points are needed to list eight equations and find eight unknown quantities of the correction transformation matrix. The perspective transformation effect is shown in Figure 4.

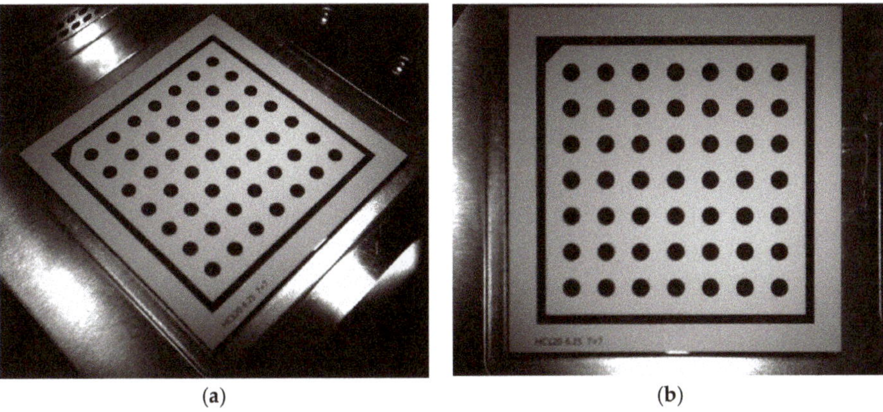

(a)   (b)

Figure 4. The perspective transformation effect: (**a**) Perspective transformation before. (**b**) After the perspective transformation.

3.2. The Steps of Defect Extraction

This article mainly describes the increased defects in the cladding layer and the stripe defects, the specific steps can be divided into three steps.

Step 1: Eliminate the effects of light.

It can be seen from Figure 3 that the image of the powdered surface is affected by the LED light, resulting in uneven illumination of the image, so measures should be taken to eliminate the influence of illumination and avoid increasing the difficulty of image processing in the later stage. Specific steps to eliminate the influence of light:

(1) Calculate the average gray value of the entire image.
(2) Perform average filter on the image, mainly using a 200×200 filter template.
(3) Subtract the original image from the average filtered image and add the gray average value of the entire image.

The above steps can be expressed as

$$G_{\text{out}} = G_{\text{in}} - G_{mean} + \text{v} \tag{3}$$

In the equation, G_{out} is the output image, G_{in} is the input image, G_{mean} is the average filtered image, and v is the average of the entire image.

For example, enter G_{in} a 5 by 5 random pixel value matrix.

$$G_{\text{in}} = \begin{bmatrix} 1 & 2 & 1 & 4 & 3 \\ 1 & 2 & 2 & 3 & 4 \\ 5 & 7 & 6 & 8 & 9 \\ 5 & 7 & 6 & 8 & 8 \\ 5 & 6 & 7 & 8 & 9 \end{bmatrix} \tag{4}$$

First, take the average of the graphs, and the formula is

$$\text{v} = \frac{\sum\limits_{i=1}^{n} x_i}{n} \tag{5}$$

In the equation, n is the total number of pixels in the image, and x_i is the original pixel value. It is concluded that v = 5 (integer).

Then, average filtering means that the pixel at the center of the graph is the average value of all pixel values, and the formula is

$$G_{\text{mean}} = g(x,y) = \frac{1}{M}\sum f(x,y) \tag{6}$$

In the equation, $g(x,y)$ is the pixel value after filtering, $f(x,y)$ is the pixel value of the original image, and M is the square matrix selected during filtering (generally 3×3).

$$f(x,y) = \begin{bmatrix} 1 & 2 & 1 & 4 & 3 \\ 1 & 2 & 2 & 3 & 4 \\ 5 & 7 & 6 & 8 & 9 \\ 5 & 7 & 6 & 8 & 8 \\ 5 & 6 & 7 & 8 & 9 \end{bmatrix} \Rightarrow g(x,y) = \begin{bmatrix} 1 & 2 & 1 & 4 & 3 \\ 1 & 3 & 4 & 4 & 4 \\ 5 & 4 & 5 & 6 & 9 \\ 5 & 6 & 7 & 8 & 8 \\ 5 & 6 & 7 & 8 & 9 \end{bmatrix} \tag{7}$$

Finally, according to Equation (3), subtract $f(x,y)$ from $g(x,y)$ and add v to get G_{out}.

$$G_{out} = \begin{bmatrix} 5 & 5 & 5 & 5 & 5 \\ 5 & 4 & 3 & 4 & 5 \\ 5 & 8 & 6 & 7 & 5 \\ 5 & 6 & 4 & 5 & 5 \\ 5 & 5 & 5 & 5 & 5 \end{bmatrix} \tag{8}$$

Step 2: Threshold segmentation.

It can be seen from Figure 3 that the brightness of the two defects is inconsistent. The defects increased by the cladding layer appear brighter in the image and the stripe defects appear darker in the image. Therefore, a high threshold is set to segment the lighter part of the paving defect, and a low threshold is set to segment the darker part of the laying defect. The threshold segmentation equation is shown as below:

$$G'(i,j) = \begin{cases} 255 & G(i,j) \geq H \ or \ G\ (i,j) \leq L \\ 0 & L < G(i,j) < H \end{cases} \tag{9}$$

In the equation, $G'(i,j)$ is the image after threshold division, $G(i,j)$ is the image after uneven illumination processing, H is the high threshold, and L is the low threshold.

In order to select the appropriate threshold under different lighting environments, this paper sets the high threshold as $u + k_1\sigma$ and the low threshold as $u - k_2\sigma$, where u is the gray average value of the image, σ is the gray standard deviation, and k is the amplification factor.

Step 3: Image denoising.

Because SLM equipment is printed with fine metal powder, the metal powder laid on the powder bed is in a loose state, so the reflection state to the light is different, which will lead to a lot of noise in the powder image. In order to reduce the difficulty of image processing, it is necessary to denoise the powdering image. The image denoising methods used in this paper mainly include morphological filter and connected domain filter. Small noise can be removed by the open operation of morphological filter, and large noise can be removed by connected domain filter.

For the defects increased in the cladding layer, the open operation of image morphology processing is carried out for denoising, that is necessary to use 3×3 structural elements to perform the opening operation, so that the pixel area containing 3×3 can be retained, and finally the area smaller than 30 pixels is deleted through the connected domain filter to extract this defect.

For the stripe defects, it is necessary to perform open operations on the image using 3×3 structural elements and 1×5 structural elements, and then merge the two open operation images, and then use 2×20 structural elements to perform expansion operations. Finally, the region with area less than 1500 pixels or width less than 150 pixels should be deleted through connected domain filter, so that the defect can be extracted.

The flow chart of powder defect extraction is shown in Figure 5. Taking Figure 3 as an example, the defect extraction effect is shown in Figure 6.

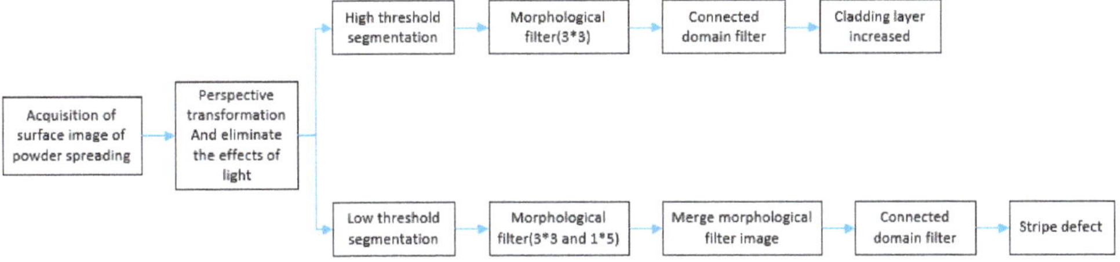

Figure 5. The flow chart of powder defect extraction.

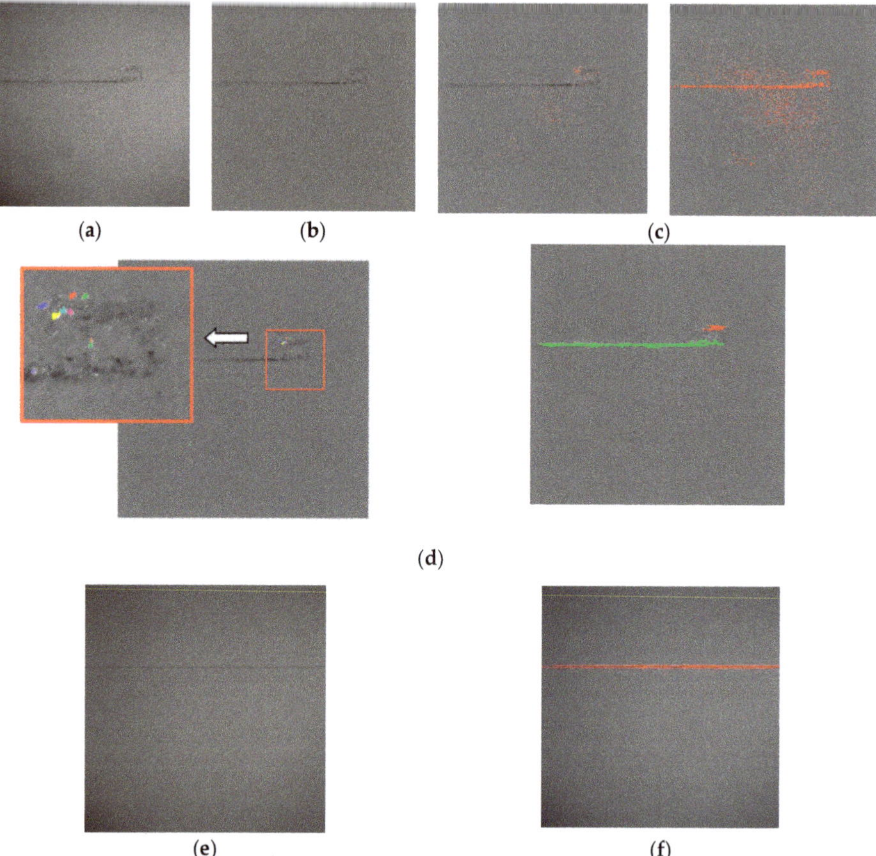

Figure 6. The defect extraction effect: (**a**) Original image. (**b**) Step 1. (**c**) Step 2: The image on the left is high threshold, and the image on the right is low threshold. (**d**) Step 3: The image on the left is cladding defect, and the image on the right is stripe defect. (**e**) Original image. (**f**) The extraction results.

4. Stripe Defect Classification

As the stripe defects include impurity defects and scraper damage defects, these two types of defects need to be classified. The Halcon machine vision software package provides two classifiers with fast detection speed: the multilayer perceptron and support vector machine. The two classifiers will be analyzed below.

4.1. The Multilayer Perceptron (MLP)

The multilayer perceptron (MLP) classifier is mainly based on an artificial neural network and is composed of multiple single-layer perceptron. Generally, the backpropagation algorithm is used to train the neural network. The classifier has excellent nonlinear mapping ability, adaptive ability, generalization ability, and fault tolerance ability, but the convergence speed of the whole algorithm is slow, leading to a long training time.

In this paper, the minimum pixel value, pixel standard deviation, defect height, and defect roundness value of stripe defects are used as feature vectors. Therefore, the number of neurons in the input layer of the multilayer perceptron network structure is set to four, and the number of neurons in the output layer is set to two. The maximum-minimum method is used for data normalization, and the excitation function is selected as softmax. At present, there is no unified theoretical basis for determining the optimal solution for the number of hidden layer neurons in a multilayer perceptron. In this paper, the number of neurons in the hidden layer is determined as two by empirical equation five.

$$H = \log_2 n \tag{10}$$

In the equation, H is the number of hidden layer neurons and n is the number of input layer neurons.

In order to verify the classification effect of the multilayer perceptron, 40 images of impurity defects and 40 images of scraper damage defects were used as training samples to train MLP. After the training, stripe defects were classified. The test samples contained 20 impurity defect samples and 20 scraped damage defect samples. At the same time, in order to verify the effect of defect-free and cladding layer increased defects, another 20 qualified samples and 20 cladding layer increased defect samples were added. The confusion matrix of MLP classification results are shown in Table 1.

Table 1. The confusion matrix of the MLP classification results.

Actual Class	Predicted Class			
	Qualified	Cladding Layer Increased Defects	Impurity Defects	Scraper Damage Defects
Qualified	19	1	0	0
cladding layer increased defects	0	20	0	0
Impurity defects	0	0	39	1
Scraper damage defects	0	0	0	40

4.2. Support Vector Machine (SVM)

Support vector machine (SVM) classifier using structural risk minimization principle can effectively solve the problem of small batch sample of learning, The idea is to find a partition hyperplane that can distinguish different samples in the training sample space. For nonlinear data, it can be mapped into a high-dimensional space through a kernel function, and finally find the support vector and seek the maximum interval hyperplane. The final decision function of the classifier is only determined by the key support vectors. To a certain extent, it can effectively avoid the dimensional disaster and the classification speed is fast, but the effect of SVM classifier for large-scale training samples will be greatly reduced.

The SVM is trained with the same feature vector and training samples, and the same test samples are used for classification. The confusion matrix of SVM classification results are shown in Table 2.

Table 2. The confusion matrix of the SVM classification results.

Actual Class	Predicted Class			
	Qualified	Cladding Layer Increased Defects	Impurity Defects	Scraper Damage Defects
Qualified	19	1	0	0
cladding layer increased defects	0	20	0	0
Impurity defects	0	0	38	2
Scraper damage defects	0	0	0	40

5. Comparison with Other Methods

5.1. Photodiode

Photodiode is a kind of photoelectric sensor that can convert an optical signal into an electrical signal [21]. It can output the corresponding analog electrical signal according to the illumination of the received light or realize the switch between different states in the digital circuit. The precision of photodiode data is related to the distance between the pool and the diode, incident angle, etc. Moreover, the higher the laser power is, the larger the fluctuation range of photodiode signal value is, and the worse the stability of the pool. Compared with photodiode, our method does not need to strictly adjust the distance angle, only needs to display the molding process in the camera, and is not affected by the laser power [22].

5.2. X-ray

X-ray can intuitively reflect the three-dimensional morphology and position of internal defects. Hu et al. [23] characterized the porosity and poor fusion defects by X-ray computer, predicted the fatigue life by combining with the fatigue crack propagation model, and effectively judged the fatigue threat level caused by different positions of defects. Although X-ray can directly monitor the morphology and location of internal defects in real-time, X-ray monitoring has a high cost, which is harmful to the human body and requires additional protection [24]. In particular, the cost of high-speed and high-resolution X-ray in situ observation is higher, which is generally used for checking and verifying other monitoring methods. In contrast, the use of industrial cameras has a lower cost.

5.3. Thermal Signal

Heat transfer is a driving force of the SLM, including the formation and dynamic behavior of molten pool, the liquid metal cooling and solidification, solidification layer of thermal cycle, microstructure, residual stress, and deformation of component has a direct impact, such as uniform temperature distribution form good quality components, unreasonable temperature distribution will affect the structural integrity and quality. However, at present, the difficulty of temperature detection is that the emissivity of the material is difficult to obtain, and in the SLM process, the form of the material includes the powder state, liquid state, solid state, and gas state, and the emissivity is also changing due to the difference of temperature at different positions. Therefore, it is very difficult to obtain the emissivity in SLM process [25].

5.4. Vibration Signal

The vibration signal in the SLM process can also reflect the processing state and component quality, such as penetration depth, crack, powder laying quality, etc. However, the research on SLM process monitoring based on vibration signals is few at present, and the selection, arrangement, signal collection, and processing of sensors need to be further studied [25].

6. Discussion

From Figures 3 and 4, it is not difficult to calculate that the recognition rates of MLP and SVM are 98.33% and 97.5%, respectively. Both have high recognition rate, but MLP recognition rate is better than SVM.

In addition, the classification time will increase the forming time of SLM equipment and reduce the forming efficiency. Therefore, the classification time of the two classifiers should be discussed; Figure 7 is the classification time comparison diagram of MLP and SVM. As you can see from Figure 7, the MLP classification time is shorter.

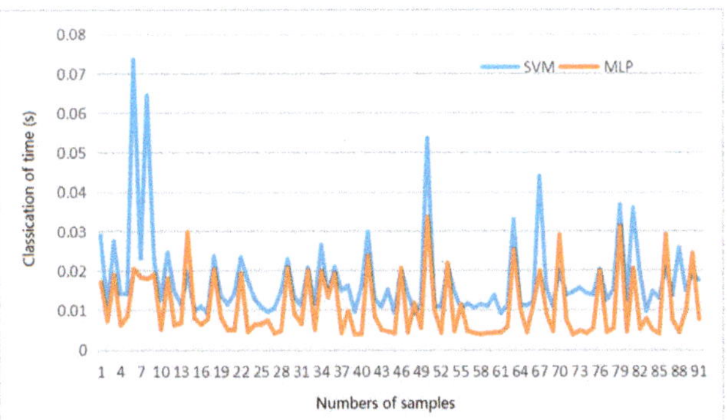

Figure 7. Classification time of MLP and SVM.

In summary, MLP has a better classification performance than SVM, so MLP is chosen to classify stripe defects.

7. Conclusions

This paper presents a method for detecting powder spreading defects and monitoring the powder spreading surface as well as providing corresponding hardware construction measures and image processing methods. Using this method, we accurately established the relationship between image features and these three defects. This method first eliminates linear distortion through perspective transformation and uses Equation (3) to eliminate lighting impact, followed by selecting a high threshold and a low threshold to extract the defects increased by the cladding layer and the stripe defects with image noise, and then filtering the image noise through the morphological filter algorithm and the connected domain filter algorithm to obtain the accurate area of the powder spreading defects. Finally, a multilayer perceptron and a support vector machine were established to classify the stripe defects. The two classifiers were compared, and the multilayer perceptron with better performance was selected as the final classifier. The detection method proposed in this paper has high recognition rate, short detection time, and good stability. Subsequently, the corresponding feedback operations can be designed according to their corresponding defect types. The study on choosing type defect resolution and recognition has achieved good effect and can be immediately applied to the industrial technology of ascension, but due to the use of industrial camera, the image resolution limit is 0.5 mm, so for defects of smaller size, establishing the rate of detection and the relationship between the image characteristics and other defects are key problems to solve in the future. Through this research, we can apply machine vision to the laser cladding of diamond, monitor the formation of the graphite interface of the diamond–substrate interface, adjust the laser spot through feedback, prevent the diamond from graphitization, and better fix the diamond; machine vision can also be applied Based on laser metal cladding technology, monitoring the thickness of the coating, feedback and adjusting the parameters to obtain a thinner effective coating to protect the base metal, and so on. It can be seen that machine vision technology can be combined with various engineering and processing methods and has a wide range of applications in the industrial field.

Author Contributions: Conceptualization, J.Z. and X.G.; methodology, Z.L.; validation, T.P., Y.L. (Yiwen Lai) and W.Z.; formal analysis, Z.L.; investigation, Z.L.; resources, J.Z.; writing—original draft preparation, Z.L.; writing—review and editing, J.Z.; supervision, Y.L. (Yuangang Liu); project administration, X.G. All authors have read and agreed to the published version of the manuscript.

Funding: This research received no external funding.

Institutional Review Board Statement: Not applicable.

Informed Consent Statement: Not applicable.

Data Availability Statement: The raw data supporting the conclusions of this article will be made available by the authors, without undue reservation.

References

1. Guo, N.; Leu, M.C. Additive Manufacturing: Technology, Applications and Research Needs. *Front. Mech. Eng.* **2013**, *8*, 215–243. [CrossRef]
2. Yakout, M.; Elbestawi, M.; Veldhuis, S.; Nangle-Smith, S. Influence of Thermal Properties on Residual Stresses in SLM of Aerospace Alloys. *Rapid Prototyp. J.* **2020**, *26*, 213–222. [CrossRef]
3. Lu, P.; Wu, M.; Liu, X.; Duan, W.; Han, J. Study on Corrosion Resistance and Bio-Tribological Behavior of Porous Structure Based on the SLM Manufactured Medical Ti6Al4V. *Met. Mater. Int.* **2020**, *26*, 1182–1191. [CrossRef]
4. Malca, C.; Santos, C.; Sena, M.; Mateus, A. Development of SLM Cellular Structures for Injection Molds Manufacturing. *Sci. Technol. Mater.* **2018**, *30*, 13–22. [CrossRef]
5. Tucho, W.M.; Lysne, V.H.; Austbø, H.; Sjolyst-Kverneland, A.; Hansen, V. Investigation of Effects of Process Parameters on Microstructure and Hardness of SLM Manufactured SS316L. *J. Alloy. Compd.* **2018**, *740*, 910–925. [CrossRef]
6. Liu, S.; Yang, W.; Shi, X.; Li, B.; Duan, S.; Guo, H.; Guo, J. Influence of Laser Process Parameters on the Densification, Microstructure, and Mechanical Properties of a Selective Laser Melted AZ61 Magnesium Alloy. *J. Alloy. Compd.* **2019**, *808*, 151160. [CrossRef]
7. Giganto, S.; Zapico, P.; Castro-Sastre, Á.; Martínez-Pellitero, S.; Leo, P.; Perulli, P. Influence of the Scanning Strategy Parameters Upon the Quality of the SLM Parts. *Procedia Manuf.* **2019**, *41*, 698–705. [CrossRef]
8. Abe, F.; Santos, E.C.; Kitamura, Y.; Osakada, K.; Shiomi, M. Influence of Forming Conditions on the Titanium Model in Rapid Prototyping with the Selective Laser Melting Process. *Proc. Inst. Mech. Eng. Part C J. Mech. Eng. Sci.* **2003**, *217*, 119–126. [CrossRef]
9. Reinarz, B.; Witt, G. Process Monitoring in the Laser Beam Melting Process-Reduction of Process Breakdowns and Defective Parts. *Proc. Mater. Sci. Technol.* **2012**, *2012*, 9–15.
10. Craeghs, T.; Clijsters, S.; Yasa, E.; Kruth, J.-P. Online Quality Control of Selective Laser Melting. In Proceedings of the 20th Solid Freeform Fabrication (SFF) Symposium, Austin, TX, USA, 8–10 August 2011; pp. 212–226.
11. Clijsters, S.; Craeghs, T.; Buls, S.; Kempen, K.; Kruth, J.-P. In Situ Quality Control of the Selective Laser Melting Process Using a High-Speed, Real-Time Melt Pool Monitoring System. *Int. J. Adv. Manuf. Technol.* **2014**, *75*, 1089–1101. [CrossRef]
12. Foster, B.; Reutzel, E.; Nassar, A.; Hall, B.; Brown, S.; Dickman, C. Optical, Layerwise Monitoring of Powder Bed Fusion. In Proceedings of the Solid Freeform Fabrication Symposium, Austin, TX, USA, 1 August 2015; pp. 10–12.
13. Xiao, G.; Li, Y.; Xia, Q.; Cheng, X.; Chen, W. Research on the on-Line Dimensional Accuracy Measurement Method of Conical Spun Workpieces Based on Machine Vision Technology. *Measurement* **2019**, *148*, 106881. [CrossRef]
14. Hsu, Q.-C.; Ngo, N.-V.; Ni, R.-H. Development of a Faster Classification System for Metal Parts Using Machine Vision under Different Lighting Environments. *Int. J. Adv. Manuf. Technol.* **2019**, *100*, 3219–3235. [CrossRef]
15. Xiang, Z.; Zhang, J.; Hu, X. Vision-Based Portable Yarn Density Measure Method and System for Basic Single Color Woven Fabrics. *J. Text. Inst.* **2018**, *109*, 1543–1553. [CrossRef]
16. Carroll, J. Machine Vision System Detects Antenna Trace Defects on Automotive Glass. *Vis. Syst. Des.* **2019**, *24*, 16.
17. Giganto, S.; Zapico, P.; Castro-Sastre, Á.; Martínez-Pellitero, S.; Leo, P.; Perulli, P. Machine Vision and Background Re-Mover-Based Approach for PCB Solder Joints Inspection. *Int. J. Prod. Res.* **2006**, *45*, 451–464.
18. Greco, S.; Gutzeit, K.; Hotz, H.; Kirsch, B.; Aurich, J.C. Selective Laser Melting (SLM) of AISI 316L—Impact of Laser Power, Layer Thickness, and Hatch Spacing on Roughness, Density, and Microhardness at Constant Input Energy Density. *Int. J. Adv. Manuf. Technol.* **2020**, *108*, 1551–1562. [CrossRef]
19. Nguyen, Q.; Luu, D.; Nai, S.; Zhu, Z.; Chen, Z.; Wei, J. The Role of Powder Layer Thickness on the Quality of SLM Printed Parts. *Arch. Civ. Mech. Eng.* **2018**, *18*, 948–955. [CrossRef]
20. Savalani, M.M.; Pizarro, J.M. Effect of Preheat and Layer Thickness on Selective Laser Melting (SLM) of Magnesium. *Rapid Prototyp. J.* **2016**, *22*, 115–122. [CrossRef]
21. Wu, S.; Dou, W.; Yang, Y. Research progress of detection technology for laser selective melting metal additive manufacturing. *Precision Forming Eng.* **2019**, *11*, 37–50.
22. Zhang, K.; Liu, T.; Liao, W.; Zhang, C.; Du, D.; Zheng, Y. Photodiode Data Collection and Processing of Molten Pool of Alumina Parts Produced through Selective Laser Melting. *Optik* **2018**, *156*, 487–497. [CrossRef]
23. Hu, Y.N.; Wu, S.C.; Withers, P.J.; Zhang, J.; Bao, H.Y.X.; Fu, Y.N.; Kang, G.Z. The Effect of Manufacturing Defects on the Fatigue Life of Selective Laser Melted Ti-6Al-4V Structures. *Mater. Des.* **2020**, *192*, 108708. [CrossRef]
24. Leung, C.L.A.; Marussi, S.; Atwood, R.C.; Towrie, M.; Withers, P.J.; Lee, P.D. In Situ X-Ray Imaging of Defect and Molten Pool Dynamics in Laser Additive Manufacturing. *Nat. Commun.* **2018**, *9*, 1355. [CrossRef] [PubMed]
25. Cao, L.; Zhou, Q.; Han, Y.; Song, B.; Nie, Z.; Xiong, Y.; Xia, L. Review on Intelligent Monitoring and Process Control of Defects in Laser Selective Melting Additive Manufacturing. *Acta Aeronaut. Astronaut. Sin.* **2021**, 1–37.

 **_materials_**

Article

Novel Calibration Strategy for Validation of Finite Element Thermal Analysis of Selective Laser Melting Process Using Bayesian Optimization

Masahiro Kusano *, Houichi Kitano and Makoto Watanabe

Research Center for Structural Materials, National Institute for Materials Science (NIMS), Sengen 1-2-1, Tsukuba, Ibaraki 305-0047, Japan; KITANO.Houichi@nims.go.jp (H.K.); WATANABE.Makoto@nims.go.jp (M.W.)
* Correspondence: KUSANO.Masahiro@nims.go.jp

Abstract: Selective laser melting (SLM) produces a near-net-shaped product by scanning a concentrated high-power laser beam over a thin layer of metal powder to melt and solidify it. During the SLM process, the material temperature cyclically and sharply rises and falls. Thermal analyses using the finite element method help to understand such a complex thermal history to affect the microstructure, material properties, and performance. This paper proposes a novel calibration strategy for the heat source model to validate the thermal analysis. First, in-situ temperature measurement by high-speed thermography was conducted for the absorptivity calibration. Then, the accurate simulation error was defined by processing the cross-sectional bead shape images by the experimental observations and simulations. In order to minimize the error, the optimal shape parameters of the heat source model were efficiently found by using Bayesian optimization. Bayesian optimization allowed us to find the optimal parameters with an error of less than 4% within 50 iterations of the thermal simulations. It demonstrated that our novel calibration strategy with Bayesian optimization can be effective to improve the accuracy of predicting the temperature field during the SLM process and to save the computational costs for the heat source model optimization.

Keywords: selective laser melting; finite element thermal analysis; model calibration; thermographic image; bead on plate test; Bayesian optimization; nickel-based superalloy

Citation: Kusano, M.; Kitano, H.; Watanabe, M. Novel Calibration Strategy for Validation of Finite Element Thermal Analysis of Selective Laser Melting Process Using Bayesian Optimization. *Materials* **2021**, *14*, 4948. https://doi.org/10.3390/ma14174948

Academic Editor: Rubén Paz

Received: 20 July 2021
Accepted: 24 August 2021
Published: 30 August 2021

Publisher's Note: MDPI stays neutral with regard to jurisdictional claims in published maps and institutional affiliations.

1. Introduction

Selective laser melting (SLM), which is classified into a powder bed fusion type in additive manufacturing methods, produces a near-net-shaped product by scanning a concentrated high-power laser beam on a thin layer of metal powder to melt and solidify it. The cyclic process of spreading a thin layer of powder and scanning the laser beam is repeated until the product is completed in the form designed by computer-aided design (CAD) software. Through the process, the material temperature periodically and rapidly rises and falls with melting and solidification with transformations. Such a thermal history affects the characteristic microstructure [1,2] and material properties [3,4], and it can also generate defects such as pores [5] and microcracks [6–8], and can cause distortion [9].

To understand the thermal history (temperature gradient, solidification rate, and cooling rate) in the SLM process, thermal analysis has been extensively performed using the finite element method (FEM) [10–16]. The FEM-based simulation solves the governing equations of heat transfer with boundary conditions, including a moving heat source model for laser scanning. Such heat source models have been originally studied for many years for welding processes, and some volumetric heat sources of various shapes and distributions have been proposed [10,14,17]. A double ellipsoidal power density model, proposed by Goldak et al. [17], can simulate shallow penetration for arc welding processes and deeper penetration for laser and electron beam processes by changing shape parameters. Accurate thermal analysis requires calibration of shape parameters, depending on material and process conditions. This model has been also applied to simulate the SLM process. The shape parameters have been calibrated so that the simulated bead shape (width, depth, and length) matches the one by the cross-sectional observation [10–14] or by in-situ monitoring with high-speed imaging [13].

In addition to the shape parameters, absorptivity in the heat source models, which means the heat efficiency of laser power in the process, depends on the material and process conditions. In the SLM process, the laser beam is reflected multiple times in the powder layer [18,19]. It also interacts with keyhole walls, a deep surface in the melting pool owing to the vapor recoil pressure of excessive laser energy [20]. Boley et al. [18] studied the absorptivity of a 1-μm-wavelength laser beam on a metal powder layer by ray-tracing calculations, and empirically measured the value by using direct calorimetric measurements. They proposed that the absorptivity of the powder layer can be predicted from that of the flat surface of its bulk metal. Trapp et al. [20] studied the effective absorptivity of a continuous 1070-nm-wavelength laser beam for some metal powders during melting. The in-situ direct calorimetric measurements revealed that the absorptivity considerably depended on laser power and scan speed (e.g., the absorptivity of 316 L stainless steel varied from 0.3 to 0.7 with the variation of laser power and scan speed). They have concluded that the measured absorptivities were different from both the one obtained by the powder layer measurements without melting and that of liquid metals estimated from the literature. Thus, the calibration of absorptivity is still necessary for the valid thermal analysis for the SLM process. However, to the best of our knowledge, no such thermal analysis using a heat source with calibrated absorptivity and shape parameters has been reported. This could be owing to the experimental difficulty of in-situ temperature measurements during the process.

Such thermal simulations are computationally expensive, and calibration is an iterative simulation process that uses different parameters to fit the simulation results with the experimental ones. Therefore, it is important to improve the efficiency of the calibration process. In short, the calibration process finds the parameters (explanatory variables x) to minimize the difference between experiment and simulation (an objective function $f(x)$). Bayesian optimization is known as a powerful tool to efficiently find minimum $f(x)$ [21,22]. A Gaussian process model of $f(x)$ is modified by a Bayesian update procedure at each new evaluation of $f(x)$. One of the features of Bayesian optimization is the use of the acquisition function to determine the next point x to be evaluated. The acquisition function can balance between sampling at the points where the modeled objective function is low and exploring the areas that have not yet been modeled, resulting in efficient parameter optimization. For example, using Bayesian optimization, Kitano et al. [7,8] have successfully found an optimal laser irradiation conditions to avoid defects and cracks in SLMed nickel superalloys. Shiraiwa et al. [23] efficiently determined the heat source parameters in welding simulation using Bayesian optimization. In order to evaluate the accuracy of the calculation results and to improve it, the SLM process simulation also needs to be iterated many times to find the optimal heat source model parameters. Thus, Bayesian optimization could be an effective technique for the calibration of a heat source model for the SLM process simulation.

This paper proposes a novel calibration strategy for the heat source model of the FEM-based thermal analysis in the SLM process. The in-situ temperature measurements were conducted using high-speed thermography to calibrate the absorptivity. The cross-sectional bead shapes were quantified to calibrate the shape parameters of the heat source model. The shape parameters were explored using Bayesian optimization, which led to the smallest difference of the melt pool shape between the FEM calculations and the experimental measurements. The optimal parameters were compared with the ones of the grid search. These iterative calibration processes were performed via two-dimensional (2D) thermal simulations for computational efficiency. Three-dimensional (3D) thermal analysis with calibrated parameters was also performed and discussed in terms of bead shape, thermal history, and cooling rate.

2. Experiments

A two-step calibration strategy for the heat source model of FEM-based thermal analysis is shown as a flowchart in Figure 1. First, the absorptivity of the heat source model was calibrated to fit the temperature field by the simulation of multiple beads to the ones measured by thermography (Section 2.2). Second, Bayesian optimization offered better shape parameters for the heat source model, minimizing the difference between single bead simulation and cross-sectional observation results (Section 2.3). Thermal simulation models and Bayesian optimization are described in Sections 3 and 4, respectively.

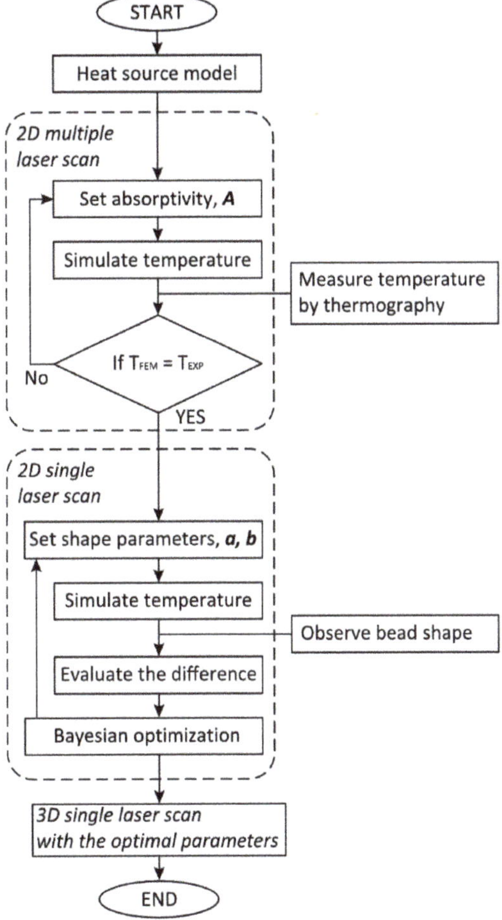

Figure 1. Flowchart of two-step of calibration strategy for the heat source model of FEM-based thermal analysis.

2.1. SLM Machine and Materials

Multiple and single laser scan tests were conducted using a commercial SLM machine, SLM 280 HL (SLM Solutions GmbH, Lübeck, Germany, Figure 2a). A thin layer of metal power on a base plate was scanned by a fiber laser beam with 1070-μm wavelength and 80-μm spot size in the chamber under an argon atmosphere. The laser power, scan velocity, and hatching space were set to 300 W, 1500 mm/s, and 100 μm, respectively. Nickel-based superalloy Hastelloy X (HX) powder (AMPERPRINT 0228 by Höganäs AB, Scania, Sweden) and 6-mm-thick baseplates (Alloy X by VDM Metals GmbH, Werdohl, Germany) were used in this study.

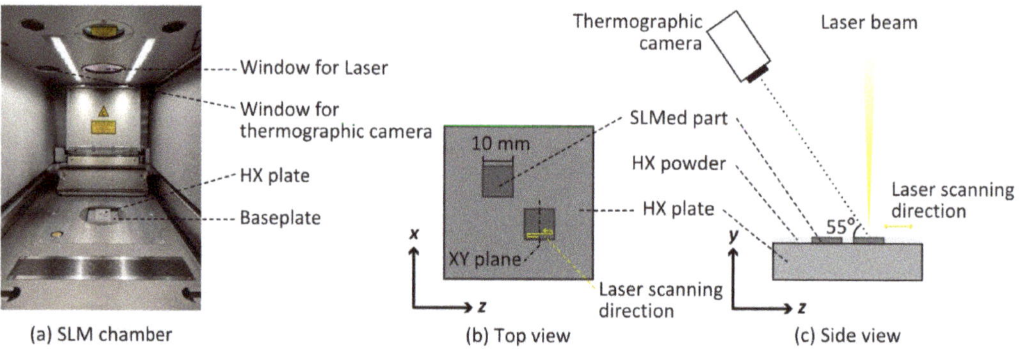

Figure 2. (**a**) A photo of the selective laser melting (SLM) chamber. Schematic images of in-situ temperature measurement of Hastelloy X (HX) by thermography: (**b**) top view and (**c**) side view.

2.2. Thermographic Measurements

The temperature distribution due to multiple laser scans was measured by a high-speed thermography camera (FAST M350, Telops Inc., Quebec, Canada) through the silicon window of the chamber. The schematic image of the experimental setup is shown in Figure 2b,c. Two rectangular components with a dimension of 10 mm × 10 mm × 0.3 mm were prefabricated on the baseplate using SLM in advance. After that, when the top surface temperature was sufficiently lowered to ambient temperature, the platform with the substrate was lowered by 30 μm, and the additional powder layer was spread on the top surface. Then, the laser beam scanned in a zigzag pattern while the thermography camera measured the surface temperature.

The thermography camera has three filters to measure various temperature ranges: (1) 0–250 °C, (2) 106–350 °C, and (3) 265–780 °C. Thus, the measurements were performed for each filter. The sampling rate and exposure time of the camera were 345 Hz and 100 μs, respectively. In the preliminary experiment, the surface radiation of the SLMed HX part was separately calibrated by comparing the temperature measured by the camera with the one by a thermocouple in a furnace. The effects of the silicon window on the SLM machine were also calibrated in advance. Since the camera was mounted on the SLM machine at an angle of approximately 55°, the resolutions of images parallel and normal to the laser scan direction were 76 and 68 μm/pixel, respectively.

2.3. Bead on Plate Test

A single 10-mm-long bead was fabricated by scanning the laser beam on the powder layer (approximate 50 μm thick) on the baseplate. The single bead was cut perpendicular to the laser scanning direction, and the three cross sections were polished. Cross-sectional images of the backscattered electron composition (BEC) mode were taken via a scanning electron microscope (SEM: JSM-6010 LA, JEOL, Tokyo, Japan).

3. Thermal Analysis

The governing equation of heat transfer of an object with a density ρ, a specific capacity C, and a thermal conductivity k is as follows:

$$\rho C \frac{\partial T}{\partial t} = \nabla(k\nabla T) + Q \tag{1}$$

where T, t, and Q are temperature, time, and heat source, respectively. The double ellipsoidal power density model [17] was applied to model the heat source by laser beam q_l, power P, and scan velocity v in the z-direction, as described as follows:

$$q_l(x, y, z, t) = \begin{cases} \frac{6\sqrt{3}A_1 P}{abc_1 \pi \sqrt{\pi}} exp\left\{-3\left[\left(\frac{x}{a}\right)^2 + \left(\frac{y}{b}\right)^2 + \left(\frac{z-vt}{c_1}\right)^2\right]\right\} & (z \geq vt) \\ \frac{6\sqrt{3}A_2 P}{abc_2 \pi \sqrt{\pi}} exp\left\{-3\left[\left(\frac{x}{a}\right)^2 + \left(\frac{y}{b}\right)^2 + \left(\frac{z-vt}{c_2}\right)^2\right]\right\} & (z < vt) \end{cases} \tag{2}$$

where A_1 and A_2 are the absorptivity forward and backward of the laser beam. The parameters a, b, c_1, and c_2 determine the shape of the volumetric heat source. In this study, it is assumed that $A_1 = A_2 = A$, $c_1 = a$, and $c_2 = 2a$. These assumptions are common in the field of welding (e.g., reference [23]) and additive manufacturing (e.g., reference [16]).

Figure 3 shows the 2D models for the single and multiple laser scanning and the 3D model for the single laser scanning. The models consist of two components corresponding to a powder layer part of 30 μm thickness and a bulk part (gray and white elements in Figure 3, respectively). As shown in Figure 1, the absorptivity A was calibrated by comparing the temperature distributions measured by the thermography camera and simulated by the 2D multiple scan model (Figure 3b). The model of 1.0 mm × 6.3 mm corresponded to the center plane of the SLMed part and HX plate in Figure 2b. In order to realize the zig-zag pattern of the laser scan, the direction of the heat source model was alternately changed by 180° for every 10 mm of movement parallel to the Z-axis, and its center was shifted in the X direction by the same distance as the hatching space (100 μm). The simulations with different A were performed up to 0.5 s after the heat source had passed.

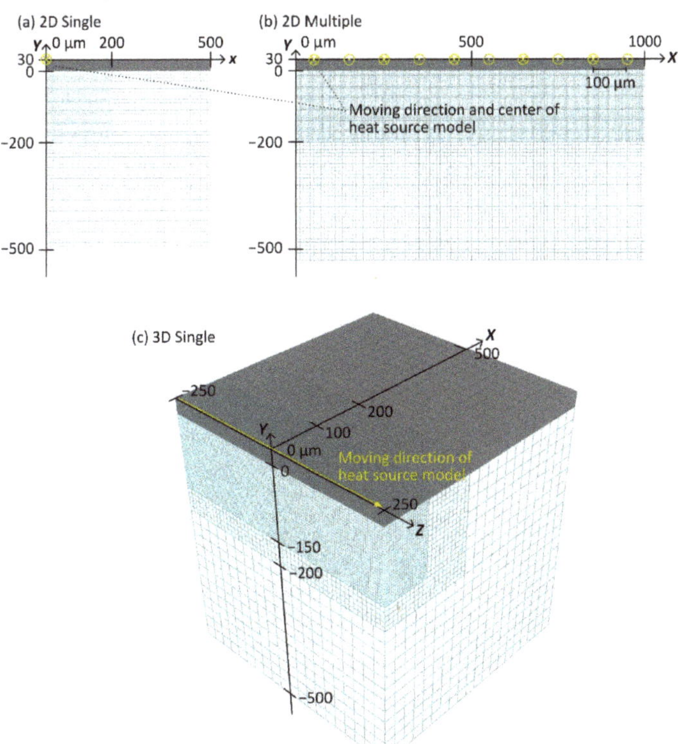

Figure 3. Thermal analysis models of (**a**) 2D single laser scan, (**b**) 2D multiple laser scan, and (**c**) 3D single laser scan.

The 2D single scan model of 500 μm × 530 μm (Figure 3a) was prepared to optimize the shape parameters a and b by comparing it with the single bead on plate test. The heat source model with the calibrated parameters was also applied to the 3D single scan model of 500 μm × 530 μm × 500 μm (Figure 3c). These simulations were performed up to 0.75 ms after the heat source passed through the model. In order to reduce the computational cost, the Y-axis for the 2D single scan model (Figure 3a) and the YZ plane for 3D single scan model (Figure 3c) were used as the symmetry wall, and the center of the heat source model passed through the point $(x, y) = (0, 30)$. The heat source model directions were indicated by symbols of ⊙ and ⊗ in Figure 3a,b, and by an arrow in Figure 3c. All boundaries except the top surface were assumed to be adiabatic. Heat loss due to natural convection and radiation, q_c and q_r on the top surface are defined as follows:

$$q_c = h_c(T - T_{amb}) \tag{3}$$

$$q_r = \sigma\varepsilon\left(T^4 - T_b^4\right) \tag{4}$$

where h_c, ε, σ, and T_{amb} are the convection coefficient, emissivity, Stefan-Boltzmann constant $(5.67 \times 10^{-8}$ (W/m^2·K^4)), and ambient temperature (30 °C), respectively.

The material properties used in the simulation are listed in Table 1. As shown in Figure 4, for the bulk part of HX, C_{solid} and k_{solid} depended on temperature [24]. For the powder part, the density ρ_{powder} and conductivity k_{powder} were assumed to be 0.5 ρ_{solid} and 0.05 k_{solid}, respectively. When the temperature exceeded the solidus temperature (T_s) of HX, the powder turned into the bulk by changing k_{powder} to k_{solid}. Additionally, when the temperature exceeded the liquidus temperature (T_l), the conductivity k_{liquid} was assumed to be 10 times larger than the one of the solid (10 $k_{solid,\ 1260\ °C} = 323$ W/(m·K)) to represent the heat transfer of the melting pool including the effect of fluidity. When the temperature was between T_s and T_l, the properties of k_{liquid} and C_{liquid} were defined as follows [25,26]:

$$\begin{cases} k = k_{solid} + \left(k_{liquid} - k_{solid}\right)\phi \\ C = C_{solid} + (C_{liquid} - C_{solid})\phi + L\frac{d\phi}{dT} \end{cases} \tag{5}$$

where ϕ represents a phase fraction defined as $(T - T_s)/(T_l - T_s)$.

Table 1. Material properties of Hastelloy X (HX) used for thermal analysis. Adapted from Ref. [24].

Material Properties	Value
Density for solid, ρ_{solid}	8240 kg/m^3
Density for powder, ρ_{powder}	4120 kg/m^3
Heat capacity for liquid, C_{liquid}	674 J/kg K
Thermal conductivity for liquid, k_{liquid}	323 W/m K
Latent heat, L	276 kJ/kg
Solidus temperature, T_s	1260 °C
Liquidus temperature, T_l	1355 °C
Heat transfer coefficient, h_c	10 W/m^2 K
Emissivity, ε	0.3

The thermal simulation was performed using the finite element code ABAQUS (ABAQUS/CAE 2019, Dassault Systems Simulia Corp., Johnston, RI, USA) with the user subroutines of the temperature and phase-dependent material properties and the moving volumetric heat source model. The 2D and 3D simulation used 4-node linear heat transfer quadrilateral and 8-node linear heat transfer brick elements (DC2D4 and DC3D8 in ABAQUS 2019), respectively. As shown in Figure 3, the mesh size was set to 5.0 μm for the part where large heat input was given by the moving heat source model, and larger sizes for other parts to save computational costs.

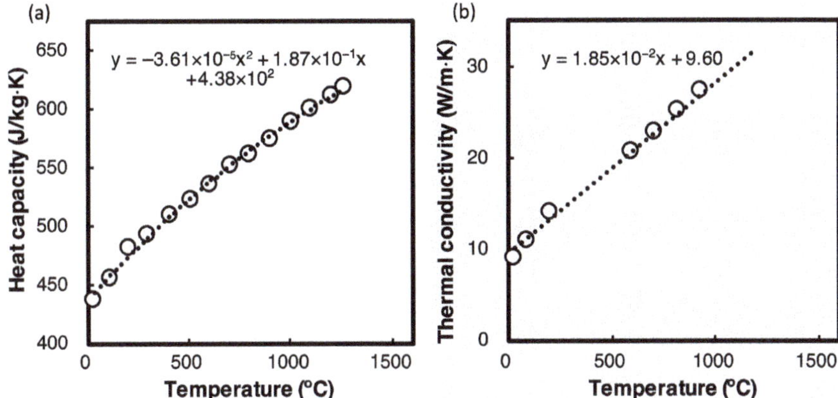

Figure 4. (**a**) Heat capacity and (**b**) thermal conductivity of HX (depending on temperature). Adapted from Ref. [24].

4. Bayesian Optimization

The shape parameters a and b of the heat source model (Equation (2)) were calibrated by comparing the shape of beads by the observation (Section 2.3) and the thermal simulation (Section 3). Here, the error between the observation and the simulation was evaluated by the following procedure (see Figure 5). First, the area containing the right half of the melted zone in the SEM image (100 μm × 150 μm, surrounded by the yellow dashed rectangle in Figure 5a) was trimmed and manually binarized so that the grayscale of the pixels was set to 1 for the melted zone and 0 for the baseplate (Figure 5b). Second, as shown in Figure 5c, the binary image was divided into blocks to match the pixel size and simulation mesh size (coarse-graining). Several coarse-grained images were prepared for different cross sections and averaged into a single image (Figure 5d). This image is the visualized matrix of the melted zone $MZ_{obs}(x, y)$, and its components range is from 0 to 1. Here, x and y are the coordinates in the width and depth directions of the melted zone, respectively. In the finite element simulation, the melted zone was defined as the part whose temperature exceeded T_s during laser scanning. As shown in Figure 5e, the melted zone was also outputted as binary values for each integral point of the elements in the corresponding area to the SEM image (100 μm × 150 μm), and the average values of four integral points in the elements were used to obtain the matrix of the melted zone $MZ_{sim}(x, y)$. Subtracting the simulated matrix $MZ_{sim}(x, y)$ (Figure 5e) from the observed matrix $MZ_{obs}(x, y)$ (Figure 5d) yields a different matrix $(diff(x, y))$ or an image, as shown in Figure 5f. Warm colors indicate that the simulated melt was smaller than the observed bead shape, and cold colors mean the opposite. Finally, the simulation error E was defined as follows:

$$E = \frac{\sum |MZ_{obs}(x, y) - MZ_{sim}(x, y)|}{\sum MZ_{obs}(x, y)} = \frac{\sum |diff(x, y)|}{\sum MZ_{obs}(x, y)} \tag{6}$$

The optimal shape parameters to minimize E were found by Bayesian optimization by using Statistics and Machine Learning Toolbox in MATLAB (R2020b, MathWorks, Natick, MA, USA). Bayesian optimization uses a Gaussian process model to find the point x that minimizes the objective function $y = f(x)$. The Gaussian process model of $f(x)$ is modified by Bayesian updates each time a new point is acquired. The Gaussian process is defined as mean $\mu(x; \theta)$ and covariance kernel function $k(x, x'; \theta)$. Here, θ is a vector of kernel parameters. The kernel function $k(x, x'; \theta)$ can significantly affect the quality of a Gaussian process regression, and the Bayesian optimization in the MATLAB toolbox uses the automatic relevance determination (ARD) Matérn 5/2 kernel, $K_{M52}(x, x')$ [22]:

$$K_{M52}(x, x') = \theta_0 \left(1 + \sqrt{5}r + \frac{5}{3}r^2 \right) exp\left(-\sqrt{5}r \right) \tag{7}$$

$$r^2 = \sum_{d=1}^{D} \frac{(x_d - x'_d)}{\theta_d^2} \tag{8}$$

where hyperparameters θ_0 and $\theta_{1:D}$ are the covariance amplitude and the length scales, respectively.

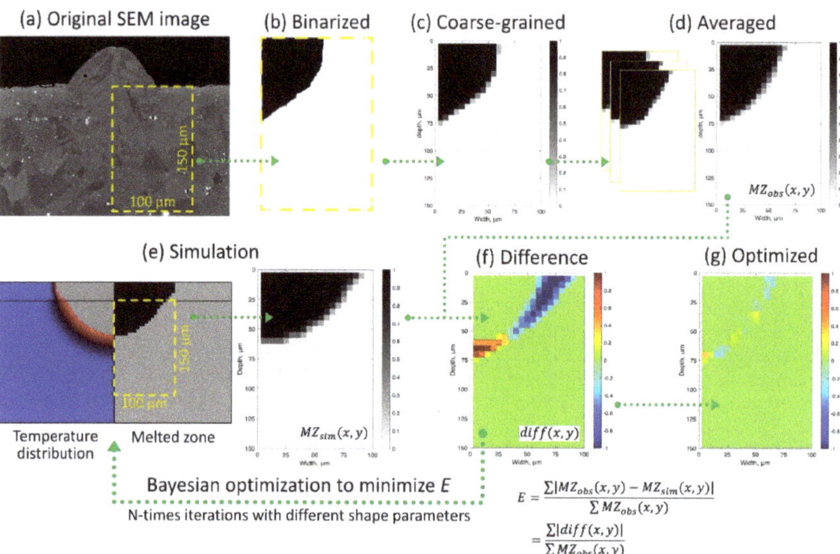

$$E = \frac{\sum |MZ_{obs}(x,y) - MZ_{sim}(x,y)|}{\sum MZ_{obs}(x,y)} = \frac{\sum |diff(x,y)|}{\sum MZ_{obs}(x,y)}$$

Figure 5. The procedure to evaluate the simulation error E. (**a**) Cross-sectional SEM image of the single bead with laser scanning, (**b**) binary image, and (**c**) coarse-grained image of the bead. (**d**) The visualization of the melted zone matrix $MZ_{obs}(x,y)$ averaged from three different cross sections. (**e**) The simulated melted zone $MZ_{sim}(x,y)$ (**right**) defined as the part exceeded T_s during the thermal simulation (**left**). (**f**) Difference in bead shapes between the observation and simulation to evaluate the error E. (**g**) The optimized difference by iterative simulations with different parameters to minimize E with Bayesian optimization. The Black and white pixels with binary values correspond to the bead and baseplate, respectively.

The next evaluation point is determined by the acquisition function $a(x)$ that evaluates the "goodness" of the point x based on the posterior distribution $Q(f|x_i, y_i \ for \ i = 1, \dots, t)$ of the Gaussian process model. The acquisition function can balance sampling points around the low objective function (exploitation) and searching for the regions that have not yet been fully explored (exploration). Thus, the efficiency of parameter retrieval by Bayesian optimization depends on the acquisition functions. In this study, three acquisition functions were examined: (1) expected improvement (EI), (2) probability of improvement (PI), and (3) lower confidence bound (LCB) [21,22,27]. The EI acquisition function evaluates the expected amount of improvement in the objective function, ignoring values that increase the objective function. EI is defined as follows:

$$EI = E\left[max\left(0, \mu_Q(x_{best})\right) - f(x)\right] \tag{9}$$

where $E(x)$ indicates the expected value of x, x_{best} is the lowest posterior mean point, and $\mu_Q(x_{best})$ is the lowest value of the posterior mean. The PI acquisition function calculates the probability that the new point x leads to a better objective function value, modified by the margin parameter m which takes the estimated noise standard deviation of the Gaussian process model:

$$PI = CDF\left(\frac{\mu_Q(x_{best}) - m - \mu_Q(x)}{\sigma_Q(x)}\right) \tag{10}$$

where CDF indicates the cumulative distribution function of the standard normal distribution, and $\sigma_Q(x)$ is the standard deviation. The LCB function minimizes the curve of $2\sigma_Q(x)$ at each point:

$$LCB = 2\sigma_Q(x) - \mu_Q(x) \tag{11}$$

In this study, the search point x corresponds to the heat source shape parameters a and b, and the objective function $y = f(x)$ is defined as the difference in bead shapes between simulations and experiments. For further details, see the literature [21,22,27]. The search areas of a and b were (10, 100 µm) and (50, 200 µm), respectively. The single bead simulation and newly selected parameters were iterated 50 times. A grid search was also performed for the same area of shape parameters for comparison. The grid sizes of both parameters a and b were 5.0 µm.

5. Results

5.1. Absorptivity Calibration

Figure 6a–c show the temperature distributions during multiple laser scans measured by the thermography camera with different filters. These images show the temperature distribution when the laser beam scans the middle of the sample. Point C in the figures indicates the sample center point. Points A and B are 1.6 and 3.2 mm away in the X-direction from point C, respectively. As shown in Figure 6d, the temperature histories with different filters were integrated into one for each point. The laser beam scanned through the sample surface (10 mm × 10 mm) at the speed of 1500 mm/s and the hatching space of 100 µm, so there was a time lag of approximately 0.1 s between the three curves. When the peak positions are aligned as shown in Figure 6e, the three curves are in good agreement, indicating that the thermal histories at these points are almost the same.

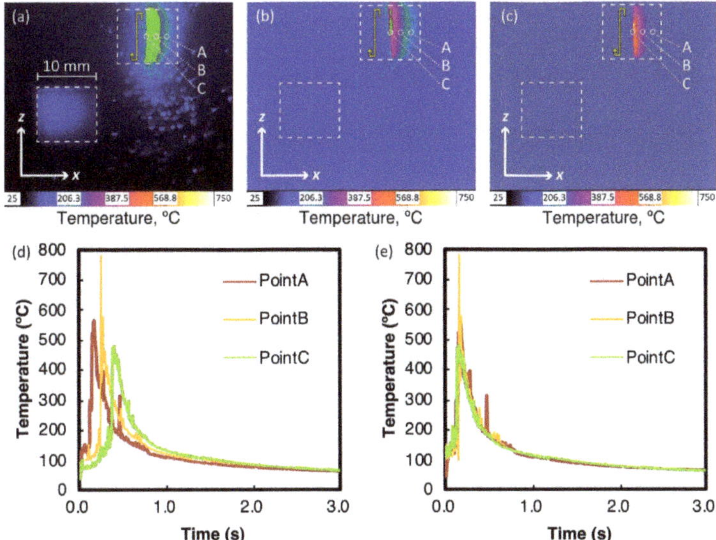

Figure 6. Temperature distributions on the top surface measured by thermography during multiple laser scanning: (**a**) 0–250 °C, (**b**) 106–350 °C, (**c**) 265–780 °C, (**d**) temperature histories at points A, B, and C, and (**e**) temperature histories at points B and C are shifted by a time lag to match the temperature history at point A.

The actual temperature at the laser-irradiated spots should exceed T_s and T_l. Therefore, another filter for the high-temperature range of 750–2500 °C was also examined. However, such high temperatures could not be measured by the thermography camera. This can be owing to the temporal and spatial resolution limitations of the camera. Zhirnov et al. [28] used an advanced thermography camera with a 20,000 frame rate and 3.07 µm/pixel resolution to measure sharp laser spots over 1900 °C. However, as discussed later, the temperature variation during cooling period below 500 °C was sufficient to calibrate the absorptivity.

Figure 7a–c show the temperature distribution by the 2D thermal simulation of multiple laser scans with $A = 0.30$, 0.50, and 0.70 when the heat source is in the center of the model. Here, it was assumed that the shape parameters a and b were 50 and 125 µm, respectively. The part whose temperature exceeds T_s is shown in gray indicating the melting pool. It was evident that the size of the melting pool increased as the absorptivity increased. The thermography camera has a spatial resolution of 68 µm, so we averaged the temperatures of the 14 nodes in the top center (yellow points in Figure 7a–c). Figure 7d,e show the average thermal histories on various axis scales. Figure 7d shows the cyclic temperature rise and fall owing to adjacent laser paths. Such heating by adjacent paths was also recognized from the asymmetry of the melting pool in Figure 7a–c.

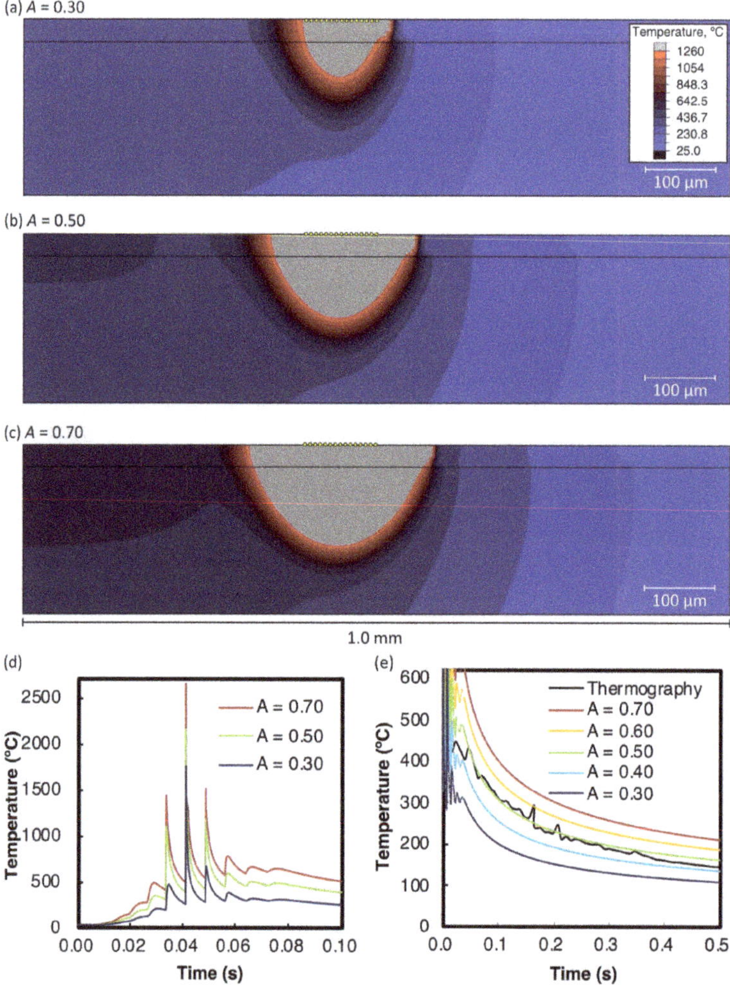

Figure 7. Temperature distributions by thermal simulation of multiple laser scanning with different absorptivities A: (**a**) 0.30, (**b**) 0.50, (**c**) 0.70, (**d**) cyclic thermal histories, and (**e**) cooling process from maximum temperature. In (**a**–**c**), the average temperature in the top center of 14 nodes is indicated in yellow. Black lines in (**e**) represent temperatures measured by the thermography camera (Figure 6e).

Figure 7e shows the cooling process from the maximum temperatures for different absorptivity assumptions. The black line in the figure indicates the experimental data measured by the thermography camera. The results indicate that "0.50" seems suitable for the absorptivity A to reproduce the surface temperature variation in the thermal analysis. As described in Section 1, the absorptivity of the heat source model includes the effects of the powder layer and the surface shape of the melting pool. There is no data available for the absorptivity of HX. For the case of 316L stainless steel powder, its absorptivity was reported as in the range from 0.3 to 0.7 for the different laser energy densities [20]. The estimated value, A ~0.50 for HX in the present study was in a similar range and was considered appropriate.

5.2. Shape Parameter Calibration of Heat Source Model

Figure 8a shows one of cross-sectional SEM images of the single bead using the SLM process at the power of 300 W and the speed of 1500 mm/s. The mean width and depth of melted zone at three different cross sections (see Figure 8a) were 60.8 ± 4.79 μm and 69.8 ± 1.22 μm, respectively. Figure 8b shows $MZ_{obs}(x,y)$ processed by the procedure described in Section 4. The gray pixels in the averaged image in Figure 8b were just observed along the boundary between the melted and non-melted zones. This indicates that the shapes of the melted zone at the three different cross sections were almost identical.

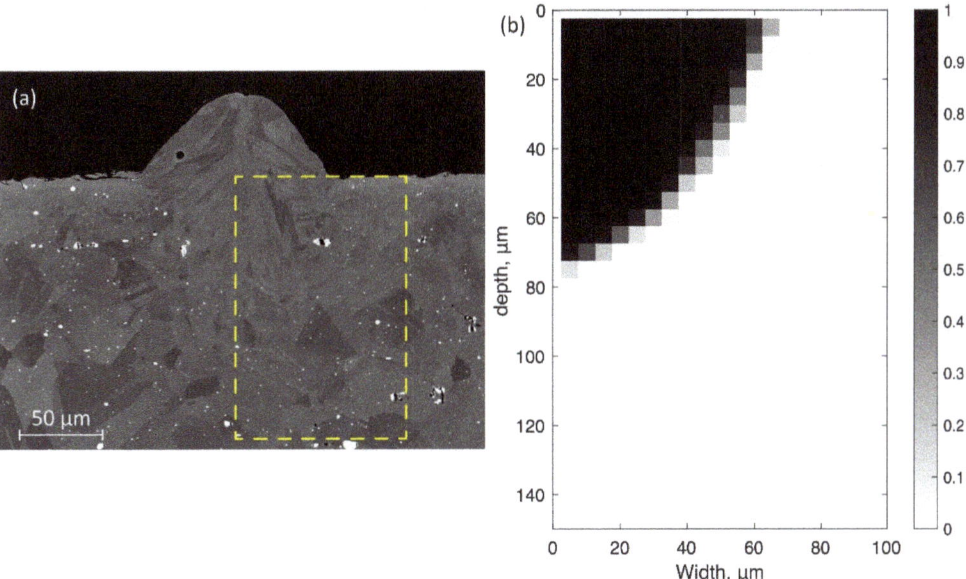

Figure 8. (**a**) Cross-sectional SEM image of the single bead with laser scanning. A yellow rectangle indicates the area to be processed. (**b**) The visualization of the melted zone matrix $MZ_{obs}(x,y)$. Black and white pixels with binary values correspond to the bead and baseplate, respectively.

Figure 9a–e show the change of temperature distribution and the formation of melted zone during the 2D thermal simulation of a single laser scan with $A = 0.50$, the shape parameters $a = 50$ μm, and $b = 125$ μm. Figure 9f shows the simulated width, depth, and area of the melted zone over time. In this simulation, the center of volumetric heat source passed through the XY plane at 0.25 ms so that the melting pool and melted zone became larger from 0.185 to 0.25 ms (see Figure 9a,b,f). After the heat source passed through, the melted zone became slightly wider at 0.255 ms (see a yellow circle in Figure 9c). Then, the melting pool gradually got small as the inputted heat was transferred to the surrounding area (Figure 9d,e).

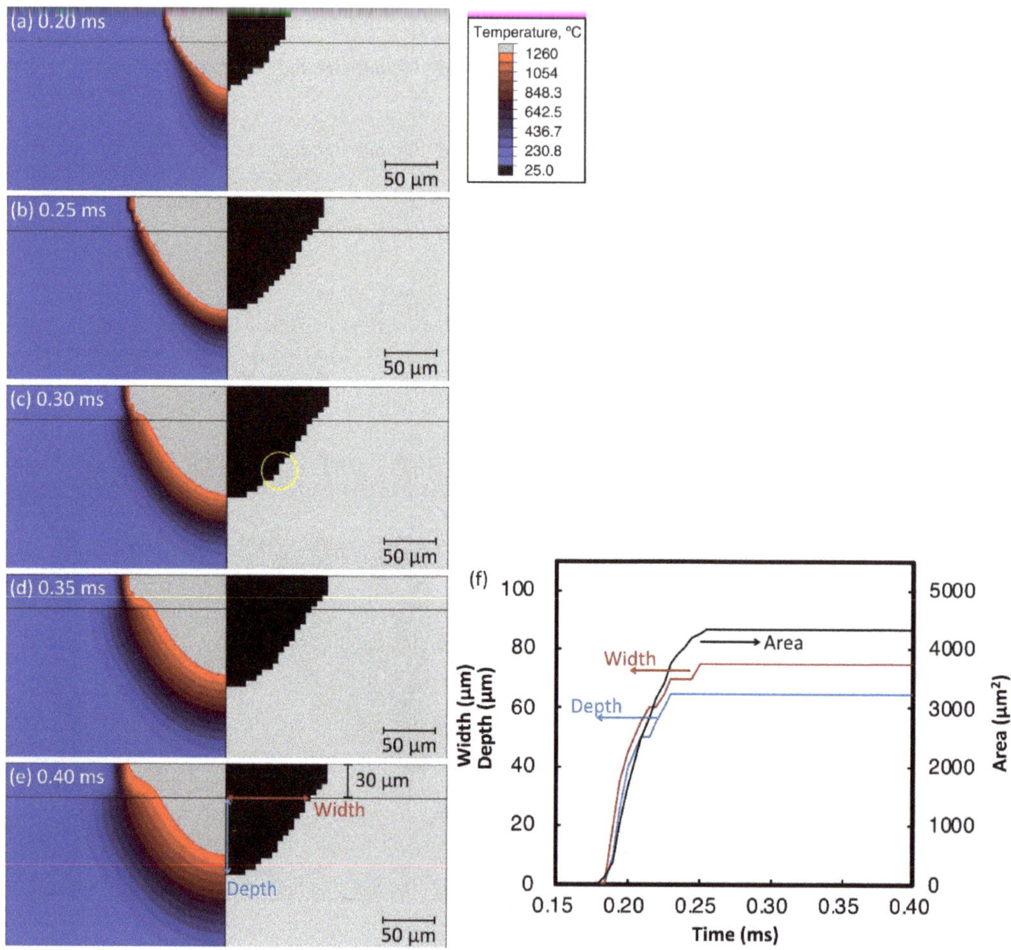

Figure 9. Temperature distribution (**left**) and melted zone (**right**) simulated by 2D thermal analysis with parameters $a = 25$ µm, $b = 125$ µm, and $A = 0.50$ at the moment of (**a**) $t = 0.20$ ms, (**b**) 0.25 ms, (**c**) 0.30 ms, (**d**), 0.35 ms, and (**e**) 0.40 ms. (**f**) The width, depth, and area of the melted zone with the time.

According to the procedure described in Section 4, the optimal shape parameters a and b to minimize error E were searched by Bayesian optimization. Figure 10b–d show contour maps of the errors searched by various acquisition functions. White circles indicate the searched parameters. Figure 11a shows the minimum errors through the search process with the number of evaluations. Figure 10a also shows the contour map of the errors by the grid search with more than 600 times simulations. The search parameters are on the grid lines in the figure. All these contour maps show that the error has one local minimum point around (55 µm, 170 µm). Instead of reducing the search for larger error areas, Bayesian optimization with all acquisition functions better exploited the objective functions around the minimum point. As a result, the optimal parameters with errors equivalent to those of grid search were achieved within 25 iterations by Bayesian optimization (Figure 11a). The optimal parameters for the grid search and each acquisition function are shown in red circles in Figure 10 and are listed in Table 2. Three acquisition functions represented similar shape parameters for minimizing E. Figure 12 shows the bead shape by simulation $BS_{sim}(x, y)$ with the optimal parameters and its difference matrix $diff(x, y)$. As shown in these figures, all the Bayesian optimizations with the three acquisition functions succeeded in efficiently detecting the shape parameters that fit the shape of the beads.

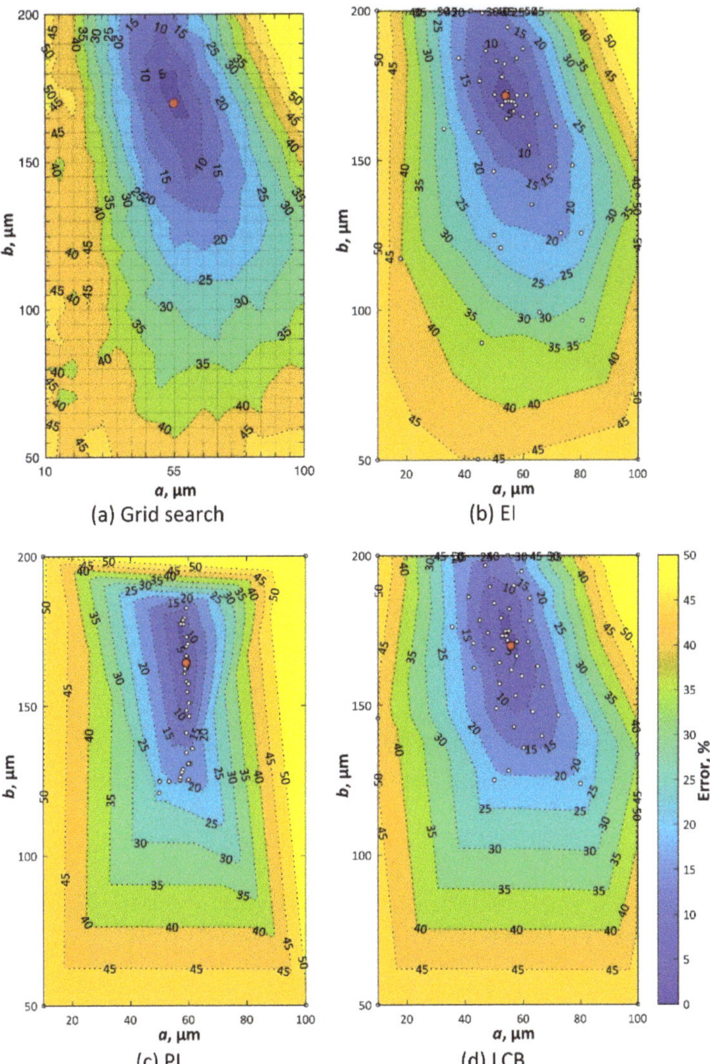

Figure 10. Contour map of error E simulated by (**a**) grid search and Bayesian optimization with the acquisition function of (**b**) EI, (**c**) PI, and (**d**) LCB. The white circles represent the searched points, and the red circles are optimal parameters to minimize E.

Table 2. Optimal parameters found by grid search and Bayesian optimization with different acquisition functions and their errors.

Search Strategy	Acquisition Function	a, μm	b, μm	Error, %
Bayesian optimization	EI	54.0	171.7	3.48
	PI	59.3	164.5	3.48
	LCB	55.7	169.8	3.68
Grid search	-	55.0	170.0	3.48

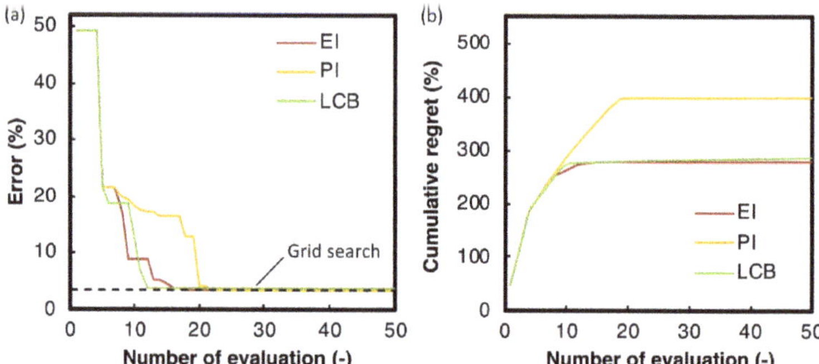

Figure 11. (**a**) Error E and (**b**) cumulative regret R with the number of evaluations.

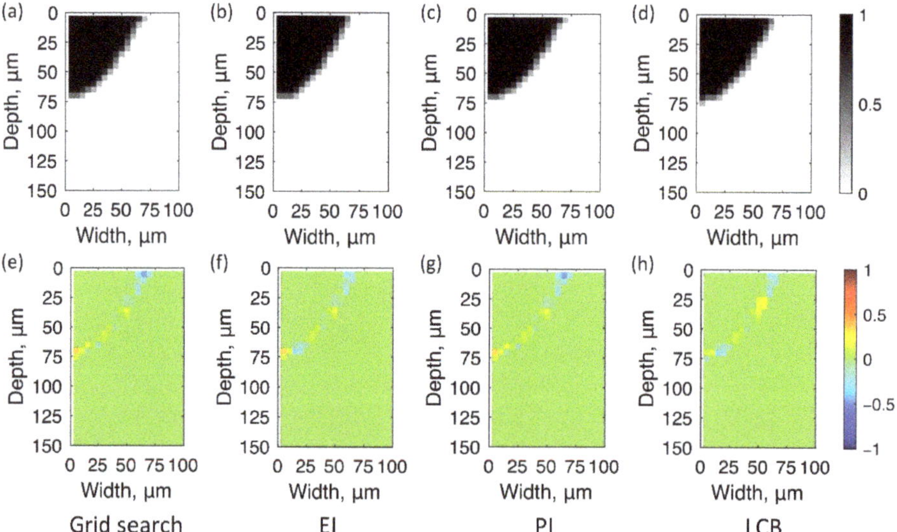

Figure 12. Bead shapes by simulation with optimal parameters (**a–d**) and their differences from the observation (**e–g**). (**a**) and (**e**) are the grid search and Bayesian optimization with the acquisition function of (**b**) and (**f**) EI, (**c**) and (**g**) PI, and (**d**) and (**h**) LCB.

The search efficiency of each acquisition function was evaluated by cumulative regret R, defined as follows [18,20]:

$$R = \sum_i (E_i - E_{min}) \tag{12}$$

where E_i is the ith error and E_{min} is the minimum error. The smaller the cumulative regret, the more efficient search in the optimization. As shown in Figure 11b, LCB and EI efficiently searched for the best parameters than PI because of the balance of exploitation and exploration. In any case, compared to the grid search, which required more than 600 calculations, Bayesian optimization with any acquisition function was much more efficient in determining optimal shape parameters.

5.3. 3D Thermal Analysis with Calibrated Parameters

The novel two-step calibration proposed above was performed using a 2D FEM-based thermal simulation model that takes computational costs into account. 3D thermal analysis requires enormous computational time and resources, as the number of meshes significantly increases with the length of the model in the Z-direction. Thus, it is not feasible to use such a 3D model for the calibration procedure with iterative simulations even though Bayesian optimization can significantly reduce the number of iterations. However, a 3D thermal analysis can be desirable to understand the temperature distribution better and combine the results of thermal simulation with other simulations (such as thermal stress analysis [9], thermal elastoplastic analysis [7,8], and grain growth simulation [2]). Consequently, a rational process is to use the 2D model to calibrate the parameters and use the calibrated parameters to simulate the 3D model. Here, the process is validated to compare 3D and 2D models with parameters calibrated in terms of bead shape, thermal history, and cooling rate.

Figure 13 shows the 2D and 3D temperature distributions of the single laser scan with the calibrated absorptivity and the shape parameters $[(A, a, b) = (0.5, 54.0, 171.7)]$. Figure 13a–d are the results of the 2D calculations and Figure 13e–p are the ones by the 3D analysis at the different times and for the different viewing angles. The 3D temperature distribution in Figure 13e–g were shown as a mirror reflection, using the symmetry in the YZ plane. Comparing Figures 13i–l and 13a–d, the temperature distribution at the XY plane in the 3D model was almost the same as that in the 2D model. Similar to the calibration procedure in Section 4, $MZ_{sim}(x, y)$ and $diff(x, y)$ were evaluated at the XY plane in the 3D model. It is clear from Figure 14 that the melted zone simulated in the 3D model was consistent with the bead shape on the cross section. The error E of the 3D simulation was 4.73% and the difference was only 1.25% comparing to the 2D simulation. In most cases, the error was small enough to justify the 3D simulation with the shape parameters optimized in the 2D model.

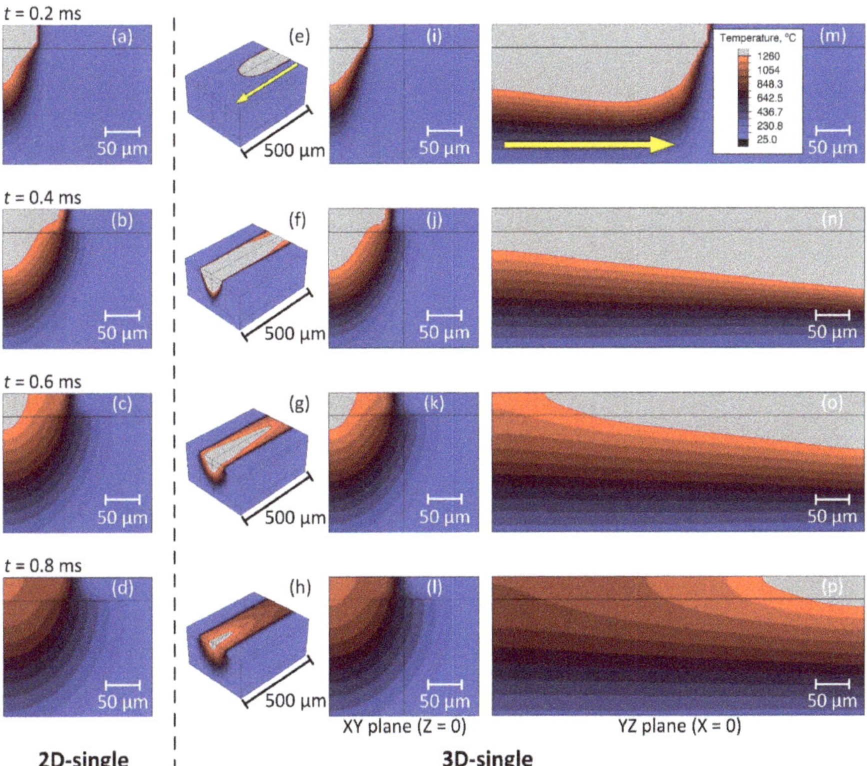

Figure 13. Temperature distributions simulated using the 2D single (**a–d**) and 3D single (**e–p**) models with calibrated parameters $(A, a, b) = (0.5, 53.95, 171.67)$ at different time steps $t = 0.2, 0.4, 0.6$, and 0.8 ms. (**i**), (**j**), (**k**), and (**l**) are XY planes at $Z = 0.0$ μm, and (**m**), (**n**), (**o**), and (**p**) are YZ planes at $X = 0.0$ μm. Yellow arrows indicate a laser scanning direction.

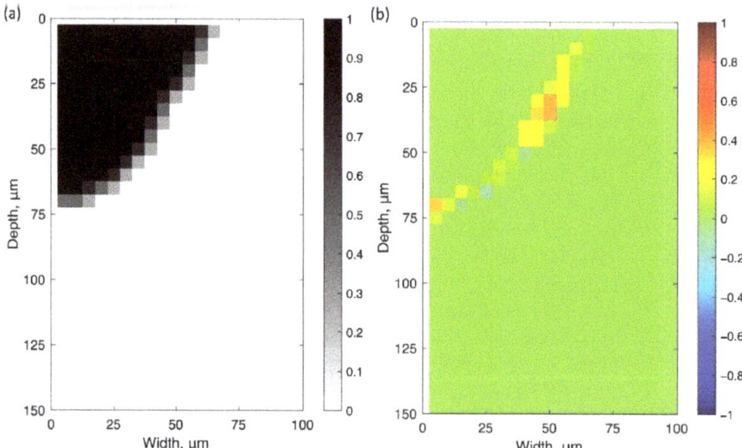

Figure 14. (**a**) Melted zone $MZ_{sim}(x, y)$ at $Z = 0$ μm in the 3D single model, and (**b**) its difference $diff(x, y)$ from the observation (Figure 8b).

Figure 15a shows thermal histories at $Y = 30, 0, -30$, and -60 μm and at $(X, Z) = (0.0, 0.0)$ in the 3D model. For comparison, the dashed lines in the figure represent the thermal histories of the 2D model at the same coordinate points. The thermal histories of both models showed almost the same behavior. Red and blue lines in Figure 15a represent the temperatures of T_l and T_s. The cooling time from T_l to T_s plotted in Figure 15b increased in the Y direction (as the depth of melting pool decreased). This indicates that shallower the melting pool, the more slowly it solidifies. The temperature difference between T_l to T_s divided by the time corresponds to an average cooling rate during the phase transformation from liquid to solid. The cooling rate decreased from 1.7×10^6 to 3.6×10^5 K/s in the Y direction. Again, the 3D and 2D models showed the same trends for the cooling rate.

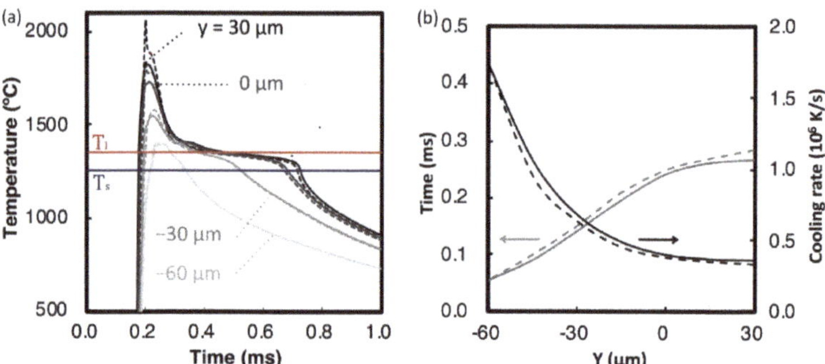

Figure 15. (**a**) Simulated thermal history at $Y = 30, 0, -30$, and -60 μm and $(X, Z) = (0.0, 0.0)$ of the 3D model and (**b**) solidification time from T_l to T_s and the cooling rate in Y direction. Solid and broken lines indicate the results of single 3D and 2D models, respectively.

6. Discussion

This section discusses the novel calibration strategy for the thermal analysis in the SLM process in terms of the computational efficiency by 2D model and Bayesian optimization, the similarity between the 2D and 3D models, the validity and limitation of the calibrated heat source model, and the advantage of the FEM-based thermal analyses over more advanced simulations for the SLM process.

In this study, the shape parameters of the heat source model were calibrated using the 2D thermal simulation models, and it was confirmed that the calibrated parameters were also valid for the 3D thermal simulation. The 2D single scan model takes less than 2 min on a desktop personal computer (PC) with an Intel Core i7 CPU (Intel Corporation, Santa Clara, CA, USA), while the 3D model takes about 2 h. Thus, as well as Bayesian optimization, the iterative simulations with the 2D single scan model could find the optimal shape parameters within 2 h (50 iterative simulations). The 3D single laser scan simulation with the calibrated parameters also showed an error of 4.73% compared to the cross-sectional observations. Thus, the novel calibration strategy effectively reduces the computational cost.

As confirmed in Section 5.3, the 2D and 3D single laser scan simulations showed almost identical melted zone, thermal history, and cooling rate. This implies that the heat input by the heat source model was mainly transferred in the X and Y directions (perpendicular to the moving direction of the heat source) rather than in the Z direction in the 3D model. Note that these simulation models were for the middle part of a sufficiently long single laser scan. At the ends of single laser scan, the cooling rate in the 2D model will be slower than in the 3D model and the experiment because the actual input heat transfers to the surrounding substrate in three dimensions.

The parameters of the double ellipsoidal power density model were calibrated using both the temperature measured by the thermography camera and the cross-sectional bead shape observed by SEM. In this study, as a first step, the parameters c_1 and c_2 in Equation (2) were assumed to be a function of a as in reference [16,23]. Some researchers reported the calibration of the shape parameters by observing the top surface of the melting pool during a laser scan using high-speed imaging [13]. Thus, in-situ observation of such melting pools helps make thermal simulations more effective. However, studying the shape of the top bead by FEM-based thermal analysis is not practical. The surface tension of the melting pool makes the shape of the top bead semi-circular. However, since the current FEM-based thermal analysis does not take such surface tension into account, the shape of the top bead becomes unrealistic (see Figure 9). Thus, if the size of the melting pool observed by high-speed imaging is used for calibration, the effects of surface tension must be considered by using computational fluid dynamics (CFD).

Absorptivity and bead shape parameters depend on laser scan conditions. As described in the Introduction section, the absorptivity is significantly dependent on the laser power and scanning speed [20]. The authors also studied the shape of a single bead of HX, which can be deep or shallow depending on the laser power and speed of the SLM process [7]. Thus, parameters calibrated for the specific laser scan conditions may not be useful for other conditions. One solution is to define the parameters as a function of laser scan conditions and calibrate the coefficients of the function by experimental measurements with various conditions. Suppose such a heat source model is developed as a function of laser scan conditions. In that case, the thermal simulation applicable to any laser scan conditions will be significantly useful for other simulations related to SLM processes such as prediction of microstructure evolution and microcracking.

One might be concerned about the effects of shape parameters on the multiple laser scan simulation for absorptivity calibration. In Section 5.1, the shape parameters a and b were assumed to be 50 and 125 µm, respectively. When the same simulations with some different values of shape parameters within the search area, the surface temperature differed by only 10 °C at most in the time range for the calibration (from 0 to 0.5 s after laser scanning). This value was sufficiently small compared to the temperature change (more than 50 °C, see Figure 7e) when the absorptivity was varied by 0.1. Thus, the two-step calibration strategy is reasonable if the first decimal place of absorptivity A is to be determined.

Fundamentally, such a heat source model with some fitting parameters is necessary because the current FEM-based thermal simulation does not directly model laser absorption and reflection in the powder layer and the fluidity of the melting pool. This can be a drawback compared with more advanced simulations such as ray-tracing simulation of laser beams on powder particles and thermal analysis with CFD. However, FEM-based thermal simulations with such a simplified heat source model can use less computational costs and resources. Again, the 3D FEM-based thermal simulation takes approximately 2 h to complete a single bead simulation on the desktop PC. On the other hand, the CFD-based thermal simulations of comparable scale take even longer, and need to be parallelized. Thus, FEM-based thermal analysis helps to efficiently understand the large-scale thermal behavior of multiple laser scans and multiple layers.

7. Conclusions

This study proposed a novel calibration strategy of the heat source model for FEM-based thermal analysis of the SLM process. Empirical data from in-situ temperature measurements by high-speed thermography and SEM observation of single bead cross sections were used to calibrate the heat source model absorptivity and shape parameters. The findings obtained in this study were as follows:

- Bayesian optimization allowed to efficiently find the optimal parameters with an error of less than 4% within 50 iterations of the thermal simulations.
- The use of the 2D models for the parameter calibrations significantly reduced the cost of iterative simulations.
- The 3D single laser scan simulation with the calibrated parameters showed almost the same results of the 2D simulation in thermal history, melted zone, and cooling rate.

Thus, our novel calibration strategy for thermal analysis is feasible to precisely and efficiently understand the thermal history of the actual SLM process.

Author Contributions: Conceptualization, methodology, validation, formal analysis, investigation, software, writing—original draft preparation, visualization, M.K.; validation, software, H.K.; writing—review and editing, project administration, supervision, M.W. All authors have read and agreed to the published version of the manuscript.

Funding: This work was supported by the Council for Science, Technology and Innovation (CSTI), Cross-ministerial Strategic Innovation Promotion Program (SIP), "Materials Integration for revolutionary design system of structural materials" (Funding agency: Japan Science and Technology (JST)).

Institutional Review Board Statement: Not applicable.

Informed Consent Statement: Not applicable.

Data Availability Statement: Data sharing not applicable.

Acknowledgments: Financial support from the Council for Science, Technology, and Innovation (CSTI), Cross-ministerial Strategic Innovation Promotion Program (SIP), "Materials Integration for revolutionary design system of structural materials" (Funding agency: JST) is gratefully acknowledged.

Conflicts of Interest: The authors declare no conflict of interest. The funders had no role in the design of the study; in the collection, analyses, or interpretation of data; in the writing of the manuscript, or in the decision to publish the results.

References

1. Li, J.; Zhou, X.; Brochu, M.; Provatas, N.; Zhao, Y.F. Solidification microstructure simulation of Ti-6Al-4V in metal additive manufacturing: A review. *Addit. Manuf.* **2020**, *31*, 100989. [CrossRef]
2. Koepf, J.A.; Gotterbarm, M.R.; Markl, M.; Körner, C. 3D multi-layer grain structure simulation of powder bed fusion additive manufacturing. *Acta Mater.* **2018**, *152*, 119–126. [CrossRef]
3. Holland, S.; Wang, X.; Chen, J.; Cai, W.; Yan, F.; Li, L. Multiscale characterization of microstructures and mechanical properties of Inconel 718 fabricated by selective laser melting. *J. Alloys Compd.* **2019**, *784*, 182–194. [CrossRef]
4. Kusano, M.; Miyazaki, S.; Watanabe, M.; Kishimoto, S.; Bulgarevich, D.S.; Ono, Y.; Yumoto, A. Tensile properties prediction by multiple linear regression analysis for selective laser melted and post heat-treated Ti-6Al-4V with microstructural quantification. *Mater. Sci. Eng. A* **2020**, *787*, 139549. [CrossRef]
5. Kasperovich, G.; Haubrich, J.; Gussone, J.; Requena, G. Correlation between porosity and processing parameters in TiAl6V4 produced by selective laser melting. *Mater. Des.* **2016**, *105*, 160–170. [CrossRef]
6. Harrison, N.J.; Todd, I.; Mumtaz, K. Reduction of micro-cracking in nickel superalloys processed by Selective Laser Melting: A fundamental alloy design approach. *Acta Mater.* **2015**, *94*, 59–68. [CrossRef]

7. Kitano, H.; Tsujii, M.; Kusano, M.; Yumoto, A.; Watanabe, M. Effect of plastic strain on the solidification cracking of Hastelloy-X in the selective laser melting process. *Addit. Manuf.* **2021**, *37*, 101742. [CrossRef]

8. Kitano, H.; Kusano, M.; Tsujii, M.; Yumoto, A.; Watanabe, M. Process Parameter Optimization Framework for the Selective Laser Melting of Hastelloy X Alloy Considering Defects and Solidification Crack Occurrence. *Crystals* **2021**, *11*, 578. [CrossRef]

9. Li, Y.; Zhou, K.; Tan, P.; Tor, S.B.; Chua, C.K.; Leong, K.F. Modeling temperature and residual stress fields in selective laser melting. *Int. J. Mech. Sci.* **2018**, *136*, 24–35. [CrossRef]

10. Zhang, Z.; Huang, Y.; Kasinathan, A.R.; Shahabad, S.I.; Ali, U.; Mahmoodkhani, Y.; Toyserkani, E. 3-Dimensional heat transfer modeling for laser powder-bed fusion additive manufacturing with volumetric heat sources based on varied thermal conductivity and absorptivity. *Opt. Laser Technol.* **2019**, *109*, 297–312. [CrossRef]

11. Shahabad, S.I.; Zhang, Z.; Keshavarzkermani, A.; Ali, U.; Mahmoodkhani, Y.; Esmaeilizadeh, R.; Bonakdar, A.; Toyserkani, E. Heat source model calibration for thermal analysis of laser powder-bed fusion. *Int. J. Adv. Manuf. Technol.* **2020**, *106*, 3367–3379. [CrossRef]

12. Kundakcıoğlu, E.; Lazoglu, I.; Poyraz, Ö.; Yasa, E.; Cizicioğlu, N. Thermal and molten pool model in selective laser melting process of Inconel 625. *Int. J. Adv. Manuf. Technol.* **2018**, *95*, 3977–3984. [CrossRef]

13. Bruna-Rosso, C.; Demir, A.; Previtali, B. Selective laser melting finite element modeling: Validation with high-speed imaging and lack of fusion defects prediction. *Mater. Des.* **2018**, *156*, 143–153. [CrossRef]

14. Ahsan, F.; Ladani, L. Temperature Profile, Bead Geometry, and Elemental Evaporation in Laser Powder Bed Fusion Additive Manufacturing Process. *JOM* **2019**, *72*, 429–439. [CrossRef]

15. Andreotta, R.; Ladani, L.; Brindley, W. Finite element simulation of laser additive melting and solidification of Inconel 718 with experimentally tested thermal properties. *Finite Elem. Anal. Des.* **2017**, *135*, 36–43. [CrossRef]

16. Zinovieva, O.; Zinoviev, A.; Ploshikhin, V. Three-dimensional modeling of the microstructure evolution during metal additive manufacturing. *Comput. Mater. Sci.* **2018**, *141*, 207–220. [CrossRef]

17. Goldak, J.; Chakravarti, A.; Bibby, M. A new finite element model for welding heat sources. *Metall. Trans. B* **1984**, *15*, 299–305. [CrossRef]

18. Boley, C.D.; Mitchell, S.C.; Rubenchik, A.M.; Wu, S.S.Q. Metal powder absorptivity: Modeling and experiment. *Appl. Opt.* **2016**, *55*, 6496–6500. [CrossRef]

19. King, W.E.; Anderson, A.T.; Ferencz, R.M.; Hodge, N.E.; Kamath, C.; Khairallah, S.A.; Rubenchik, A.M. Laser powder bed fusion additive manufacturing of metals; physics, computational, and materials challenges. *Appl. Phys. Rev.* **2015**, *2*, 041304. [CrossRef]

20. Trapp, J.; Rubenchik, A.M.; Guss, G.; Matthews, M.J. In situ absorptivity measurements of metallic powders during laser powder-bed fusion additive manufacturing. *Appl. Mater. Today* **2017**, *9*, 341–349. [CrossRef]

21. Brochu, E.; Cora, V.; de Freitas, N. A tutorial on Bayesian optimization of expensive cost functions, with application to active user modeling and hierarchical reinforcement learning. *arXiv* **2010**, arXiv:1012.2599. Available online: https://arxiv.org/abs/1012.2599 (accessed on 20 July 2021).

22. Snoek, J.; Larochelle, H.; Adams, R.P. Practical Bayesian optimization of machine learning algorithms. *Adv. Neural Inf. Process. Syst.* **2012**, *4*, 2951–2959.

23. Shiraiwa, T.; Enoki, M.; Goto, S.; Hiraide, T. Data Assimilation in the Welding Process for Analysis of Weld Toe Geometry and Heat Source Model. *ISIJ Int.* **2020**, *60*, 1301–1311. [CrossRef]

24. Mills, K.C. Ni—Hastelloy-X, Recommended Values of Thermophysical Properties for Selected Commercial Alloys. In *Woodhead Publishing Series in Metals and Surface Engineering*; Woodhead Publishing: Sawston, UK, 2002; pp. 175–180. [CrossRef]

25. Dong, L.; Correia, J.P.M.; Barth, N.; Ahzi, S. Finite element simulations of temperature distribution and of densification of a titanium powder during metal laser sintering. *Addit. Manuf.* **2017**, *13*, 37–48. [CrossRef]

26. Muhieddine, M.; Canot, E.; March, R. Various Approaches for Solving Problems in Heat Conduction with Phase Change. *Int. J. Finite Vol.* **2009**, *6*, hal-01120384. Available online: https://hal.inria.fr/hal-01120384 (accessed on 20 July 2021).

27. Select Optimal Machine Learning Hyperparameters Using Bayesian Optimization. Available online: https://www.mathworks.com/help/stats/bayesopt.html (accessed on 20 July 2021).

28. Zhirnov, I.; Mekhontsev, S.; Lane, B.; Grantham, S.; Bura, N. Accurate determination of laser spot position during laser powder bed fusion process thermography. *Manuf. Lett.* **2020**, *23*, 49–52. [CrossRef]

materials

Article

Comparison of CAD and Voxel-Based Modelling Methodologies for the Mechanical Simulation of Extrusion-Based 3D Printed Scaffolds

Gisela Vega [1,*], Rubén Paz [1,*], Andrew Gleadall [2], Mario Monzón [1] and María Elena Alemán-Domínguez [1]

[1] Mechanical Engineering Department, Campus de Tafira Baja, Universidad de Las Palmas de Gran Canaria, 35017 Las Palmas, Spain; mario.monzon@ulpgc.es (M.M.); mariaelena.aleman@ulpgc.es (M.E.A.-D.)
[2] Wolfson School of Mechanical and Manufacturing Engineering, Loughborough University, Loughborough LE11 3TU, UK; a.gleadall@lboro.ac.uk
*   Correspondence: gisela.vega@ulpgc.es (G.V.); ruben.paz@ulpgc.es (R.P.)

Abstract: Porous structures are of great importance in tissue engineering. Most scaffolds are 3D printed, but there is no single methodology to model these printed parts and to apply finite element analysis to estimate their mechanical behaviour. In this work, voxel-based and geometry-based modelling methodologies are defined and compared in terms of computational efficiency, dimensional accuracy, and mechanical behaviour prediction of printed parts. After comparing the volumes and dimensions of the models with the theoretical and experimental ones, they are more similar to the theoretical values because they do not take into account dimensional variations due to the printing temperature. This also affects the prediction of the mechanical behaviour, which is not accurate compared to reality, but it makes it possible to determine which geometry is stiffer. In terms of comparison of modelling methodologies, based on process efficiency, geometry-based modelling performs better for simple or larger parts, while voxel-based modelling is more advantageous for small and complex geometries.

Keywords: tissue engineering; scaffold; material extrusion additive manufacturing; 3D geometry modelling; finite element analysis; mechanical properties

Citation: Vega, G.; Paz, R.; Gleadall, A.; Monzón, M.; Alemán-Domínguez, M.E. Comparison of CAD and Voxel-Based Modelling Methodologies for the Mechanical Simulation of Extrusion-Based 3D Printed Scaffolds. *Materials* **2021**, *14*, 5670. https://doi.org/10.3390/ma14195670

Academic Editor: Antonino Recca

Received: 2 September 2021
Accepted: 23 September 2021
Published: 29 September 2021

Publisher's Note: MDPI stays neutral with regard to jurisdictional claims in published maps and institutional affiliations.

1. Introduction

In tissue engineering applications, porous structures are desired for promoting cell growth and tissue regeneration. The morphology, size, and distribution of the pores have a vital effect not only on the mechanical properties of the structure [1], but also on its biological performance. Pore size has an optimal value for each type of tissue. Pore size below the optimal range hinders the vascularisation of the structure and cell migration. On the other hand, large pore sizes lead to a reduced surface area (and, therefore, a limited cell adhesion) and weak structures when biodegradable polymers are used to manufacture the structure. For instance, the osseointegration process is enhanced when scaffolds with pore sizes between 150 and 500 μm are used [2–4]. Moreover, the vascularisation achieved with interconnected pores with sufficient pore size enhances the osteogenesis process [5].

In addition to its role in the biological processes that take place during in vivo regeneration of the target tissue, the porosity is relevant during the degradation process of biodegradable scaffolds [6], as it is related to the permeability of the structure and, consequently, to the removal of the degradation by-products. These substances have an autocatalytic effect; hence, their local concentration has a strong impact on the degradation rate and mechanism of the scaffold [7].

Several techniques are used in tissue engineering for scaffold fabrication, which are the conventional ones (particulate leaching, phase separation, gas-foaming, or emulsion freeze-drying), as well as electrospinning and additive manufacturing (AM) techniques. In recent years, this last method has been the most selected one. AM allows a high degree of control of porosity, and this is one of the reasons for its popularity in the tissue engineering field [8–11].

AM techniques can be grouped into seven different categories according to ISO/ASTM 52900:2015: binder jetting, directed energy deposition, material extrusion, material jetting, powder bed fusion (e.g., selective laser sintering), sheet lamination, and vat photopolymerisation (e.g., stereolithography). Among them, the most widely used technology for scaffold manufacturing is material extrusion (MEX), commonly known as extrusion-based 3D printing or fused deposition modelling (FDM). FDM presents ease and flexibility in the processing and selection of materials, as well as in the manufacturing process [8–14].

However, in the case of material extrusion AM, scaffolds are not always designed in the same way. Some authors define the solid part and then set its porosity in the printing settings of the slicer software [15–17]. In this step, the 3D printing path is determined according to the printing settings, and the resulting geometry is shown in the software. Nevertheless, this geometrical information cannot be exported and, therefore, it cannot be used for the prediction of mechanical behaviour through simulation techniques such as finite element analysis (FEA).

Other researchers preferred firstly to design the model using CAD software, representing it with filaments instead of as a solid part, and then print the scaffold [10,14,18–23]. In those cases, they were able to simulate the model behaviour and compare the results with the experimental ones or optimise the part before printing it. Despite this, the printing process of this method is not suitably defined; it could present problems depending on the slicer software chosen, as each filament is typically interpreted as a solid piece and represented as several filaments, instead of a single pass.

On the other hand, there are several valid options to simulate the mechanical performance of the printed scaffold based on its model, such as the homogenisation [24–27], the voxel-based [12,28,29], or the CAD-based modelling techniques [10,14,18–23,30,31].

The CAD-based modelling method is the most widely used. There are several ways to obtain the geometric model. One of them is the definition of a layer, which is repeated along the height, combining the direction of the filaments. This method has the limitation that it is only applicable to quadrangular prisms or cylinders with the same mesostructured layers (when the base figure is rotated 90°, it keeps its shape). Another way is to model the deposited part by computer-aided software based on the G-code information. This method can reproduce simple geometric objects that are defined by their boundaries (vertices, edges, and loops). However, it could raise problems if it is applied to non-Eulerian solids or biomorphic porous structures. Moreover, it could also fail if there are gaps or overlaps, and its manipulation may be difficult when objects have fine internal architectures or are too large [32,33].

To solve the computational limitations of CAD models, unit cells are used. These are basic building blocks that represent the scaffold microarchitectures. There are unit cell libraries available, which, combined with the homogenisation technique (application of the unit cell properties to the entire solid), make the computational method more efficient [32,33].

On the other side of modelling methods, the image-based approach is also used and is compatible with computed tomography (CT) and magnetic resonance imaging (MRI). It allows the internal architecture to be controlled by the intersection between 3D binary images which define the voxel values (solid or void). However, this method presents the limitations of resolution and dataset dimensions and the need for a reference image [32,33].

Recent studies have introduced another way of modelling porous structures using natural geometries as reference (e.g., bones). This modelling technique takes into account the irregularities and the spatial evolution of the reference porous structure. This is the top-down design through the Voronoi tessellation method. It considers the irregularity of bones and controls the distribution and shape of the pores, in addition to the gradient interconnected porosity. The main challenge of this method is to find a suitable manufacturing technology, which seems to be the selective laser melting (SLM) one [34,35].

The development of a modelling method combined with a simulation tool to analyse the mechanical behaviour of different pore sizes and structures would allow reducing the experimental work needed when developing a new design for a scaffold to be used in tissue engineering. Therefore, this tool would improve the cost-efficiency of the development process.

This is the first study to evaluate and compare several modelling techniques, particularly geometry-based and voxel-based ones. This work developed a new automated modelling technique, which is the first to define the methodology for modelling any CAD part from the G-code file, from simple parts to more complex ones. Furthermore, the new technique was evaluated in comparison to the only other modelling approach that is able to simulate microscale geometry in practical-sized structures, the VOLCO software [28]. On the other hand, a methodology is defined to make use of the models obtained to estimate the mechanical behaviour of the parts before printing them, which would allow their optimisation. As a case study, FEA was applied to 3D printed scaffolds (material extrusion) in order to compare their capabilities and limitations in the modelling and simulation process. Several scaffolds were designed by defining different porosities and infill patterns, and the modelling and FEA simulation were carried out to compare the results between both methodologies. Moreover, some samples were also manufactured and tested to compare the simulations with the experimental results.

2. Materials and Methods

2.1. Materials

The printing material employed for manufacturing the scaffolds used in this experiment was polycaprolactone (PCL), a biodegradable polyester with suitable properties for tissue engineering applications. The material properties are shown in Table 1.

Table 1. Material Properties of PCL.

Material Properties of PCL-Regemat 3D	
Molecular weight	50,000 g/mol
Density	1.1 g/cm^3
Tensile strength	45 MPa
Elongation at yield	15%
Tensile modulus	350 MPa
IZOD impact strength (notched)	8 kJ/m^2
Shore D hardness	46
Heat deflection temperature (0.45 MPa)	57 °C

2.2. Geometries and Manufacturing Parameters

A solid, cylindrical scaffold (Ø10 × 7 mm) was used for this study. This solid part was introduced in the slicer software (Slic3r 1.3.0) (Alessandro Ranellucci, Rome, Italy), and three different printing settings were selected to obtain three parts with different fill patterns (rectilinear and gyroid) and fill densities (40% and 50% density, corresponding to 60% and 50% porosity, respectively). Table 2 shows the printing parameters of the three configurations studied: a scaffold defined by a rectilinear pattern and a porosity of 50% (50_rect), a scaffold similar to the first one but with a porosity of 60% (60_rect), and another one with a gyroid fill pattern and a porosity of 50% (50_gyr).

Table 2. Printing Parameters of Scaffolds.

Print Settings		
Common Settings to All the Parts		
External perimeter extrusion width		0.44 mm (0.39 mm^3/s)
Perimeter extrusion width		0.48 mm (0.88 mm^3/s)
Infill extrusion width		0.48 mm (2.51 mm^3/s)
Filament diameter		1.75 mm
Infill speed		20 mm/s
Nozzle diameter		0.4 mm
Temperature		190 °C
First layer height		0.35 mm
Fill angle		90°
Particular Settings		
50_rect	Fill density	50%
	Fill pattern	Rectilinear
60_rect	Fill density	40%
	Fill pattern	Rectilinear
50_gyr	Fill density	50%
	Fill pattern	Gyroid

After setting the scaffold characteristics, the G-code file for printing and modelling was obtained with Slic3r 1.3.0.

2.3. Modelling Methods

To be able to apply FEA to the final printed part, a modelling strategy is needed to reproduce the deposited part. This work compares two different approaches to obtain the 3D CAD part from the G-code file: voxel-based and geometry-based modelling techniques.

2.3.1. G-Code Conditioning

A Matlab 2021a script processes the G-code and conditions it to be used in whichever of the modelling techniques. The first step is the conversion of the G-code into a coordinate file. This is a spreadsheet (.xls) which consists of four columns that contain all the movement coordinates and deposition guidelines. From the first to the third column, the file contains the *x*-, *y*-, and *z*-coordinates, respectively, of each point of the path. The fourth column indicates, in a binary format, if that point is the first one of the movement (0) or an intermediate or final point of the path (1).

The coordinates are extracted from the G-code reading it and looking for its specific denomination, *X*, *Y*, or *Z*, followed by a number. In the spreadsheet, only the corresponding number is collected.

To fill the fourth column, comments are required in the G-code; thus, the verbose option must be selected during the G-code generation. Some key comments for starting a new path are detected to set a "0" in the first point of that path. This step is automatically done by the Matlab script that obtains the coordinates from the G-code file. This script was configured to be able to manage different slicer software, which uses different comments to indicate the positioning movements. Some examples of these movements are "move to first infill point" or "move to next layer", specific to the file obtained in Slic3r 1.3.0 or, in the case of Simplify 3D 4.1.1., "layer <number>", "feature infill", "feature outer perimeter", or "feature solid layer".

2.3.2. Voxel-Based Modelling (VOLCO)

In the voxel-based methodology, the tool developed at the University of Nottingham, Volume Conserving Model for 3D Printing (VOLCO) (run in Matlab R2021a), was used [28]. It is executed by macros of a datasheet file, called "Setup.xls", which needs the introduction of the coordinates of the part. These coordinates can be obtained from a G-code file with the Matlab script described above. The scheme of the voxel-based modelling process from the G-code to the results of the mechanical behaviour is shown in Figure 1.

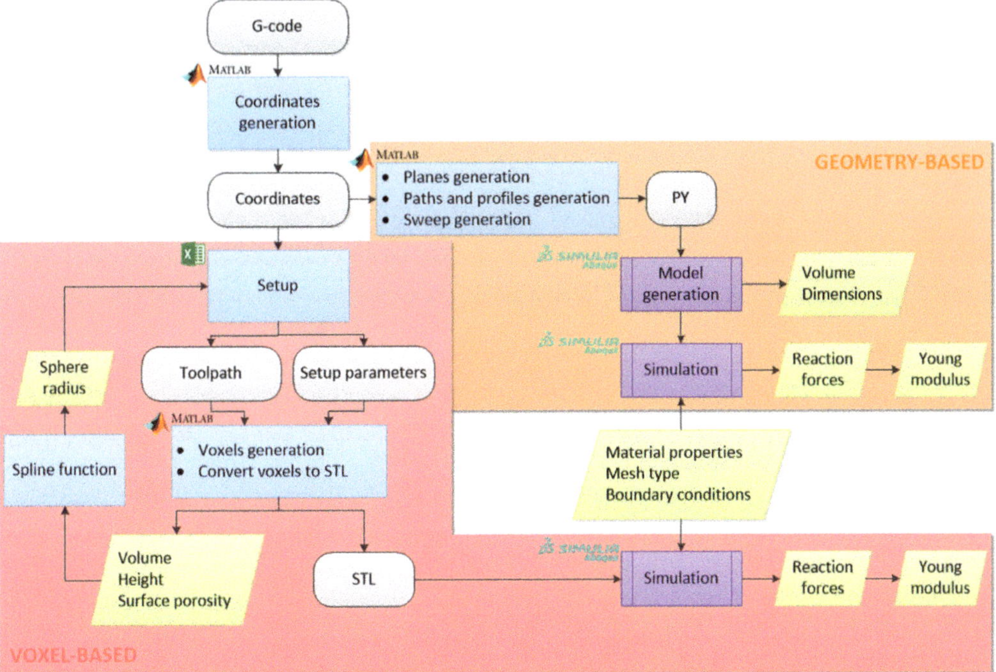

Figure 1. Flowchart of the modelling methodologies.

VOLCO simulates the material extrusion during the manufacturing process and generates a voxelised 3D-geometry model of the predicted microarchitecture. The software deposits voxelised spheres, placed one after the other, to simulate the filaments, and, when they interact, there is an expansion of the material in the available space, keeping the volume constant. This predictive model starts from the premise that the deposited filaments are not perfect cylinders, but that the molten material interacts with those obstacles that are encountered in the deposition process [28].

Once VOLCO has been executed, a voxelised model is obtained and represented in an STL file. Additionally, the dimensional and porosity data are also depicted, i.e., the volume, height, and porosity of the part. The STL can be imported into Abaqus/CAE 6.14-1 (Dassault Systèmes Simulia Corp., Providence, RI, USA) and, after the FEA, the reaction forces can be determined and, therefore, the Young's modulus of the part can be deduced.

The procedure previously described is the basic one, but an additional step can be applied to improve the accuracy of the resulting voxelised volume. This consists of an iterative process to equalise the volume of the voxelised part and the theoretical one, according to the G-code. It is done through the spline function of Matlab R2021a (The Mathworks Inc., Natick, MA, USA). In an iterative process, this function is manually applied with the available data (sphere radius applied and volume of the model obtained) to estimate, by cubic spline interpolation, the correct sphere radius in the VOLCO setup ("OriginalSphereRadius" in the Setup.xls) that will lead to a voxelised model with the theoretical volume of the G-code.

2.3.3. Geometry-Based Modelling

The second method is the automated sweep CAD modeller (run in MATLAB R2021a) of the extrusion-based G-code (DECODE). This consists of a Matlab script that reads the previously mentioned coordinates file. This information is used to automatically write a Python 3.9.6 (Python Software Foundation, Beaverton, OR, USA) file with the code that generates the planes that represent each layer, the sketches of the filament paths and profiles (sections), and the sweep operations to model the deposited part in Abaqus/CAE 6.14-1. This Python file can be subsequently run in the FEA software (Abaqus/CAE 6.14-1) to obtain and visualise the CAD model.

As mentioned above, a plane is defined at each printed layer height, where the filament is deposited. To sweep each layer, it is necessary to have drawn the sketches of the path and the profiles (section of the deposited filament). The sketches of the profiles are positioned at the beginning of the path, perpendicular to the first line direction. The shape of the profile is the combination of a rectangle and two semicircles, as represented in Figure 2.

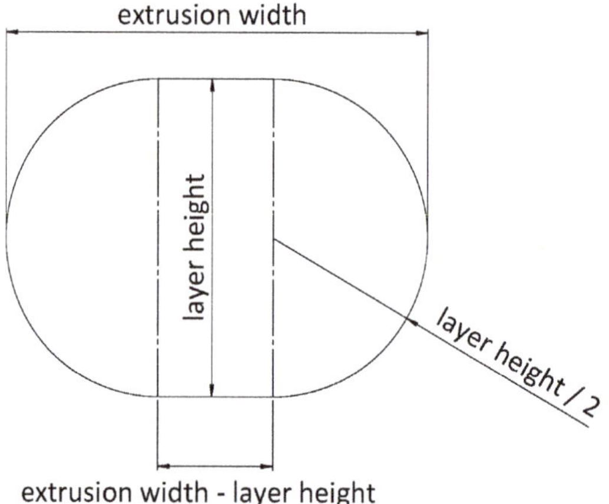

Figure 2. Filament profile shape and dimension.

Several authors have defined how the profile of a printed filament should be represented. Some of them affirmed that the cross-section must be an ellipse, such as Bellehumeur or Rupal [30,36]. However, the solution presented in this work is more similar to the elliptic–rectangular cross-section proposed by Park and Rosen [37] and supported by Gleadall [28]. This shape was assumed by taking into account the deformation of the fused filament when it is deposed on a plane layer and the effect of the extruder nozzle running through the top layer.

After running the Python file, the model is obtained in Abaqus/CAE 6.14-1. After these steps, the volume and dimensions of the part can be obtained. The model is then prepared for the simulation, taking into account the material properties, mesh type, and boundary conditions. Once all the steps are completed, the mechanical behaviour such as displacements or reaction forces can be measured. The scheme of the complete process is shown in Figure 1

During the development of the geometry-based modelling technique, some CAD modelling or meshing issues were observed. For these reasons, several strategies were developed and tested to obtain a model without problems in Abaqus/CAE 6.14-1. These strategies were "whole-part modelling", "line-by-line modelling", "line-by-line modelling with corner revolutions", and "modelling by sections".

To test the effectiveness of these strategies, a simple part, test specimen ($40 \times 5 \times 2$ mm^3), and a complex part, the gyroid scaffold, were modelled.

1. Whole-Part Modelling

The first and simpler methodology was "whole-part modelling". In this strategy, each path corresponds to a continuous deposition made by the extruder. In case that there is a discontinuity in the path of the same layer, each part of the path is represented through a separate sweep feature. The sketches of the paths are represented by straight lines that consecutively join each point of the coordinates file between 0 and 1 in the fourth column, which correspond to all the consecutive coordinates of continuous deposition. These coordinates are located in all the direction changes of the printhead during the deposition.

This type of model presents sharp corners in very acute angles, and, in the case of a close sweep, Abaqus/CAE 6.14-1 gives an error in the modelling process. This last problem can be solved by automatically checking this condition (automated modelling). If the first and the last points of a sweep are too close (distance < extrusion_width/2), it is considered a close sweep, and, as a consequence, a division is applied by separating the last line of the sweep into a new sweep, as shown in Figure 3. Otherwise (open path), the sweep does not need to be divided.

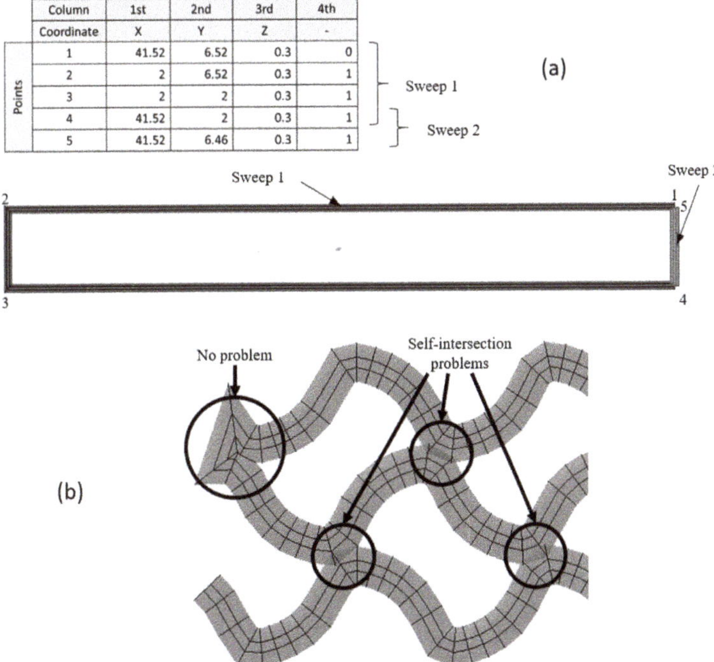

Figure 3. Division of close sweep (**a**) and self-intersection of a gyroid fill pattern (**b**).

Although this additional checking solved the modelling error, when this modelling strategy is applied to more complex parts such as a scaffold with a gyroid fill pattern, the model presents self-intersections that often result in meshing problems in Abaqus/CAE 6.14-1 (overlapped volumes). An example of a self-intersection is shown in Figure 3.

2. Line-by-Line Modelling

To avoid the problems of "whole-part modelling", another strategy called "line-by-line modelling" was developed, which consists of creating a sweep per line, thus resulting in empty corners at the joints (Figure 4).

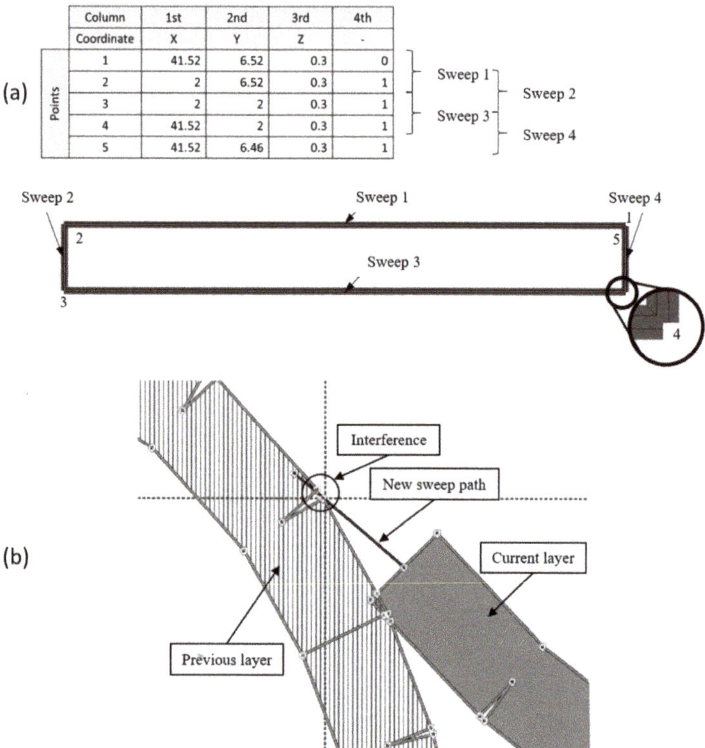

Figure 4. Line-by-line modelling (**a**) and its error (**b**).

This new strategy led to new problems previously undetected in Abaqus/CAE 6.14-1. The error in the modelling process appears while trying to sweep certain lines of the path that interfere with points present in previous layers, corresponding to the corners created in the empty joints. The error is shown in Figure 4.

3. Line-by-Line Modelling with Corner Revolutions

The previously mentioned error was solved by filling the joints with revolutions. This led to a new strategy called "line-by-line modelling with corner revolutions". The result of this technique is shown in Figure 5. This strategy requires the creation of planes perpendicular to the path at each joint to create the corresponding sketches (the same filament profile depicted in Figure 4) for the revolution feature with respect to the vertical symmetric axis. As a consequence, each line and corner would require several operations, which would slow down the modelling process. For this reason, it was decided to find another strategy that does not require the separation of each line from the path but avoids self-intersections, which is the "modelling by sections".

4. Modelling by Sections

The last strategy proven was the "modelling by sections", where the sweeps that present self-intersection are divided to avoid the previous error. Although the model presents sharp corners as in the "whole-part modelling" strategy, it was later proven that their influence is not too relevant from the mechanical point of view, at least for the prediction of the stiffness in static conditions.

The "modelling by sections" strategy identifies the peaks of each path (relative maximums and minimums) in x- and y-coordinates. Then, it compares the distance between all the relative maximums and minimums of the x-coordinate on the one hand and, on the other hand, all the relative maximums and minimums of the y-coordinate. If there is any distance lower than the extrusion width and there are at least three points between the two to be compared, there is a self-intersection in that coordinate (x or y). The reason for excluding points that are located close is that, in those cases, there would be no self-intersection in modelling, since the self-intersection problem only arises when the path separates from the previous path and intersects again (Figure 6).

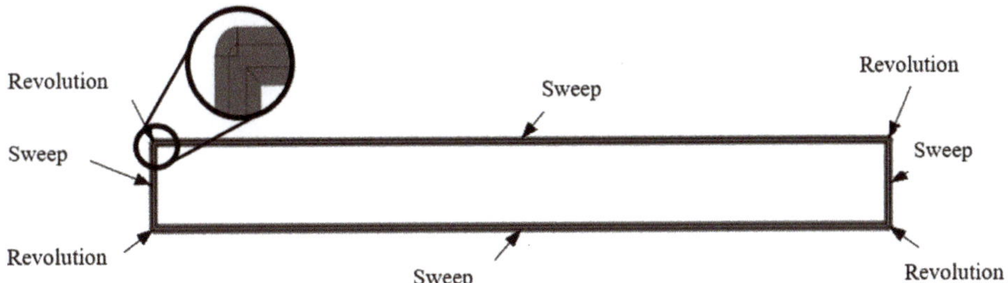

Figure 5. Line-by-line modelling with corner revolutions.

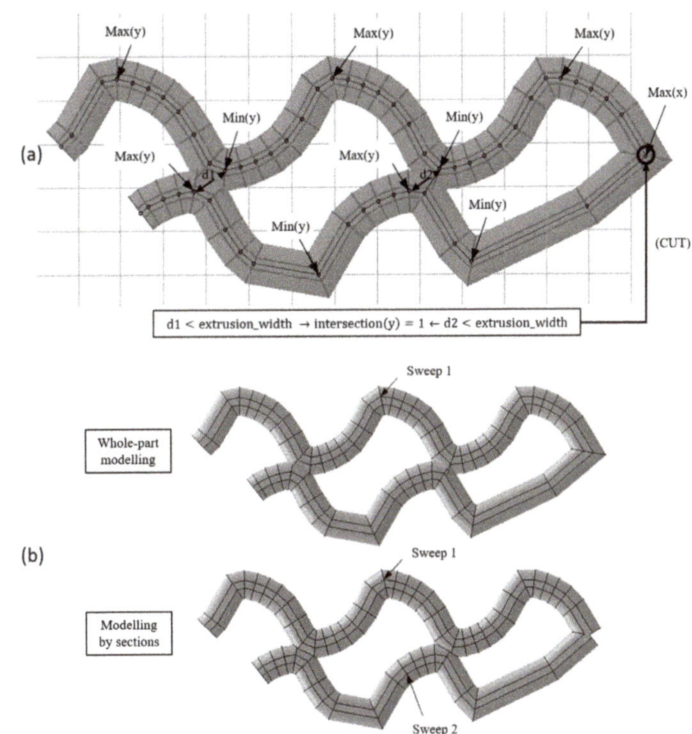

Figure 6. Self-intersection identification and division process (**a**) and its result (**b**).

If there is a self-intersection in x-coordinate, the path is divided into the y-coordinate peaks, and vice versa. This cut is represented in the coordinate matrix as "-1" in the fourth column. It works differently than "0" as it is the last point of the previous sweep and the first point of the next one.

An example of a division process is shown in Figure 6, in which the distances between peaks are higher than the extrusion width, except for d1 and d2. These self-intersections are located at the peaks of the y-coordinate; therefore, a cut at the x-coordinate peak is assigned "-1" in the fourth column.

However, there is also the possibility that there is a self-intersection in the sweep, but that it does not exactly coincide with the peaks. Therefore, if two points are near (the distance is greater than the extrusion width but smaller than the 150% of this width), it also compares the maximum and its two previous and following points with the minimum and its two previous and following points.

The diagram of the whole process and conditions for section divisions is shown in Figure 7.

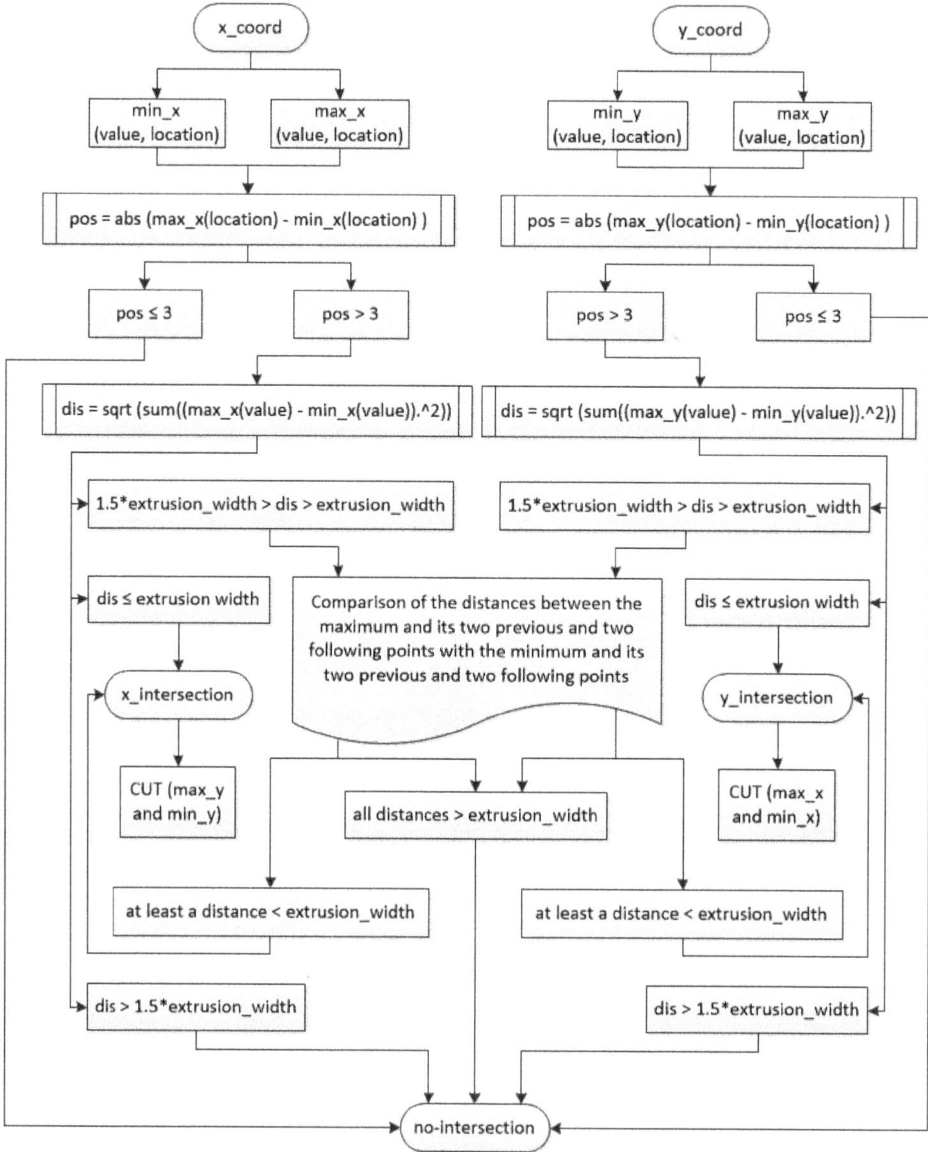

Figure 7. Modelling by sections diagram.

As another example, in Figure 8, the identification of relative maximums and minimums of a sweep path is shown. Some of them fulfil the condition of being at a distance smaller than the extrusion width. However, an example of a maximum and a minimum that are close is shown, but the points that make the sweep auto-intersect are not the peaks at points 2 and 7, 1 and 6, or 6 and 7.

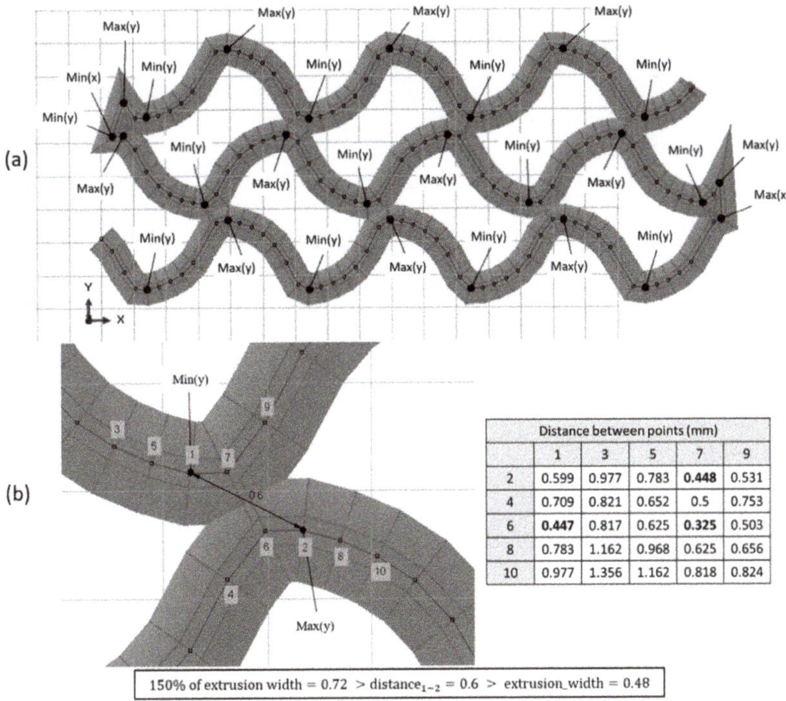

Distance between points (mm)					
	1	3	5	7	9
2	0.599	0.977	0.783	**0.448**	0.531
4	0.709	0.821	0.652	0.5	0.753
6	**0.447**	0.817	0.625	**0.325**	0.503
8	0.783	1.162	0.968	0.625	0.656
10	0.977	1.356	1.162	0.818	0.824

Figure 8. Peak identification in a complex geometry (**a**) and distance comparison of peaks and their previous and following points (**b**).

The "−1" value is also used to indicate the cut of a close sweep, applying the same process as in the "whole-part modelling" to detect it, as shown in Figure 9. It should be noted that, if there are no self-intersections in the model, the resulting model will be the same as that obtained with "whole-part modelling". Thus, this last modelling strategy is suitable for complex geometries, while "whole-part modelling" is useful only for simple parts.

5. Comparison of Modelling Strategies

The resulting shape differences between the developed modelling strategies are shown in Figure 10 for simple and complex parts (with self-intersection). Empty corners are found in the division of the last section of "whole-part modelling" and "modelling by sections", in the separation of self-intersecting paths in "modelling by sections", and between each section of "line-by-line modelling". Sharp corners can be found in both "whole-part modelling" and "modelling by sections". Lastly, it can be observed that, in complex parts, "line-by-line modelling", "line-by-line modelling with corner revolutions", and "modelling by sections" avoid self-intersections.

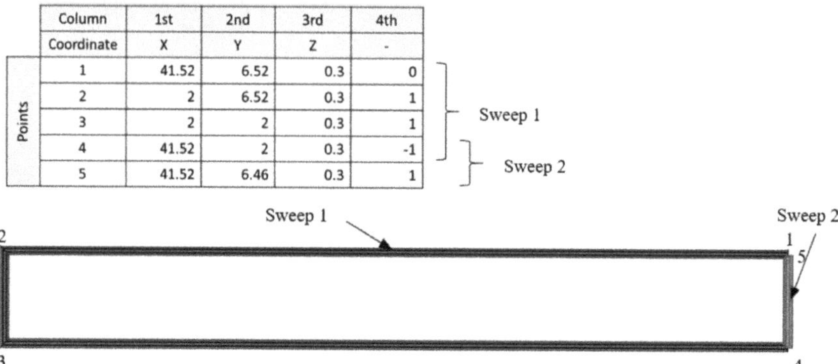

	Column	1st	2nd	3rd	4th
	Coordinate	X	Y	Z	-
Points	1	41.52	6.52	0.3	0
	2	2	6.52	0.3	1
	3	2	2	0.3	1
	4	41.52	2	0.3	-1
	5	41.52	6.46	0.3	1

Figure 9. Close sweep in modelling by sections.

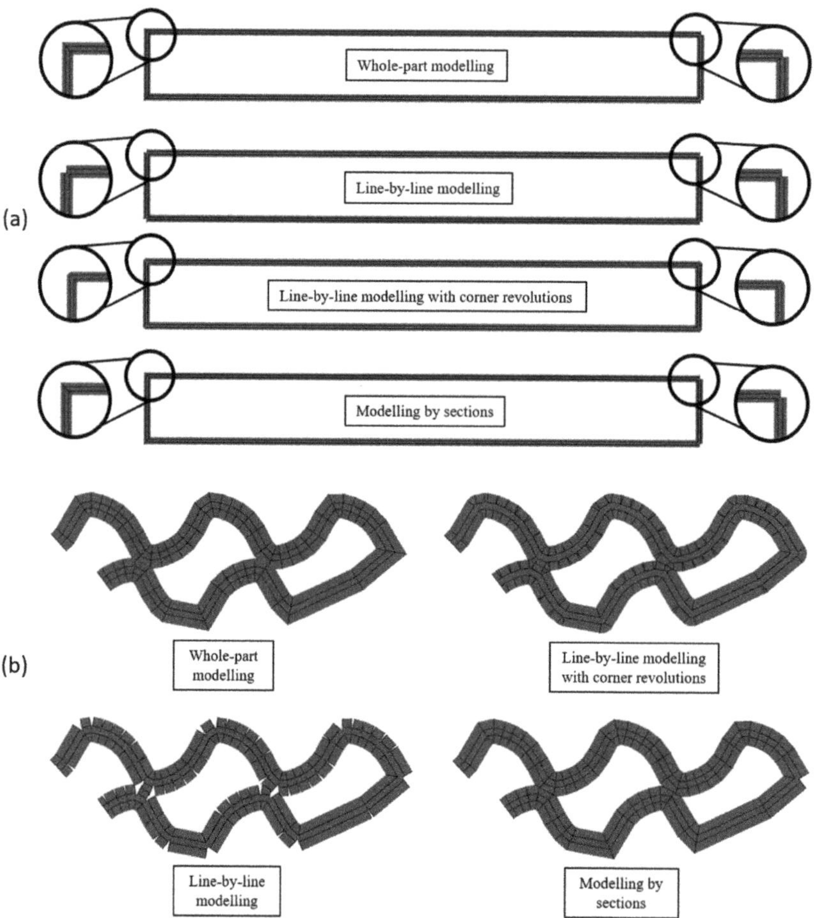

Figure 10. Comparison of modelling strategies in simple parts (**a**) and complex parts (**b**).

To validate the modelling strategies, the test specimen previously defined was modelled following the DECODE methodology steps. To obtain the G-code file, the solid part was introduced in Slic3r 1.3.0, and the printing parameters listed in Table 3 were set.

Table 3. Printing Parameters of Test Specimen.

Print Settings	
External perimeter extrusion width	0.48 mm (1.87 mm^3/s)
Perimeter extrusion width	0.48 mm (3.74 mm^3/s)
Infill extrusion width	0.48 mm (3.74 mm^3/s)
Filament diameter	1.75 mm
Infill speed	30 mm/s
Nozzle diameter	0.4 mm
Temperature	200 °C
Layer height	0.3 mm
First layer height	0.3 mm
Number of perimeters/solid layers	1
Fill angle	0°
Fill density	20%
Fill pattern	Concentric

The modelling strategies used were "whole-part modelling", "line-by-line modelling", "line-by-line modelling with corner revolutions", and "modelling by sections". All the models obtained were simulated with a three-point flexural test in Abaqus/CAE 6.14-1, as shown in Figure 11. For this purpose, three rigid cylindrical parts with a radius and a height of 5 mm were designed, and surface-to-surface interaction was defined between the test specimen and each cylinder. A friction coefficient of 0.15 was assigned to the tangential behaviour of the interaction, and the normal behaviour was defined as a hard contact. The models were meshed with a second-order tetrahedral mesh and a seed size of 1 mm. The supports were encastred, and a downward vertical displacement of 1 mm was applied to the crosshead. After the definition of all the above conditions, the reaction forces obtained in the different cases were as shown in Table 4.

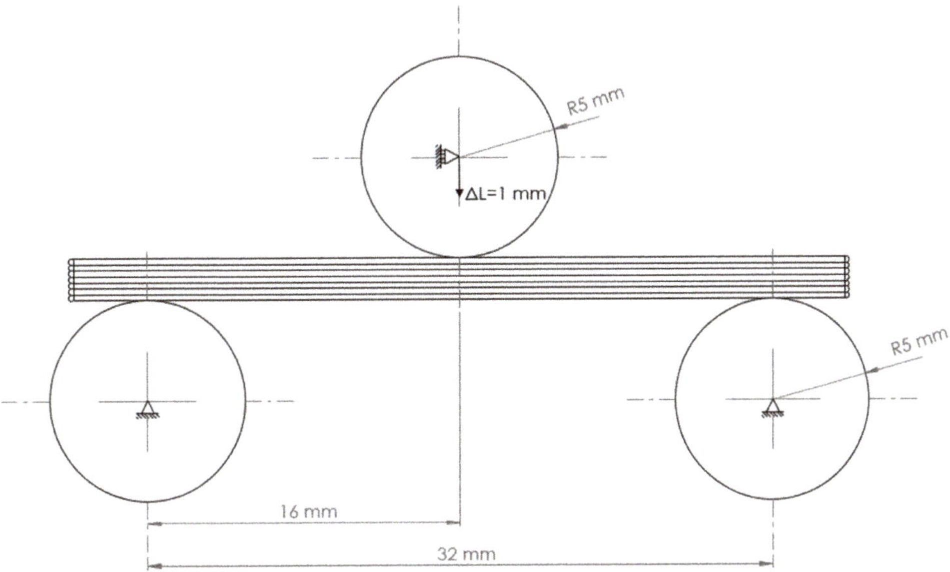

Figure 11. Flexural test of test specimen.

Table 4. Comparison of Simulation Results.

Modelling Strategy	Reaction Force (N)
Whole-part modelling	13.4898
Line-by-line modelling	13.4721
Line-by-line modelling with corner revolutions	13.4759
Modelling by sections	13.4898

The resultant reaction forces did not vary significantly among the different modelling strategies, and it was found that they were the same in "whole-part modelling" and "modelling by sections", since, in parts without self-intersection, their structures were the same. For this reason, it was decided to choose "modelling by sections" as the best strategy, since fewer modelling errors arose, and the modelling strategy is relatively easy to automate.

2.3.4. Compression Simulation by Finite Element Analysis

In order to obtain the stiffness and elastic modulus of the scaffolds models, they were simulated in Abaqus/CAE 6.14-1 by FEA. In particular, a compression test was applied with the following boundary conditions: the vertical movement was restricted at the base of the scaffold, two points of this base were encastred to avoid displacement in the horizontal plane, and a vertical displacement of 0.2 mm was applied on the top layer to simulate the compression movement of the compression plate.

Another important part of the simulation conditioning is the material properties. The selection of the model that defines the material behaviour may have a great influence on the results. In this case, linear behaviour of the elastic material (isotropic) was assumed (based on previous mechanical tests) and nonlinear effects of large deformations and displacement were included. Considering the selected material behaviour, the density and tensile elastic modulus were introduced according to the datasheet of the PCL, presented in Table 1, and the Poisson's ratio was defined as 0.46, according to some studies [38,39]. Although several physical parameters of the material may have an effect on the mechanical behaviour, the most relevant properties are the elastic modulus and Poisson's ratio. However, other models could be applied depending on the availability in the FEA software (hyperelasticity, plasticity, nonlinear behaviour, anisotropy, etc.). Each of these models would require different physical parameters.

On the other hand, regarding the mesh type definition, geometry-based models were meshed with C3D10 (10-node quadratic tetrahedron) and C3D4 (four-node linear tetrahedron) element types, and the voxel-based models were meshed only with C3D4, the default element type for converting voxels to tetrahedrons. This difference was due to the fact that the voxelised model does not allow meshing with second-order elements. The seed size of the mesh was varied depending on the part in order to avoid distorted elements, as well as the application of virtual topology. The meshing conditions for each model are shown in Table 5. Moreover, rigid joints between layers (a single solid) were assumed.

During the model conditioning step in Abaqus/CAE 6.14-1, it was not possible to mesh the CAD model of the 50_gyr scaffold with quadratic elements.

After the simulation, the reaction forces were obtained in the base layer and, therefore, the Young's modulus was evaluated according to Equation (1), extracted from ISO 604:2002 (plastics—determination of compressive properties). Note that the equivalent area corresponds to the area of the base as if it were a solid part.

$$E = \frac{\sigma}{\varepsilon} = \frac{F/A}{\Delta L/L_0} = \frac{F \cdot L_0}{A \cdot \Delta L},$$

(1)

where E is the Young's modulus or modulus of elasticity (MPa), σ is the tensile stress (MPa), ε is the deformation, F is the reaction force (N), A is the equivalent area (mm^2), ΔL is the displacement (mm), and L_0 is the original height (mm).

Table 5. Mesh Assignment.

Part	Modelling Technique	Mesh Type	Seed Size (mm)	Virtual Topology
50_rect	DECODE	C3D10 C3D4	0.3 0.4	Yes No
	VOLCO	C3D4	0.05	No
60_rect	DECODE	C3D10 C3D4	0.1 0.1	No No
	VOLCO	C3D4	0.05	No
50_gyr	DECODE	C3D10 C3D4	- 0.05	- No
	VOLCO	C3D4	0.05	No

2.4. Scaffold Manufacturing and Compression Tests

The real scaffolds were manufactured by FDM in a BQ Hephestos 2 3D printer. The extrusion temperature was set to 190 °C.

Four replicas per group were subjected to mechanical characterisation under compression load. The mechanical testing was carried out in a LIYI (LI-1065, Dongguan Liyi Environmental Technology Co., Ltd., Dongguan, China) testing machine in displacement control mode. The crosshead speed was set to 1 mm/min, and the compression modulus was calculated as the slope of the initial segment in the stress–strain graph.

2.5. Morphological Characterisation

Printed scaffolds were geometrically compared with models by microscope imaging to determine which representation was closer to reality. Concretely, the Olympus BX51 microscope (Olympus Co., Ltd., Tokyo, Japan) with a 2× magnification factor was used to collect images of the top of the three types of scaffolds (50_rect, 60_rect, and 50_gyr).

3. Results and Discussion

3.1. Modelling Efficiency

3.1.1. Scaffold Modelling Efficiency

The modelling efficiency was estimated from the time and CPU required for modelling and simulation processes. The times required to generate the coordinates and to follow the modelling and simulation steps in each scaffold configuration and both modelling techniques are collected in Table 6, as well as the minimum memory required for simulation.

The computer used for modelling and simulation was an Intel® Core™ i9-9820X CPU @3.30GHz, with 64.0 GB installed RAM and a 64-bit operating system, x64-based processor.

The use of C3D4 meshes saved memory in the simulation of the models, but not always time. In general, modelling and simulations using DECODE were more efficient, although they were very similar to VOLCO in complex geometries such as the gyroid scaffold, which also presented the difficulty of not being able to be meshed with second-order elements with the available software and hardware.

Table 6. CPU Time and Memory Required to Model and Simulate the Scaffolds.

Part	Modelling Technique	Mesh Type	Coordinates Generation (s)	Modelling Process (s)	Simulation Process (s)	Total Time (h:mm:ss)	Minimum Memory Required (MB)
50_rect	DECODE	C3D10	10	55	255	0:05:10	1925
		C3D4			477	0:08:52	326
	VOLCO	C3D4		2339	4967	2:01:46	9549
60_rect	DECODE	C3D10	7	33	1559	0:26:32	7491
		C3D4			319	0:05:52	1238
	VOLCO	C3D4		959	2955	1:05:14	7985
50_gyr	DECODE	C3D10	7	338	-	-	-
		C3D4			2970	0:55:08	6905
	VOLCO	C3D4		920	2280	0:53:20	6901

The models generated through the different modelling techniques are presented in Figure 12.

3.1.2. Modelling Limitations

Voxel-based and geometry-based modelling and simulation methodologies were studied in terms of applicability to the full range of FDM structures. In the particular case of dimensional and shape adequacy, limitations were found depending on the software and hardware used, since, for large or complex parts, Abaqus/CAE 6.14-1 cannot support simulation, meshing, or modelling. For example, for the large part presented in Figure S12, modelling was not possible from the python file. On the other hand, the 50_gyr scaffold and the corner part shown in Figure S13 were modelled but not meshed by Abaqus/CAE 6.14-1 due to complex geometry and large memory required to mesh the part, respectively. Lastly, in other cases, parts could be modelled and meshed but not simulated because the minimum memory required for FEA was higher than that available.

On the other hand, in both modelling techniques, Microsoft Excel was used to record the coordinates obtained from the G-code. In the most favourable case, using Microsoft Excel 64 bit, the worksheet was limited to 1,048,576 rows by 16,384 columns, but not to the maximum use of 2 GB of RAM as in the 32 bit version. In short, as each coordinate is saved in a new row, there would be a limit of 1,048,576 coordinates in VOLCO. In DECODE, this would not be a limitation as it allows the reading of several worksheets continuously, as well as in the coordinate matrix generated in Matlab R2021a, which would only present memory limitations due to the available hardware.

The computer used for modelling had 64.0 GB of installed RAM, but the memory available for all arrays was 43.7 GB. In addition, each element of the matrix required approximately 8 bytes of memory. Thus, the number of elements was limited to 5.86×10^9.

On the one hand, the array limitations and the four columns needed to define a coordinate set of 1.465×10^9 as the maximum number of coordinates in the matrix.

On the other hand, this affected VOLCO in the construction of the voxel matrix, which would need as many elements as voxels (defined in a binary way) needed to define the part. This number depends on the dimensions and voxel size, as defined in Equation (2).

$$\text{No. voxels} = \frac{(X_{end} - X_{start}) \cdot (Y_{end} - Y_{start}) \cdot (Z_{end} - Z_{start})}{\text{VoxelSize}^3}, \tag{2}$$

where X_{end}, Y_{end}, Z_{end} are the maximum coordinates (mm), X_{start}, Y_{start}, Z_{start} are the minimum coordinates (mm), and VoxelSize is the set voxel size (mm).

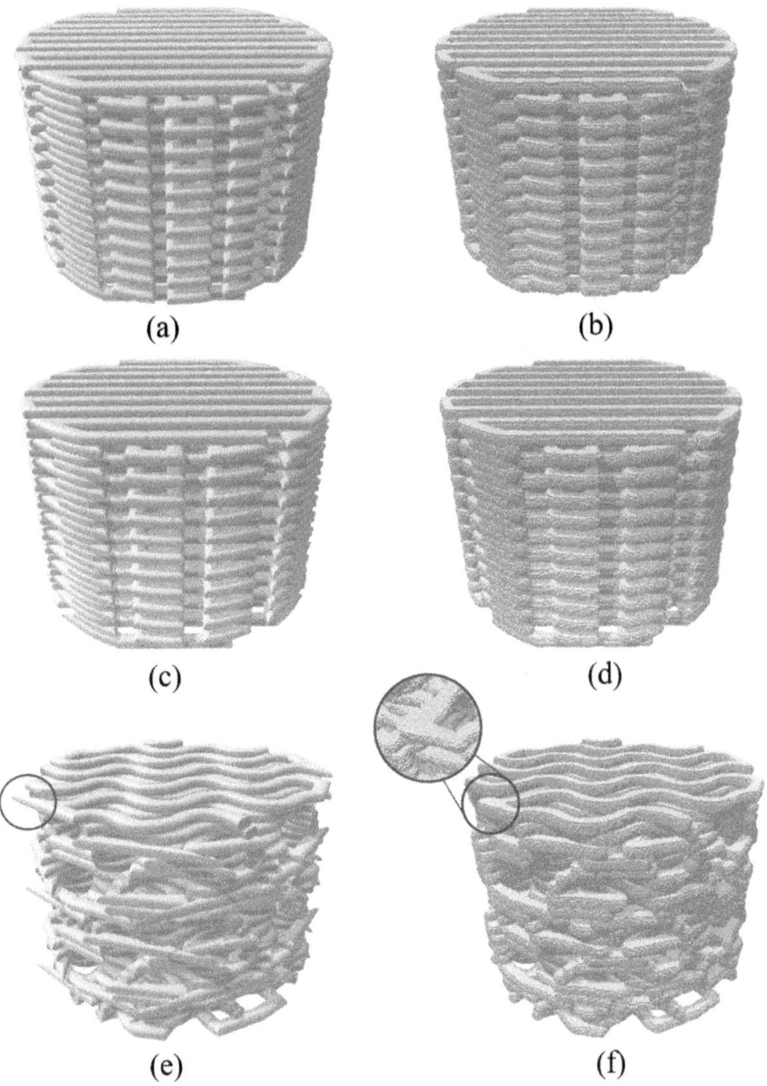

Figure 12. Models: (**a**) 50_rect DECODE (**b**) 50_rect VOLCO, (**c**) 60_rect DECODE, (**d**) 60_rect VOLCO, (**e**) 50_gyr DECODE and (**f**) and 50_gyr VOLCO.

Several examples are shown in Table 7 to determine the practical-sized structures that can be modelled by VOLCO. Those that require more memory than available cannot be modelled. It should be noted that a maximum voxel size of 100 μm was set to allow proper filament resolution, as the nozzle tip used was 0.4 mm, which allowed a maximum layer height of 0.3 mm.

Table 7. Number of voxels according to dimensions and set voxel size.

Type	Dimensions (mm)	Voxel Size (μm)	Min. No. Voxels (Memory Required) vs. Max. No. Elements (Memory Available)
Specimen	$80 \times 10 \times 4$	100	3.2×10^6 (24.41 MB) < 5.86×10^9 (44,713 MB)
Specimen	$80 \times 10 \times 4$	5	2.56×10^{10} (190.73 GB) > 5.86×10^9 (43.66 GB)
Large part	$200 \times 250 \times 500$	100	2.5×10^{10} (186.26 GB) > 5.86×10^9 (43.66 GB)
Corner part	$181 \times 181 \times 181$	100	5.93×10^{10} (44.18 GB) > 5.86×10^9 (43.66 GB)

In summary, Table 8 presents the steps that are possible or not for each part, using each modelling methodology, with the available hardware and software.

Table 8. Part Simulation Feasibility.

Part	VOLCO				DECODE			
	STL Generation	Abaqus/CAE 6.14-1			PY Generation	Abaqus/CAE 6.14-1		
		Importation	Meshing	FEA		Importation	Meshing	FEA
Specimen	✓	✓	✓	✓	✓	✓	✓	✓
Large part					✓			
Corner part					✓	✓		
Rectilinear scaffolds	✓	✓	✓	✓	✓	✓	✓	✓
Gyroid scaffold	✓	✓	✓	✓	✓	✓		
Mini specimen	✓	✓	✓	✓	✓	✓	✓	✓

In all cases, the limitation for modelling, meshing, or FEA was the available memory, except for the meshing of the geometry-based model of the gyroid scaffold and the STL file generation of the large part by VOLCO. In the case of the scaffold, this was due to the limitations for meshing complex parts in Abaqus/CAE 6.14-1. In the case of VOLCO, the large part had 7,196,287 coordinates that could not be recorded in a single Microsoft Excel worksheet.

3.2. Dimensional Accuracy

3.2.1. Volume

One of the most important parameters to compare the geometrical adjustment is the volume. It can be measured in the models through Abaqus/CAE 6.14-1 query tools, but VOLCO also generates an output with this information. The theoretical volume was extracted from the G-code file, which specifies the amount of filament used.

All the data collected are compared in Table 9. The deviation between the theoretical and the VOLCO model volumes was almost 0% because of the application of the spline function previously mentioned.

Table 9. Comparison of Volumes.

Part	Theoretical Volume (mm^3)	Experimental Volume (mm^3)	Modelling Technique	Model Volume (mm^3)	Theoretical Deviation	Experimental Deviation
	Vt	Ve		Vm	$\lvert Vm - Vt \rvert /Vt \times 100$	$\lvert Vm - Ve \rvert /Ve \times 100$
50_rect	278.8	274.91	DECODE	278.46	0.12%	1.29%
			VOLCO	278.8	0%	1.42%
60_rect	231.16	224.18	DECODE	230.95	0.09%	3.02%
			VOLCO	231.15	0%	3.11%
50_gyr	188.83	183.64	DECODE	188.81	0%	2.82%
			VOLCO	188.83	0%	2.83%

Compared with the theoretical data, the differences in volume were negligible. On the other hand, taking into account the deviations from the experimental results, between 1.29% and 3.11%, it could be assumed that the dimensional accuracy of the models is acceptable. This last statement is reinforced by the premise that the selected printing parameters and the temperature have a considerable influence on the final dimensions of the printed part, which do not always respect the theoretical measurements [40].

3.2.2. Dimensions and Shape

Another important comparison involves the dimensions of the models compared with the initial solid design (theoretical) and the printed scaffolds (real), as shown in Table 10.

Table 10. Comparison of Dimensions.

Part	Dimension	Theoretical (mm)	Real (mm)	Modelling Technique	Model (mm)	Theoretical Deviation	Real Deviation
		T	R		M	$\lvert M - T \rvert/T \times 100$	$\lvert M - R \rvert/R \times 100$
50_rect	Diameter	10	7.06	DECODE	10.104	1.04%	43.12%
				VOLCO	10.025	0.25%	42.00%
	Height	7	6.04	DECODE	6.95	0.71%	15.07%
				VOLCO	6.95	0.71%	15.07%
60_rect	Diameter	10	6.83	DECODE	10.098	0.98%	47.85%
				VOLCO	10.05	0.50%	47.14%
	Height	7	5.91	DECODE	6.95	0.71%	17.60%
				VOLCO	6.95	0.71%	17.60%
50_gyr	Diameter	10	7.42	DECODE	10.832	8.32%	45.98%
				VOLCO	10.075	0.75%	35.78%
	Height	7	5.63	DECODE	6.954	0.66%	23.52%
				VOLCO	6.95	0.71%	23.45%

The differences between the theoretical scaffold and the models lie in the modelling techniques limitations, as shown in Table 8. In the case of the DECODE modelling, the sweep operation creates sharp tips in very acute angles. On the other hand, VOLCO modelling expands the filament in joints. However, another determining factor in the dimensional inaccuracy of FDM printed parts is nonuniform temperature gradients [41]. Furthermore, there are physicochemical characteristics of the materials such as glass transition temperature or free volume that cannot be considered in the modelling techniques, as both methods (geometry-based and voxel-based) utilise the G-code file to generate the geometry. This limitation could be relevant in other technologies such as electrospinning or rotary jet spinning [42]; however, in the case of material extrusion, the effect could be negligible.

The comparison of filaments shape and width between the models and printed scaffolds was done for 50_rect, 60_rect, and 50_gyr types, as shown in Figure 13. The images and measurements of the printed scaffolds were obtained by microscopy imaging. As is shown in Figure 13, the voxelised filaments did not have constant width as in reality, in contrast with the CAD model. However, the voxelised filaments had smaller widths than the other scaffolds.

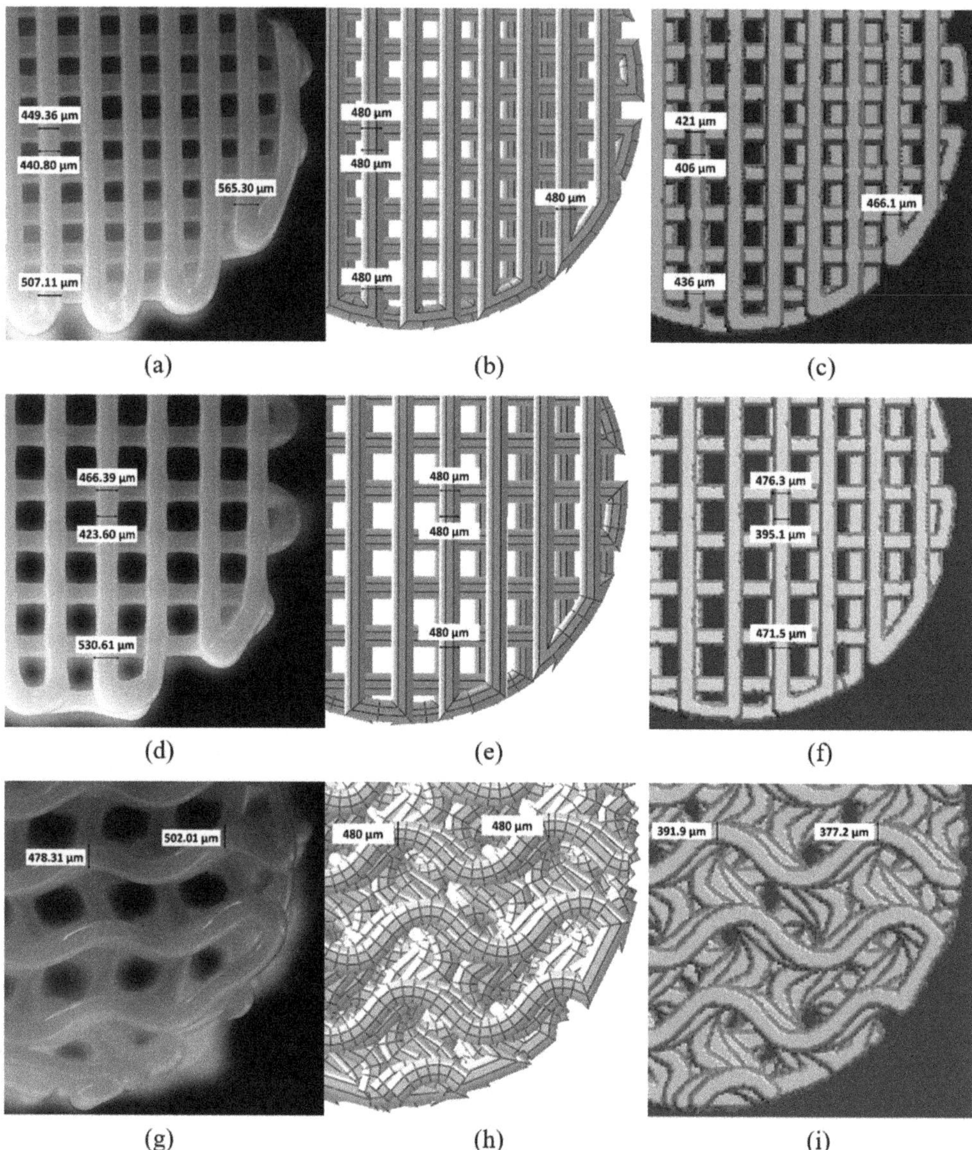

Figure 13. Filament width comparison of: (**a**) 50_rect microscopy image of printed scaffold, (**b**) 50_rect CAD model, (**c**) 50_rect voxelised model, (**d**) 60_rect microscopy image of printed scaffold, (**e**) 60_rect CAD model, (**f**) 60_rect voxelised model, (**g**) 50_gyr microscopy image of printed scaffold, (**h**) 50_gyr CAD model and (**i**) 50_gyr voxelised model.

Therefore, the DECODE model could be an approximation of the printed part, representing filaments with non-variable width but with an average value similar to that along the path. On the other hand, the VOLCO model presented a final shape closer to the real scaffold, although the value of the filament width did not necessarily coincide.

3.3. Equivalent Elastic Modulus

The models for the different scaffold configurations were simulated with Simulia Abaqus/CAE 6.14-1. A compression test was applied to each model; accordingly, the reaction forces, which appear in the base layer, and the Young's moduli were obtained. The results of the simulations and the experimental tests are collected in Table 11.

Table 11. Comparison of Elastic Modulus.

Part	Experimental Young Modulus (MPa)	Modelling Technique	Mesh Type	Model Young Modulus (MPa)	Deviation
	Er			Em	$\|Em - Er\|/Er \times 100$
50_rect	58.88	DECODE	C3D10	49.16	16.52%
			C3D4	85.19	44.68%
		VOLCO	C3D4	103.4	75.61%
60_rect	55.58	DECODE	C3D10	28.93	47.95%
			C3D4	40.56	27.03%
		VOLCO	C3D4	71.09	27.91%
50_gyr	39.21	DECODE	C3D10	-	-
			C3D4	11.88	69.70%
		VOLCO	C3D4	41.71	6.37%

During the simulation step, it was not possible to mesh the CAD model of the 50_gyr scaffold, but it was possible to obtain a meshed 50_gyr model with linear elements, trying multiple seed sizes until finding that which did not generate distorted elements. The meshing problem is one of the most recurrent issues in geometry-based modelling methodology, especially in complex geometries. In these cases, voxel-based modelling is more suitable.

The comparison of the results of both modelling techniques shows that the voxel-based model was stiffer than the geometry-based one. This is because VOLCO modelling takes into account the intersection of the filaments, expanding the material at the joints. This leads to larger sections of material in the contact between layers, consequently increasing the stiffness of the part under compression compared to DECODE modelling.

The mesh type used also influenced the stiffness of the part, as can be seen in the different geometry-based models. Linear elements (C3D4) were less precise and, therefore, tended to increase the stiffness of the model.

Another conclusion that can be extracted from the elastic modulus results is that related to the prediction of the part stiffness. As previously demonstrated and presented in Table 11, there is a large dimensional deviation between the printed parts and the theoretical values. Moreover, during the deposition process, voids may appear, leading to a higher porosity and consequently, affecting the mechanical properties. This effect depends on the materials used, especially in composite materials [43]. Therefore, both the dimensional deviation and the voids arising lead to differences between the simulated and the experimental mechanical results. Using the current procedure, it is not possible to predict the exact value of the Young's modulus, whether with the VOLCO or DECODE modelling techniques, but it is possible to accurately compare different geometries. In other words, the modulus variations are not proportional between the experimental and simulation results, but the most rigid model according to the simulations corresponds with the most rigid part.

4. Conclusions

This is the first study to consider modelling the printed parts before manufacturing to apply FEA and to compare which modelling technique is more suitable, practical, and efficient for a specific case. Furthermore, a new CAD-based modelling technique based on sweep operations was developed to obtain any part from a G-code file, without geometry limitations, in an automated way and with capabilities to obtain the 3D model in a few seconds.

One of the conclusions that can be extracted from the results is that the models obtained cannot be considered to be accurate representations of the printed parts, as they do not take into account dimensional variations. On the other hand, the models are not comparable, as they provide very different results depending on the methodology or type of mesh used. VOLCO modelling gives stiffer models due to the expansion of the material at the joints. However, the elastic modulus of the models follows the same pattern as the experimental one. In other words, even though they are not proportional, the most rigid model corresponds to the most rigid printed part in both methodologies. Therefore, it is possible to determine which will be the stiffest geometry.

Although geometry-based modelling is faster and more computationally efficient in simple geometries, it also presents more meshing problems than the voxel-based one. These meshing problems can be solved by reducing the seed size or the application of "virtual topology"; however, in certain cases, this does not work either.

Related to the dimensional accuracy, the volume differences between the models and the theoretical reference are negligible. In the CAD model, the volume is very close to the one expected in the G-code file. On the other hand, in the voxelised model, a manual adjustment of the volume through the spline function is required.

For all these reasons, if the objective is to find the fastest and easiest modelling technique to design and optimise a part to be printed with FDM technology, the new automatic CAD-based modelling is the preferred option. However, for small and complex models, voxel-based modelling is the most suitable option.

Following this research, the next steps need to be focused on optimising 3D printed parts before manufacturing, using the DECODE modelling technique (or VOLCO for small models) and FEA to drive the optimisation, thereby reducing the experimental work and improving the cost-efficiency of the process.

Supplementary Materials: The following are available online at https://www.mdpi.com/article/10.3390/ma14195670/s1: Figure S1. FEA result of 50_gyr scaffold (base reaction forces and deformation) modelled by DECODE with C3D4 mesh type; Figure S2. FEA result of 50_gyr scaffold (base reaction forces and deformation) modelled by VOLCO with C3D4 mesh type; Figure S3. FEA result of 50_rect scaffold (base reaction forces and deformation) modelled by DECODE with C3D4 mesh type; Figure S4. FEA result of 50_rect scaffold (base reaction forces and deformation) modelled by DECODE with C3D10 mesh type; Figure S5. FEA result of 50_rect scaffold (base reaction forces and deformation) modelled by VOLCO with C3D4 mesh type; Figure S6. FEA result of 60_rect scaffold (base reaction forces and deformation) modelled by DECODE with C3D4 mesh type; Figure S7. FEA result of 60_rect scaffold (base reaction forces and deformation) modelled by DECODE with C3D10 mesh type; Figure S8. FEA result of 60_rect scaffold (base reaction forces and deformation) modelled by VOLCO with C3D4 mesh type; Figure S9. FEA result (stress and deformation) of test specimen; Figure S10. Specimen modelling and FEA process by VOLCO; Figure S11. Specimen modelling and FEA process by DECODE; Figure S12. G-code generation of large part; Figure S13. Corner part modelling process.

Author Contributions: Conceptualization, G.V., R.P., M.M. and M.E.A.-D.; methodology, G.V., R.P. and A.G.; software, G.V. and A.G.; investigation, G.V., R.P., A.G. and M.E.A.-D.; resources, M.M.; data curation, G.V. and M.E.A.-D.; writing—original draft preparation, G.V., R.P. and M.E.A.-D.; writing—review and editing, R.P. and M.M.; visualization, A.G., M.E.A.-D. and M.M.; supervision, M.M. and R.P.; project administration, M.M.; funding acquisition, M.M. All authors have read and agreed to the published version of the manuscript.

Funding: This research was funded by the BioAM project (Improvement of the biofunctionality of polymeric scaffolds obtained by additive manufacturing, DPI2017-88465-R) from the Ministerio de Ciencia, Innovación, y Universidades, the PhD Grant Program of the University of Las Palmas de Gran Canaria (PIFULPGC-2019-ING-ARQ-1), and the BAMOS project (Biomaterials and additive manufacturing: osteochondral scaffold innovation applied to osteoarthritis, H2020-MSCA-RISE-2016-734156) from the European Union's Horizon 2020 research and innovation programme.

Institutional Review Board Statement: Not applicable.

Informed Consent Statement: Not applicable.

Data Availability Statement: The data presented in this study are available within the article.

Conflicts of Interest: The authors declare no conflict of interest.

References

1. Perez, R.A.; Mestres, G. Role of pore size and morphology in musculo-skeletal tissue regeneration. *Mater. Sci. Eng. C* **2016**, *61*, 922–939. [CrossRef]
2. Chuenjitkuntaworn, B.; Inrung, W.; Damrongsri, D.; Mekaapiruk, K.; Supaphol, P.; Pavasant, P. Polycaprolactone/Hydroxyapatite composite scaffolds: Preparation, characterization, and in vitro and in vivo biological responses of human primary bone cells. *J. Biomed. Mater. Res. Part A* **2010**, *10*. [CrossRef]
3. Gómez-Lizárraga, K.K.; Flores-Morales, C.; Del Prado-Audelo, M.L.; Álvarez-Pérez, M.A. Polycaprolactone-and polycaprolactone/ceramic-based 3D-bioplotted porous scaffolds for bone regeneration: A comparative study. *Mater. Sci. Eng. C* **2017**, *79*, 326–335. [CrossRef] [PubMed]
4. Chen, S.; Guo, Y.; Liu, R.; Wu, S.; Fang, J.; Huang, B.; Li, Z.; Chen, Z.; Chen, Z. Tuning surface properties of bone biomaterials to manipulate osteoblastic cell adhesion and the signaling pathways for the enhancement of early osseointegration. *Colloids Surfaces B Biointerfaces* **2018**, *164*, 58–69. [CrossRef] [PubMed]
5. Karageorgiou, V.; Kaplan, D. Porosity of 3D biomaterial scaffolds and osteogenesis. *Biomaterials* **2005**, *26*, 5474–5491. [CrossRef] [PubMed]
6. Agrawal, C.M.; Mckinney, J.S.; Lanctot, D.; Athanasiou, K.A. Effects of fluid flow on the in vitro degradation kinetics of biodegradable scaffolds for tissue engineering. *Biomaterials* **2000**, *21*, 2443–2452. [CrossRef]
7. Bosworth, L.A.; Downes, S. Physicochemical characterisation of degrading polycaprolactone scaffolds. *Polym. Degrad. Stab.* **2010**, *95*, 2269–2276. [CrossRef]
8. Hendrikson, W.J.; van Blitterswijk, C.A.; Rouwkema, J.; Moroni, L. The Use of Finite Element Analyses to Design and Fabricate Three-Dimensional Scaffolds for Skeletal Tissue engineering. *Front. Bioeng. Biotechnol.* **2017**, *5*, 1–13. [CrossRef] [PubMed]
9. Velasco, M.A.; Narváez-Tovar, C.A.; Garzón-Alvarado, D.A. Design, Materials, and Mechanobiology of Biodegradable Scaffolds for Bone Tissue Engineering. *Biomed Res. Int.* **2015**, *2015*. [CrossRef] [PubMed]
10. Naghieh, S.; Karamooz Ravari, M.R.; Badrossamay, M.; Foroozmehr, E.; Kadkhodaei, M. Numerical investigation of the mechanical properties of the additive manufactured bone scaffolds fabricated by FDM: The effect of layer penetration and post-heating. *J. Mech. Behav. Biomed. Mater.* **2016**, *59*, 241–250. [CrossRef]
11. Seol, Y.-J.; Kang, H.-W.; Lee, S.J.; Atala, A.; Yoo, J.J. Bioprinting technology and its applications. *Eur. J. Cardio-Thoracic Surg.* **2014**, *46*, 342–348. [CrossRef]
12. Ribeiro, J.F.M.; Oliveira, S.M.; Alves, J.L.; Pedro, A.J.; Reis, R.L.; Fernandes, E.M.; Mano, J.F. Structural monitoring and modeling of the mechanical deformation of three-dimensional printed poly (ε-caprolactone) scaffolds. *Biofabrication* **2017**, *9*, 025015. [CrossRef]
13. Qu, H. Additive manufacturing for bone tissue engineering scaffolds. *Mater. Today Commun.* **2020**, *24*, 1–16. [CrossRef]
14. Saygili, E.; Dogan-Gurbuz, A.A.; Yesil-Celiktas, O.; Draz, M.S. 3D bioprinting: A powerful tool to leverage tissue engineering and microbial systems. *Bioprinting* **2020**, *18*, e00071. [CrossRef]
15. Singh, D.; Babbar, A.; Jain, V.; Gupta, D.; Saxena, S.; Dwibedi, V. Synthesis, characterization, and bioactivity investigation of biomimetic biodegradable PLA scaffold fabricated by fused filament fabrication process. *J. Brazilian Soc. Mech. Sci. Eng.* **2019**, *41*, 1–13. [CrossRef]
16. Temple, J.P.; Hutton, D.L.; Hung, B.P.; Huri, P.Y.; Cook, C.A.; Kondragunta, R.; Jia, X.; Grayson, W.L. Engineering anatomically shaped vascularized bone grafts with hASCs and 3D-printed PCL scaffolds. *J. Biomed. Mater. Res. Part A* **2014**, *102*, 4317–4325. [CrossRef] [PubMed]
17. Roopavath, U.K.; Malferrari, S.; Van Haver, A.; Verstreken, F.; Rath, S.N.; Kalaskar, D.M. Optimization of extrusion based ceramic 3D printing process for complex bony designs. *Mater. Des.* **2019**, *162*, 263–270. [CrossRef]
18. Sahai, N.; Saxena, K.K.; Gogoi, M. Modelling and simulation for fabrication of 3D printed polymeric porous tissue scaffolds. *Adv. Mater. Process. Technol.* **2020**, 1–10. [CrossRef]
19. Souness, A.; Zamboni, F.; Walker, G.M.; Collins, M.N. Influence of scaffold design on 3D printed cell constructs. *J. Biomed. Mater. Res. Part B* **2018**, *106B*, 533–545. [CrossRef]
20. Soufivand, A.A.; Abolfathi, N.; Hashemi, S.A.; Lee, S.J. Prediction of mechanical behavior of 3D bioprinted tissue-engineered scaffolds using finite element method (FEM) analysis. *Addit. Manuf.* **2020**, *33*, 2214–8604. [CrossRef]
21. Schipani, R.; Nolan, D.R.; Lally, C.; Kelly, D.J. Integrating finite element modelling and 3D printing to engineer biomimetic polymeric scaffolds for tissue engineering. *Connect. Tissue Res.* **2020**, *61*, 174–189. [CrossRef]
22. Miranda, P.; Pajares, A.; Guiberteau, F. Finite element modeling as a tool for predicting the fracture behavior of robocast scaffolds. *Acta Biomater.* **2008**, *4*, 1715–1724. [CrossRef]

23. Entezari, A.; Fang, J.; Sue, A.; Zhang, Z.; Swain, M.V.; Li, Q. Yielding behaviors of polymeric scaffolds with implications to tissue engineering. *Mater. Lett.* **2016**, *184*, 108–111. [CrossRef]
24. Melancon, D.; Bagheri, Z.S.; Johnston, R.B.; Liu, L.; Tanzer, M.; Pasini, D. Mechanical characterization of structurally porous biomaterials built via additive manufacturing: Experiments, predictive models, and design maps for load-bearing bone replacement implants. *Acta Biomater.* **2017**, *63*, 350–368. [CrossRef]
25. Kadkhodapour, J.; Montazerian, H.; Darabi, A.C.; Anaraki, A.P.; Ahmadi, S.M.; Zadpoor, A.A.; Schmauder, S. Failure mechanisms of additively manufactured porous biomaterials: Effects of porosity and type of unit cell. *J. Machanical Behav. Biomed. Mater.* **2015**, *50*, 180–191. [CrossRef] [PubMed]
26. Campoli, G.; Borleffs, M.S.; Amin Yavari, S.; Wauthle, R.; Weinans, H.; Zadpoor, A.A. Mechanical properties of open-cell metallic biomaterials manufactured using additive manufacturing. *Mater. Des.* **2013**, *49*, 957–965. [CrossRef]
27. Ravari, M.R.K.; Kadkhodaei, M.; Badrossamay, M.; Rezaei, R. Numerical investigation on mechanical properties of cellular lattice structures fabricated by fused deposition modeling. *Int. J. Mech. Sci.* **2014**, *88*, 154–161. [CrossRef]
28. Gleadall, A.; Ashcroft, I.; Segal, J. VOLCO: A predictive model for 3D printed microarchitecture. *Addit. Manuf.* **2018**, *21*, 605–618. [CrossRef]
29. Sanz-Herrera, J.A.; García-Aznar, J.M.; Doblaré, M. On scaffold designing for bone regeneration: A computational multiscale approach. *Acta* **2009**, *5*, 219–229. [CrossRef]
30. Rupal, B.S.; Mostafa, K.G.; Wang, Y.; Qureshi, A.J. A Reverse CAD Approach for Estimating Geometric and Mechanical A Reverse CAD Approach for Estimating Geometric and Mechanical Behavior of FDM Conference Printed Parts Behavior of FDM Printed Parts Costing models for capacity optimization Trade-off between. *Procedia Manuf.* **2019**, *34*, 535–544. [CrossRef]
31. Sukindar, N.A.; Dahan, A.A.A.; Halim, N.F.H.A.; Kamaruddin, S.; Sulaiman, M.H.; Ariffin, M.K.A.M. The Effect of Printing Parameters on Scaffold Structure Using Low-cost 3D Printer. *Test Eng. Manag.* **2020**, *82*, 8255–8263.
32. Hollister, S.J. Porous scaffold design for tissue engineering. *Nat. Mater.* **2005**, *4*, 518–524. [CrossRef]
33. Giannitelli, S.M.; Accoto, D.; Trombetta, M.; Rainer, A. Current trends in the design of scaffolds for computer-aided tissue engineering. *Acta Biomater.* **2014**, *10*, 580–594. [CrossRef] [PubMed]
34. Gómez, S.; Vlad, M.D.; López, J.; Fernández, E. Design and properties of 3D scaffolds for bone tissue engineering. *Acta Biomater.* **2016**, *42*, 341–350. [CrossRef] [PubMed]
35. Wang, G.; Shen, L.; Zhao, J.; Liang, H.; Xie, D.; Tian, Z.; Wang, C. Design and Compressive Behavior of Controllable Irregular Porous Scaffolds: Based on Voronoi-Tessellation and for Additive Manufacturing. *ACS Biomater. Sci. Eng.* **2018**, *4*, 719–727. [CrossRef] [PubMed]
36. Bellehumeur, C.; Li, L.; Sun, Q.; Gu, P. Modeling of bond formation between polymer filaments in the fused deposition modeling process. *J. Manuf. Process.* **2004**, *6*, 170–178. [CrossRef]
37. Park, S.; Rosen, D.W. Quantifying effects of material extrusion additive manufacturing process on mechanical properties of lattice structures using as-fabricated voxel modeling. *Addit. Manuf.* **2016**, *12*, 265–273. [CrossRef]
38. Wan, C.; Chen, B. Reinforcement and interphase of polymer/graphene oxide nanocomposites. *J. Mater. Chem.* **2012**, *22*, 3637–3646. [CrossRef]
39. Iannace, S.; De Luca, N.; Nicolais, L.; Carfagna, C.; Huang, S.J. Physical characterization of incompatible blends of polymethylmethacrylate and polycaprolactone. *J. Appl. Polym. Sci.* **1990**, *41*, 2691–2704. [CrossRef]
40. Northcutt, L.A.; Orski, S.V.; Migler, K.B.; Kotula, A.P. Effect of processing conditions on crystallization kinetics during materials extrusion additive manufacturing. *Polymer* **2018**, *154*, 182–187. [CrossRef]
41. Sood, A.K.; Ohdar, R.K.; Mahapatra, S.S. Improving dimensional accuracy of Fused Deposition Modelling processed part using grey Taguchi method. *Mater. Des.* **2009**, *30*, 4243–4252. [CrossRef]
42. Sebe, I.; Kállai-Szabó, B.; Oldal, I.; Zsidai, L.; Zelkó, R. Development of laboratory-scale high-speed rotary devices for a potential pharmaceutical microfibre drug delivery platform. *Int. J. Pharm.* **2020**, *588*, 119740. [CrossRef] [PubMed]
43. Savandaiah, C.; Maurer, J.; Gall, M.; Haider, A.; Steinbichler, G.; Sapkota, J. Impact of processing conditions and sizing on the thermomechanical and morphological properties of polypropylene/carbon fiber composites fabricated by material extrusion additive manufacturing. *J. Appl. Polym. Sci.* **2021**, *138*. [CrossRef]

 *materials*

Article

Manufacturing of Closed Impeller for Mechanically Pump Fluid Loop Systems Using Selective Laser Melting Additive Manufacturing Technology

Alexandra Adiaconitei [1,*], Ionut Sebastian Vintila [1], Radu Mihalache [1], Alexandru Paraschiv [2], Tiberius Florian Frigioescu [2], Ionut Florian Popa [1] and Laurent Pambaguian [3]

[1] Satellites and Space Equipment Department, Romanian Research and Development Institute for Gas Turbines (COMOTI), 061126 Bucharest, Romania; sebastian.vintila@comoti.ro (I.S.V.); radu.mihalache@comoti.ro (R.M.); ionut.popa@comoti.ro (I.F.P.)

[2] Gas Turbine Special Equipment, Physics and Mechanical Testing Laboratory, Romanian Research and Development Institute for Gas Turbines (COMOTI), 061126 Bucharest, Romania; alexandru.paraschiv@comoti.ro (A.P.); tiberius.frigioescu@comoti.ro (T.F.F.)

[3] European Space Research and Technology Centre (ESA-ESTEC), Mechanical Department, European Space Agency, 2200 AG Noordwijk, The Netherlands; laurent.pambaguian@esa.int

* Correspondence: alexandra.adiaconitei@comoti.ro; Tel.: +40-754-598-130

Citation: Adiaconitei, A.; Vintila, I.S.; Mihalache, R.; Paraschiv, A.; Frigioescu, T.F.; Popa, I.F.; Pambaguian, L. Manufacturing of Closed Impeller for Mechanically Pump Fluid Loop Systems Using Selective Laser Melting Additive Manufacturing Technology. *Materials* **2021**, *14*, 5908. https://doi.org/10.3390/ma14205908

Academic Editors: Rubén Paz and Thomas Niendorf

Received: 28 July 2021
Accepted: 6 October 2021
Published: 9 October 2021

Abstract: In the space industry, the market demand for high-pressure mechanically pumped fluid loop (MPFL) systems has increased the interest for integrating advanced technologies in the manufacturing process of critical components with complex geometries. The conventional manufacturing process of a closed impeller encounters different technical challenges, but using additive manufacturing (AM) technology, the small component is printed, fulfilling the quality requirements. This paper presents the Laser Powder Bed Fusion (LPBF) process of a closed impeller designed for a centrifugal pump integrated in an MPFL system with the objective of defining a complete manufacturing process. A set of three closed impellers was manufactured, and each closed impeller was subjected to dimensional accuracy analysis, before and after applying an iterative finishing process for the internal surface area. One of the impellers was validated through non-destructive testing (NDT) activities, and finally, a preliminary balancing was performed for the G2.5 class. The process setup (building orientation and support structure) defined in the current study for a pre-existing geometry of the closed impeller takes full advantages of LPBF technology and represents an important step in the development of complex structural components, increasing the technological readiness level of the AM process for space applications.

Keywords: additive manufacturing; selective laser melting; closed impeller; MPFL pumps; balancing; non-destructive testing

1. Introduction

Additive manufacturing technology has gained a large amount of interest due to its manufacturing advantages in obtaining components and structures with complex shapes. The use of its applications in the space industry is significantly advancing, for example, the Juno (Jupiter Near-polar Orbiter New Frontiers 2) spacecraft (launched in 2011) was equipped with additively manufactured brackets, and Aerojet Rocketdyne Company (El Segundo, CA, USA) uses this type of manufacturing for LOX/H2 rocket engine injectors and thruster systems for CubeSats and other small satellites [1]. Additive manufacturing has the potential to re-design the space system architectures with a long-term impact to reduce costs and increase performances, taking into consideration design and structural requirements. The competitive environment in the space industry regarding the use of AM is also maintained by the global market dynamics, where the AM industry expanded by 7.5%, reaching near to USD 12.8 billion in 2020 [2].

One of the critical technology areas of the NASA Space Technology Roadmap is the Thermal Management Systems, where technology development is needed to enable space exploration and to integrate advanced and additive manufacturing technology. The active thermal systems, where a liquid coolant is circulated in a closed loop under the action of a pump, are known as Fluid Loop Systems, and these are able to maintain the functionality of the spacecraft during the extreme temperature differences, specific for the space environment. The MPFLs present high interest for Europe Large System Integrators (LSI) as a continuous demand for communication satellite platforms that use electric power of around 25 kW [3]. In MPFL, the pump is considered the weakest component and the most likely to fail during operation, and the most used type of centrifugal pump is the one with a canned rotor. The hydraulic capacity and pump performances rely on the impeller, a critical component with a very small and complex shape; therefore, the advantages of AM technology could be used in the manufacturing process in order to achieve a simplified process, mass and costs savings and high performances. Having a low-volume, complex design that formative or subtractive methods are unable to produce, the AM technology fulfils the requirements for manufacturing the closed impeller.

The manufacturing process of the closed impeller using different AM technologies was investigated by Allison et al. [4], and the results showed that the tested Direct Metal Laser Sintering (DMLS) impellers possess acceptable mechanical characteristics, even when some localized material yielding was experienced during speed testing. The hydraulic performance of an AM closed impeller was the major concern of Fernandez's study [5]. The authors concluded that the inherent roughness of the Fused Deposition Modelling (FDM) process did not limit the head-flow curve results of the pump, and by using a chemical post-treatment, assures a more stable behavior in the high flow operating range of the pump. As the AM technology permits the fabrication of pump impellers with a significantly reduced lead-time compared with conventional processes (casting and machining), Rettberg et al. [6] focused on a new additive manufacturing approach for closed impellers. Sulzer (Winterthur, Switzerland) is developing an impeller manufacturing process, which combines Laser Metal Deposition (LMD) with subtractive 5 Axis CNC Milling. Additionally, Sulzer presented different orientations and support structure designs in order to avoid material deposition in inaccessible areas for a LPBF closed impeller [7]. Consequently, efforts were made by Yanghi et al. [8] towards optimizing the manufacturing process of a closed impeller using Laser Powder Bed Fusion (L-PBF), in order to mitigate the distortions of the part by applying the finite element method (FEM). The validated FEM predicted distortion was used to compensate for distortion at the design stage by numerical reverse engineering, and a new impeller was produced following the same AM and post-machining procedures, resulting in a distortion-compensated impeller with mitigated distortion.

The current study focused on a specific AM LPBF method for manufacturing a closed impeller out of Inconel 625. Through this process, the part was fabricated by the sequential addition of material. Although it is known as a near-net-shape technique, as the part is directly built based on a computer aided design (CAD) model and no tools are required to initiate the manufacturing process (except the ones required by post-processing), not all technically feasible approaches to additive manufacturing parts are adaptable for space applications. A detailed investigation for space application demanding requirements, such as dimensional accuracy, quality of the surface or material characterization, is required for a better understanding of additive manufacturing processes. Multiple research studies have aimed to perfect and understand the various additive manufacturing techniques, expand the type of materials and parts that are used, and explore the manufacture of a complex, integrated system. Huber et al. [9] described the LPBF manufacturing of an already existing closed impeller design at the lower limit of castability, and the purpose was to find the optimal building direction as well as to design a proper support structure. The results show that LPBF processed prototype closed impeller, obtained through a heuristic and iterative process, fulfilled all geometrical requirements. Thomas [10] developed a set of design rules to achieve more predictable and reliable results regarding LPBF parts. The geometric limitations of LPBF were evaluated through a quantitative cyclic experimental methodology, and an important conclusion was presented regarding the self-supporting surfaces where alternative orientations can eliminate the need for support and can improve the surface quality of the down-facing surfaces. The same subject of surface quality on different sides of Inconel 625 parts was studied by Mumtaz [11], and it was concluded that the parameters that aid a reduction in both top and side surface roughness are achieved through the use of a higher peak power due to the flattening/smoothing of the melt pool surface (due to increased recoil pressure). Additionally, Yang et al. [12] studied the influence of process parameters on the vertical surface roughness of AlSi10Mg parts and concluded that the surface roughness was reduced to 4 µm from 15 µm when a proper linear energy density was used, improving the surface roughness by more than 70%.

Particular attention is paid to the accuracy of AM parts, and the geometrical accuracy tolerances were found to be ±20–50 µm, even reaching 100 µm [13]; nonetheless, the part-quality may vary due to the nature of layer-by-layer processing. Kamarudin et al. [14] highlighted the dimensional accuracy analysis of the SML benchmark model. Positive deviations of 11.66% (maximum) were identified for the cylinder part and a maximum negative deviation of −3.30% for a rectangular slot. Therefore, the dimensional accuracy may vary depending on the geometry of the printed part and process parameters, as demonstrated by Wang et al. [15]. Overall, the LPBF technology can produce geometrical features, such as sharp corners and cylinders.

As most alloys (Al-Cu-Mg-Sc-Si, Ni-Cr-based super alloys or Ti6Al-4V) used in the AM field to create complex geometries have been developed for traditional manufacturing processes, such as casting and forging, the performances of the AM part could be severely limited if the influence of the LPBF process on the microstructure and mechanical properties of the material is not properly assessed. In this context, several studies were conducted for a deeper understanding control of the effects that currently limit the fidelity of LPBF as a microstructure, residual stress, micro-roughness and porosity of AM materials. Simonelli et al. [16] investigated the influence of Fe on the microstructural development of Ti-6Al-4V used for LPBF, and Gussone et al. [17] demonstrated the feasibility of Ti-Fe alloys used for LPBF with ultrafine microstructures and mechanical strength for structural application. In [18] and [19], the authors provide a brief overview of alloy design strategies, highlighting the potential for alloys to match to the unique processing conditions encountered during the AM process. In the work performed by Shi et al. [20], the effects of laser beam shape on the temporal evolution of the melt pool geometry were investigated, while Roehling et al. [21] identified different strategies to control microstructures locally and to tailor the mechanical performance of additively manufactured parts.

All the ongoing research activities prove the increasing interest in the AM field not only for the general advantages related to complex shapes, but also for the role of alloys tailored for LPBF.

The activities performed in this study follow a preliminary analysis [22], where an investigation of three different building orientations was conducted (0, 32 and 45°) for an Inconel 625 closed impeller. The impeller orientation on the building plate was selected based on two main criteria. First, the major concern was to avoid as much as possible the deposition of the support structure on the internal surface area. Due to the very small dimension of the impeller (φ 22.2 mm internal diameter), it would have been impossible to remove it. Second, it is well known that an orientation around 45° is best suited for a lower roughness and high dimensional accuracy. Having a self-supporting surface inside of the impeller (the shroud), it was important to achieve the minimum roughness from the printing process. The high roughness of the down-facing surfaces is a common disadvantage of the AM technology. The printed closed impellers were subjected to dimensional accuracy and surface quality evaluations, and it was observed that by increasing the printing angle, a better dimensional stability was obtained for both the exterior regions as well as for the blade surface accuracy. On the suction side of the 32° oriented closed impeller, the deviations were between −0.238 and 0.140 mm. Increasing the printing angle to 45°, the deviations were considerably reduced to the range −0.056 to +0.010 mm. The same observation was applicable for the pressure side of the blades: deviation of the 32°-oriented closed impeller was between −0.153 and +0.204 mm, but for the 45°-oriented part, the deviation was between −0.092 and +0.111 mm. Additionally, post-processing activities were preliminary evaluated; however, the finishing process was not uniform on the entire length of the blade, as the suction sides of each blade remained unfinished.

Considering the preliminary work mentioned above, the current study proposes an optimized manufacturing process of closed impellers with a geometry that is difficult to achieve through traditional methods (casting or welding), considering LPBF technology. The AM closed impeller was subjected to detailed non-destructive analyses, such as X-ray computer tomography (CT) scans, liquid penetration evaluation, dimensional accuracy, surface finishing and quality analysis. Additionally, the closed impeller was subjected to balancing activities in the G2.5 balancing class, as a preliminary step, into developing an AM rotary component for MPFL systems. The manufacturing process is detailed below, which analyzed from the CAD model to the complete post-processed closed impeller.

2. Materials and Methods

2.1. Design Approach

While the level of part complexity that metal printing is able to produce exceeds that of traditional manufacturing techniques, the primary challenge for AM space products is the fulfilment of qualification requirements and the guarantee that all batches of parts have the expected mechanical properties and the same high quality.

The baseline model of the closed impeller and the AM model were designed using Solid Edge (version 2019, Siemens PLM Software, Cologne, Germany), following the AM recommendations and constrains [23–29]. The closed impeller design for AM does not have any holes (keyway and thrust balancing holes) to prevent the retention of the metal powder and the deposition of the support structure. Additionally, the closed impeller has an offset material on the outer surface for post-processing operations. No additional material was added on the internal surfaces. The outside diameter of the baseline model is 42.6 mm, and the height is 22 mm, but for the AM closed impeller, the outer diameter is 44.6 mm and the height is 25.5 mm, as seen in Figure 1.

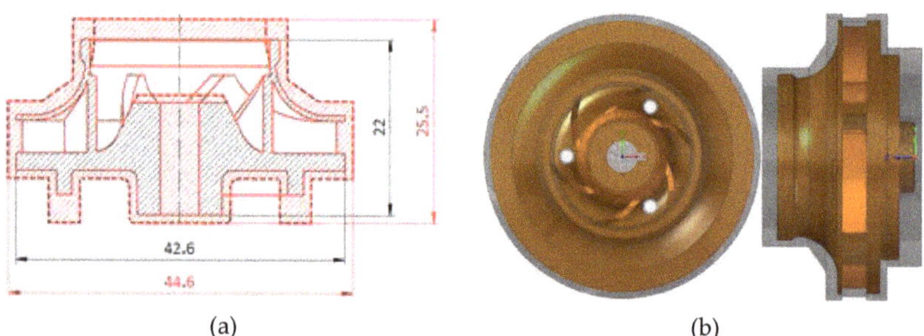

Figure 1. Schematic representation of (**a**) the closed impeller, where black area represents baseline model and red area represents AM model; top and side view of (**b**) the baseline model (orange) with offset material (gray).

2.2. Printing Process Parameters

The closed impellers were manufactured out of Inconel 625 (purchased from LPW Technology Ltd., Runcom, UK), using a Lasertec 30SLM facility (DMG MORI, Bielefeld, Germany) with a building volume of 300 × 300 × 300 mm (L × W × H). The chemical composition of Inconel 625 powder is presented in Table 1 and printing parameters are presented in Table 2. The in-depth material characterization of the Ni-Alloy was performed, and the performances and capabilities of Lasertec 30 SLM (supplier DMG MORI, Bielefeld, Germany) were analyzed in order to define the optimized process parameters to produce high density material. Specimens were manufactured using variable process parameters and were subjected to density and porosity measurements in order to define the most appropriate workspace that generates material with higher relative densities as compared to the theoretical density of the IN 625 alloy. Additionally, the influence of process parameters on the specimen surface roughness and material hardness was assessed. The main conclusion was that for 250 W laser power, 700–800 mm/s scan speed, and layer thicknesses in the range of 30–50 μm, the relative densities achieved are over 99.5%, as highlighted by the authors in [30–32]. However, during the manufacturing of the closed impeller and due to the appearance of the adherent dross on the interior side of the impeller, subjected to analysis in [33], the laser power was decreased to 200 W, which was found to be the best corrective measure.

Table 1. Chemical composition of metal powder.

Chemical Comp.	Al	C	Co	Cr	Fe	Mo	Nb	Si	Ti	Ni
Lot (%wt)	0.09	0.02	<0.1	21.2	4	9	3.62	<0.05	<0.05	Bal.

Table 2. LPBF process parameters [33].

Building Orientation	Laser Power (W)	Scanning Speed (mm/s)	Layer Thickness (μm)	Hatch Distance (mm)	Laser Focus (μm)
B + 60°	200	750	50	0.11	70–120

2.3. Post-Processing Operations

The as-printed closed impeller was subjected to heat treatment using an electric air furnace (Nabertherm LH 30/14 GmbH, Lilienthal/Bremen, Germany) that involves stress relief heat treatment (heating with 10 °C/min up to 870 °C, held for 1 h, followed by air cooling) and annealing heat treatment (heating up to 1000 °C, held for 1 h, followed by fast cooling and oil quenching). Post-processing operations were performed for both the interior and exterior surfaces of the closed impeller, in three separate steps. The first step comprises removing support material from the closed impeller, followed by an interior post-processing operation, and finally machining the exterior of the closed impeller to its final dimensions, as defined by the baseline model in Figure 1.

The removal of support material and machining of the exterior surfaces was performed on a conventional lathe turning machine. Abrasive Flow Machining (AFM) was performed on the closed impellers' interior surface, at Extrude Hone GmbH, Remscheid, Germany, using a VECTOR 6 AFM system, which is ideal for polishing and deburring the internal surface with a small and complex geometry. This technology uses a chemically inactive or non-corrosive media to enhance the roughness and edge conditions. The abrasive particles in the media grind away rather than shear off the unwanted material. Turning operations for the external surface area follow the AFM process as the final machining process of the closed impeller to its final dimensions, as a cost-effective and in-house process.

2.4. Verification Plan

As the AM closed impeller must be free of internal defects, contamination, cracks, lack of fusion or inclusions, and respect the imposed geometrical accuracies and roughness requirements, a verification plan was considered by performing X-ray CT, dye Liquid Penetrant Inspection (LPI), dimensional control and roughness measurements, and the flow-chart is presented in Figure 2.

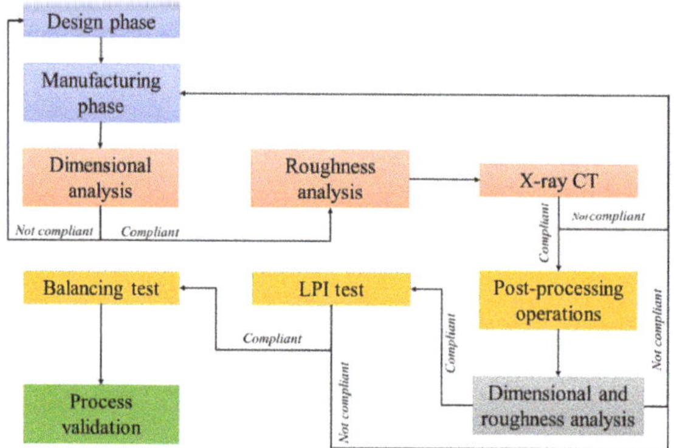

Figure 2. Verification plan—flow chart.

2.4.1. Non-Destructive Tests

An X-ray CT scan was performed on a Micro CT System (Diondo GmbH, Hattingen, Germany) at Dynamic Instruments (Bucharest, Romania) with a resolution of 20 μm and a dimension of 8 Voxels for investigating defects/porosities (Voxel size of 0.041 mm on all three directions). LPI was performed using MR 71 Cleaner, MR 68 NF Dye Penetrant and MR 70 Developer (MR Chemie GmbH, Unna, Germany) to observe any defects that may appear during or after machining the external surfaces of the AM closed impeller. Roughness was measured using a Mahr Surf PS10 instrument (Mahr GmbH, Gottingen, Germany) before and after post-processing operations for process validation. Length of measurement was considered 0.8×10 mm with a 1.0 mm/s speed with respect to impellers dimensions. Dimensional accuracy analysis was performed using a 3D laser surface scanning ATOS Compact Scan 5M machine, integrated with GOM's software for scanning and inspection with $2 \times 5 \times 106$ pixels and measuring point distance between 0.017 and 0.481 mm. The correlation between the measured model of the closed impeller and the CAD model was performed by means of three-point alignment. The three alignment points were: (i) the closed impeller axis of rotation; (ii) interior top disc surface; (iii) thrust balancing holes axis, as presented in Figure 3.

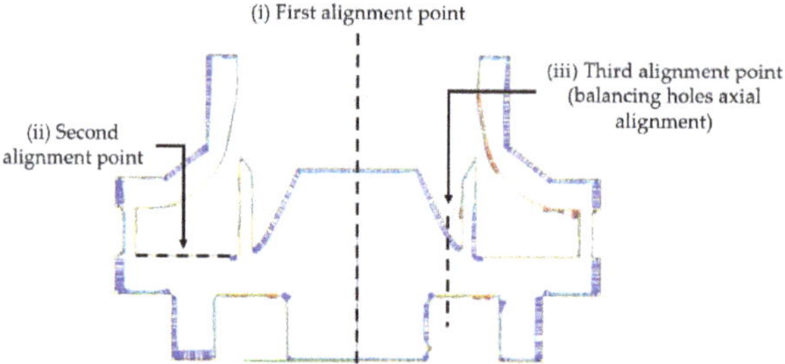

Figure 3. Alignment procedure for the measured model of the closed impeller and the CAD model.

2.4.2. Balancing Operations

The balancing activity is a mandatory step in the development process of this closed impeller in order to ensure a proper operation and the lifetime requirements for an MPFL pump. The balancing procedure was conducted following ISO standard 1940-1: 2003 (E) [34], where due to the small dimensions and mass of the closed impeller, the minimum acceptable class is G2.5. The balancing operations were performed in a single correction plane, with a Passio 5 balancing machine (SCHENCK RoTec GmbH, Darmstadt, Germany) at Aeroteh SA (Bucharest, Romania). The unbalance measurements were conducted at 2200 rot/min.

3. Results

The morphology of the virgin IN 625 powder used in the current study presents a typical morphology for gas atomized powder consisting of relatively small particles (size range 10–45 µm), and mainly spherical particles with satellites joined during solidification. Smooth, spherical powder particles were observed as well as elongated particles. Representative SEM images can be observed in Figure 4.

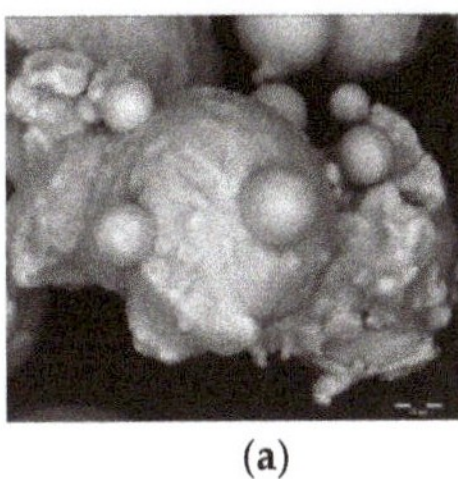
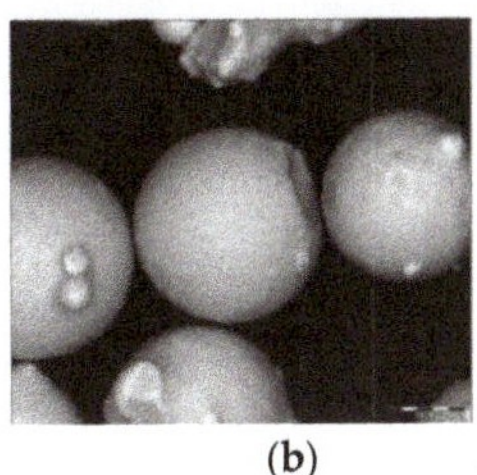

(a) (b)

Figure 4. Powder particle shape and morphology: (**a**) particles with satellites joined during solidification; (**b**) smooth, round particles.

Using the results of the analysis performed by the authors in [33] with respect to an inherent defect of LPBF parts, more precisely, an adhered dross on overhanging structures, it was concluded that the adherent dross can be minimized using the orientation of B + 60° for the closed impeller and a maximum laser power of 200 W. Consequently, a set of three closed impellers at B + 60° orientation were manufactured (Figure 5), for which dimensional evaluation, post-processing operations and a balancing test were performed. The terminology for building orientations is in line with standard terminology for additive manufacturing [35]. The three closed impellers were subjected to heat treatment and the removal of support material before further investigations. Dimensional stability evaluation is presented in Figure 6. The red color on the top of the impeller represents a small area with a high roughness due to the support material still being attached to the shroud. The highest deviation was found on the exterior of the impeller (−0.2 mm), but it can be considered negligible as the offset was set to 1 mm.

Blade positioning was found to have a maximum deviation of ±0.09 mm, showing that the printing process follows the geometrical constraints of ±0.1 mm on the blade positioning and tolerances, with respect to the CAD model.

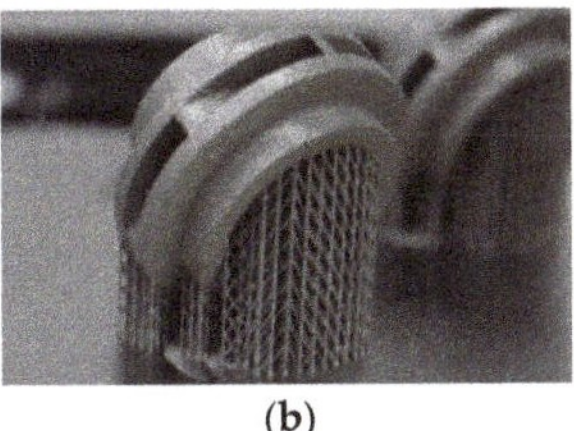

(a) (b)

Figure 5. Closed impellers built at B + 60° orientation: (**a**) front view; (**b**) the back view of closed impeller.

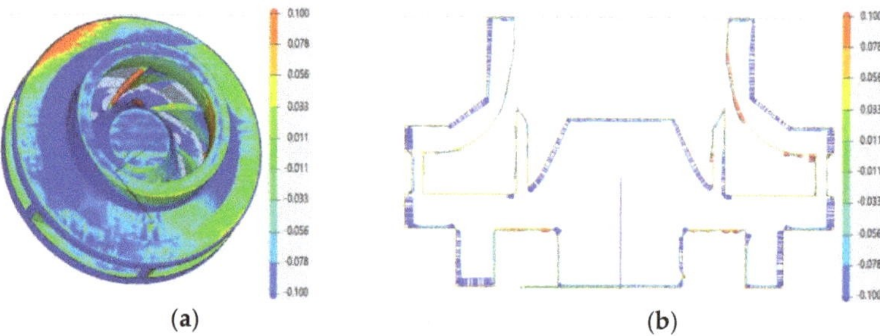

(a)   (b)

Figure 6. Dimensional evaluation: (**a**) top view; (**b**) section view.

An X-ray CT scan (voxel size of 0.041 mm on all three directions) was performed on all closed impellers to detect any internal defects (pores, lack of fusion, cracks or inclusions), before the post-processing activities. Some representative X-ray CT results are presented in Figure 7 for one of the closed impellers. Small voids in the material were identified, both in depth of the material and on the added offset material. As the added offset material was to be machined, only the defects found internally were analyzed more carefully. The identified defects were measured using the MyVGL software (Volume Graphics GmbH, Heidelberg, Germany), where pores with diameters between 0.04 and 0.09 and superficial cracks with lengths between 0.26 and 0.61 mm were found. However, due to the small size and low volume of the voids identified by the CT scan, it can be concluded that they do not affect the mechanical properties of the closed impeller. This conclusion is also supported by the results of the tensile tests, presented in Table 3, and the obtained relative density (average between 99.4 and 99.5%) and porosity (average between 0.5 and 0.6%), higher than imposed values (min. 99.3% for relative density and max. 0.7% for porosity). Tensile test specimens of 3 mm gauge diameter were machined from 3.5 mm diameter printed coupons. A tensile test was performed on Instron 3369 equipment (Instron, Norwood, MA, USA) with a ± 50 kN cell force. The strain rate was set to $\dot{e}Lc = 0.00025$ s^{-1} until the detection of yield strength, then the strain rate was changed to $\dot{e}Lc = 0.0067$ s^{-1}, in accordance with the ISO 68921-1:2009 standard [36].

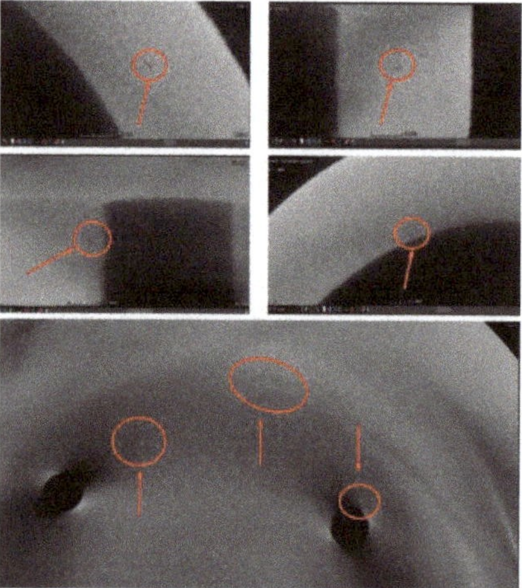

Figure 7. Representative X-ray CT scans for one closed impeller built at B + 60° orientation.

Table 3. Tensile test results of 3 mm gauge diameter specimens machined from 3.5 mm diameter printed.

Specimen No.	Specimen 1	Specimen 2	Specimen 3
d_0 [mm]	2.984	3.014	2.982
Rm [MPa]	776	769	769
$Rp_{0.2}$ [MPa]	472	465	465
RA [%]	51	51	50
EL [%]	50	50	50

where d0—measured diameter of the test specimen; Rm—e tensile strength of the specimen; Rp0.2—yield strength; RA—reduction in area; EL—elongation.

Based on tensile test results presented in Table 3, it was concluded that the tensile properties of IN 625 specimens built with 200 W laser power and 750 mm/s scanning speed meet the requirements of [37].

A preliminary AFM process was investigated in a previous study [22], showing that the finishing process could be applied to the AM closed impeller, with such small dimensions. However, the process required different optimization approaches, such as the use of optimized tooling, and different media with higher viscosity, pressure and number of cycles. For the current study, the post-processing activities were performed on two closed impellers. Special tooling was required to aid the orientation of the abrasive media, as presented in Figure 8. AFM process parameters are presented in Table 4. The dark-grey color of the closed impeller is due to heat treatment.

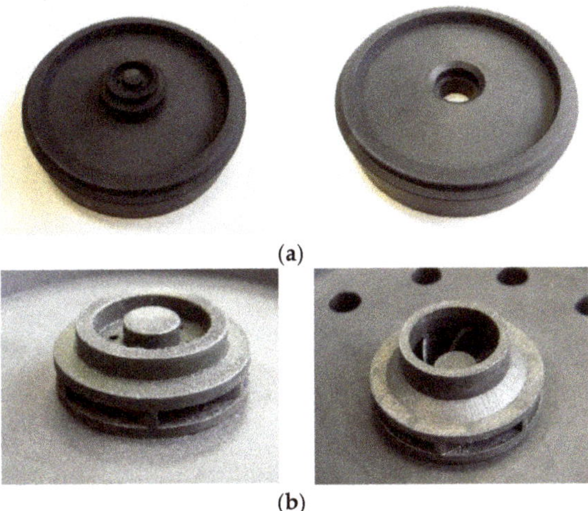

(a)

(b)

Figure 8. Illustration of the (**a**) customized tool required for performing the AFM finishing process and (**b**) the closed impeller mounted on the customized tool.

Table 4. AFM process parameters.

Trial	Media Type	Pressure [bar]	Volume [m^3]	No. of Cycles
Closed Impeller 1 (tooling change)	649 Z1 BC	35	0.00819	10
Closed Impeller 2	EM 25048	50	0.00819	8

After a visual inspection performed on the first closed impeller, the AFM finishing process showed good results in terms of surface roughness improvement. Nevertheless, small areas at the suction side of each blade tail remained unfinished. This resulted from the impeller blades' geometry, which does not guide the media along with the full extent of the blade. Details of the internal surface area after the AFM process are presented in Figure 9. The second impeller presents a more uniform surface, including the suction side of each blade, and the dimensions (edges and channels) are less affected. Details of the second impeller subjected to the AFM process are presented in Figure 10.

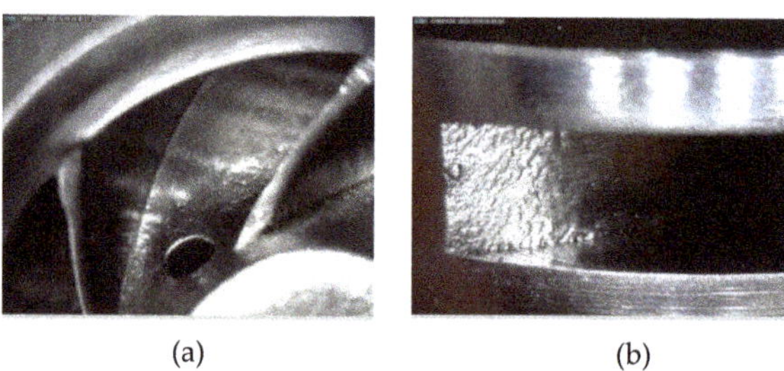

(a)   (b)

Figure 9. Details of first impeller (after tooling optimization) showing (**a**) interior surface and (**b**) pressure side.

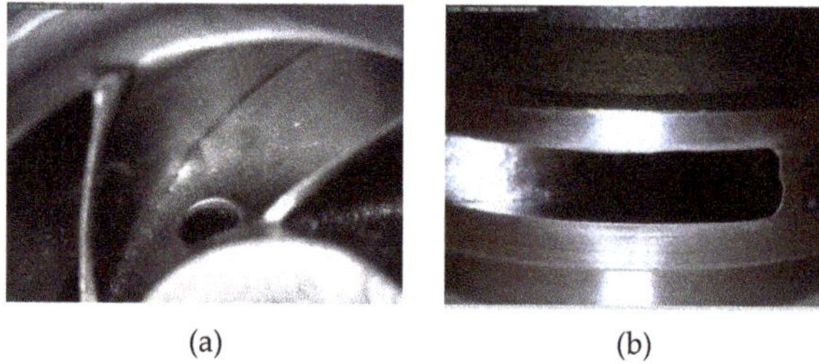

(a)   (b)

Figure 10. Details of Impeller 2 (after media change) showing (**a**) interior surface and (**b**) pressure side.

After the AFM process, the closed impeller was mechanically post-processed by means of turning operation, to evaluate the dimensional stability of the outer surfaces. A comparison between the as-printed closed impeller and finished one is presented in Figure 11.

Both impellers were halved (Figure 11) in order to better observe and analyze the interior finishing, as well as the blade thickness. It should be noted that a total of 2.7 and 2.96 g of material was removed during the first and second AFM processes, respectively (average mass of the closed impeller after heat treatment and removal of support material is of 122 g).

As can be observed in Figures 12 and 13 after the printing process, all six blades' thicknesses were under the imposed tolerances of ±0.1 mm. As the AFM process affects the blade thickness, both the media and the parameters were modified, according to Table 4. After the second trial, the closed impeller showed better results in terms of blade thickness (Table 5). The media gliding on the blade's profile induces the differences of thicknesses when compared to the imposed tolerances.

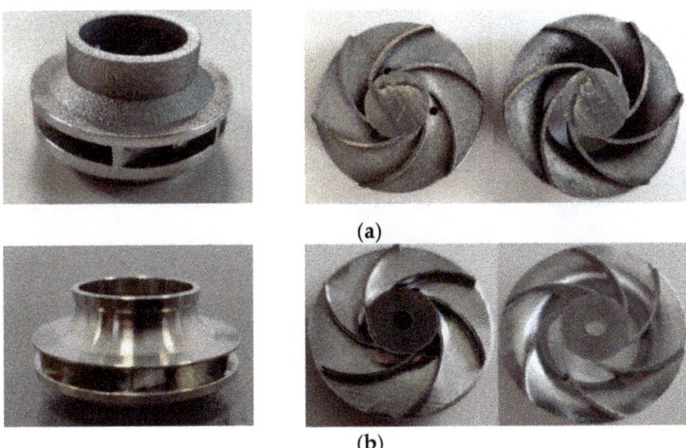

Figure 11. AM closed impeller: (**a**) before post-processing (no heat treatment applied); (**b**) after post-processing all surfaces.

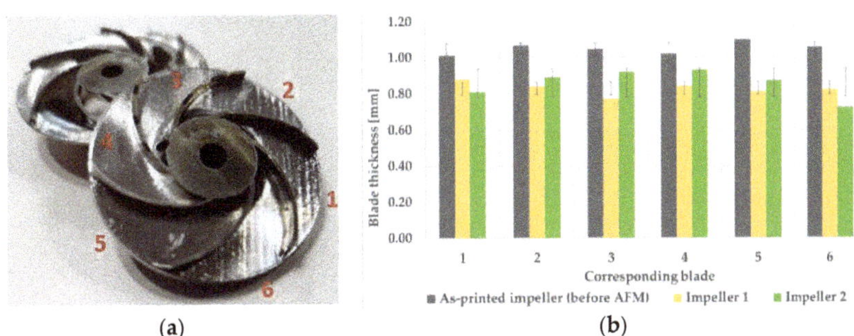

Figure 12. Blade thickness evaluation after AFM process: (**a**) halved AM closed impeller; (**b**) blade thickness values measured using a calibrated equipment with an average of three values with respect to as-printed impeller.

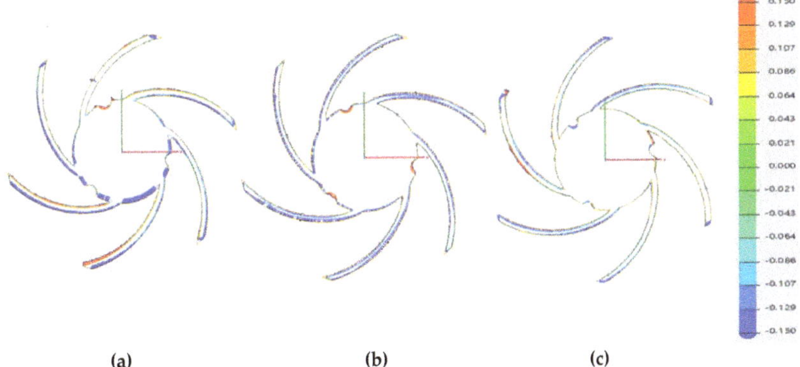

Figure 13. Blade thickness evaluation using 3D measurements for: (**a**) as-printed impeller (before AFM), (**b**) Impeller 1, and (**c**) Impeller 2.

Table 5. Average blade thickness after AFM finishing process.

	Blade 1 [mm]	Blade 2 [mm]	Blade 3 [mm]	Blade 4 [mm]	Blade 5 [mm]	Blade 6 [mm]
As-printed	1.01	1.07	1.05	1.02	1.10	1.05
Std. dev.	0.06	0.04	0.03	0.06	0.04	0.09
Impeller 1	0.88	0.84	0.77	0.84	0.81	0.82
Std. dev.	0.017	0.062	0.045	0.04	0.065	0.075
Impeller 2	0.81	0.89	0.92	0.93	0.87	0.72
Std. dev.	0.06	0.10	0.01	0.06	0.03	0.03

A roughness evaluation (Table 6) was also performed for the two finished halves of the impeller, and a mean value for the shroud was found at Ra 3.81 µm and for the disc at Ra 0.87 µm, as compared to as-printed values of Ra 7.86 µm for the shroud and Ra 8.13 µm for the disc.

Table 6. Roughness evaluation.

Average Values	Disc		Shroud	
	Ra [µm]	Rz [µm]	Ra [µm]	Rz [µm]
As-printed impeller	7.8681	60.9075	8.138	45.898
Std. dev.	1.019	9.686	0.784	3.478
Impeller 1	0.532	3.843	3.813	33.573
Std. dev.	0.093	0.491	1.602	7.280
Impeller 2	0.907	6.076	2.758	30.085
Std. dev.	0.234	1.598	1.546	13.619

However, in order to evaluate how this affects the overall performances of the pump, a comparison is foreseen by the authors among a conventional manufactured closed impeller, an as-printed closed impeller (mechanically post-process on the outer surfaces), and a fully post-processed AM closed impeller. For this reason, the balancing study was performed on the AM closed impeller only on the exterior surfaces. The third closed impeller was machined to its final dimensions and prepared for Liquid Penetrant Inspection and balancing operations. After performing the LP test on the machined external surfaces, it was found that the closed impeller does not present any visible defects.

The balancing was performed considering a one-correction plane analysis to verify if the AM closed impeller could be balanced, taking into consideration the possible geometrical deviations that may occur due to the manufacturing and post-processing operations. The closed impeller after being fitted to the dynamic balancing machine is rotated on high speed (2200 rot/min) in order to determine any possible unbalances. The positioning of the closed impeller with the shaft of the balancing machine is pre-sented in Figure 14. The analysis presented in Figure 15 was recorded during the bal-ancing procedure. As mentioned previously, it should be noted that the closed impeller subjected to this balancing trial was not exposed to AFM surface finishing post-processing, as the outcome of this trial was to validate the balancing procedure for such small additively manufactured closed impellers, using only a correction plane. The closed impeller mounted on the balancing machine is illustrated in Figure 16.

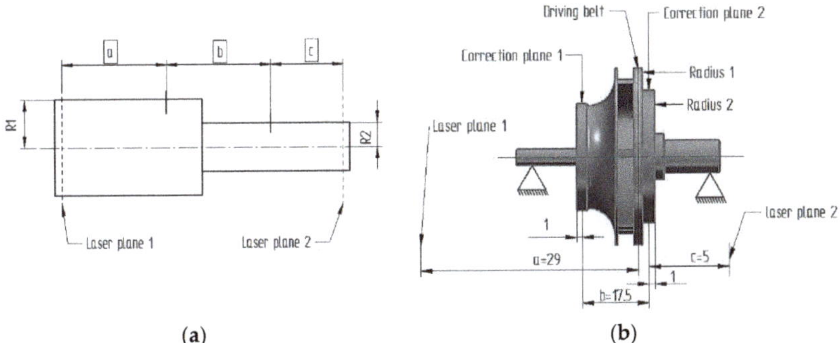

Figure 14. Positioning of the closed impeller on the balancing machine: (**a**) correction plane positioning with corresponding radii; (**b**) closed impeller and shaft mounted on balancing machine.

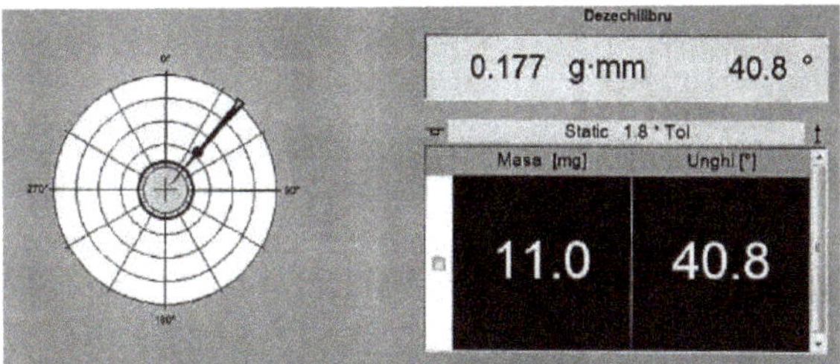

Figure 15. Unbalanced results after balancing operation.

Figure 16. Closed impeller mounted on the balancing machine.

After the first balancing procedure, a mass of 72.8 mg was removed at the corresponding 90° angle, after which the closed impeller was retested, concluding that a secondary mass removal (results presented in Figure 16) is not necessary as the unbalance is within the tolerance defined by the G2.5 balancing class (0.1 g·mm). The balancing results are presented in Table 7. After mass removal, it was concluded, based on the obtained results, that the AM closed impeller can be balanced using a single correction plane.

Table 7. Closed Impeller balancing test.

Measuring speed for first test	2231 rot/min	
Correction plane	1.09 g·mm	90.0°
Correction plane—Mass removal	72.8 mg	90.0°
Measuring speed for second test	2208 rot/min	
Correction plane	0.177 g·mm	40.8°
Correction plane—Mass removal	11.0 mg	40.8°

4. Discussion

The high interest in the space industry for the MPFL systems is reflected in the development processes of individual components, where the additive manufacturing was successfully integrated. The main objective of the present paper was to define a complete manufacturing process for a closed impeller by means of LPBF technology, with respect to dimensions accuracy and surface quality. The geometry of the closed impeller presented a challenge for the LPBF technology with respect to the deposition of support material in an unreachable area.

As the closed impeller is designed for a centrifugal pump that shall be further integrated in MPFL systems, the dimensional accuracy and roughness of the internal surfaces of the closed impeller have a major impact on the pump's lifetime. Therefore, the current research study focused on presenting an evaluation process for a closed impeller in terms of geometrical and dimensional stability, post-processing activities and a balancing activity for such small AM rotary parts. More precisely, with respect to previous investigations on the manufacturing process of small closed impellers [22,33], the present paper started with the closed impellers manufactured using B + 60° as the building orientation with a laser power of 200W in order to avoid any defects, such as adherent dross on the shroud, being a self-supporting structure.

Before the 3D measurements of closed impellers, the support material was removed by a turning operation. Figure 6 presents the dimensional accuracy for the as-printed part, highlighting the dimensional accuracy of the printing process (geometrical constrains of ±0.1 mm on the blades positioning and tolerances), with respect to the CAD model. Considering the offset material, the geometrical deviations from the back of the closed impeller are not considered, as the turning process removed between 0.1 and 0.2 mm of material during support structure removal.

The finishing AFM process was investigated using a new type of medium and adjusting the process parameters. The optimization process focused on reducing the impact over the dimensions of the closed impeller. Consistent results were achieved regarding the roughness of the internal surface area. Compared with [22] where the roughness after AFM process for the shroud was found at Ra 3.85 μm and for the disc at Ra 0.66 μm, after optimization approaches applied in the current study, the roughness for the shroud was improved (Ra 2.7 μm), and it was slightly increased for the disc Ra 0.9 μm; however, a better protection was found for the blade geometry of the closed impeller after process optimization.

The balancing investigation of the closed impeller aimed to achieve a dynamically balanced rotary component that, when installed on the MPFL system, induces an acceptable magnitude of vibration.

This paper presents not only the advantages of LPBF technology, but also the challenges of the manufacturing process, in this case, the surface quality. Depending on the applicability of the AM component, a compromise was made during the manufacturing process between surface quality and dimensional accuracy. Further investigations on the AM closed impeller will be conducted with a focus on the efficiency of the AM part compared to the cast or welded counterpart.

5. Conclusions

This study represents a new step in understanding the complexity of additive manufacturing technology applied for the design of metallic components for space applications to increase the technological readiness level. In addition, a customized post-processing method for the interior finishing of complex geometries, such as closed components, was studied and presented.

Within this study, a full fabrication process of an Inconel 625 closed impeller for MPFL systems was investigated, by means of LPBF technology. The AM closed impeller was built at B + 60° orientation. X-ray CT scans were conducted to analyze possible defects that may occur during fabrication (porosity, cavitation, voids, inclusions, etc.), showing a very small void content that did not affect the material properties or the performance of the closed impeller. Post-processing operations showed good results in terms of roughness and dimensional stability; however, the AFM process could be further enhanced by using more adaptable abrasive media and process parameters in order to achieve a homogenous finishing process over the blades.

A balancing study was performed on the closed impeller at a balancing class of G2.5 using a single correction plane, bringing us one step closer to integrating the AM closed impeller into the MPFL system and testing its performance under relevant conditions, in comparison to a conventionally made closed impeller.

Author Contributions: Conceptualization, A.A., I.S.V. and R.M.; methodology, A.A., I.S.V., R.M., A.P. and T.F.F.; software, I.S.V., A.P. and T.F.F.; validation, L.P., R.M., A.A. and I.S.V.; formal analysis, A.A. and R.M.; investigation, A.A., I.S.V. and T.F.F.; resources, R.M.; data curation, A.A., I.S.V. and R.M.; writing—original draft, I.S.V., A.A, I.F.P. and A.P.; writing—review and editing, A.A., I.S.V., I.F.P. and L.P.; visualization, A.A. and I.S.V.; supervision, L.P. and R.M.; project administration, R.M.; funding acquisition, R.M. All authors have read and agreed to the published version of the manuscript.

Funding: This work was performed under the framework of AM Process Development for Manufacturing a Closed Pump Impeller, contract number 4000129552/19/NL/AR/ig, funded by the European Space Agency (ESA). The APC was funded by the "NUCLEU" Program TURBO 2020+, grant no. 2N/2019, supported by the Romanian Ministry of Research, Innovation and Digitalization.

Institutional Review Board Statement: Not applicable.

Informed Consent Statement: Not applicable.

Data Availability Statement: The data presented in this study are available on request from the corresponding author. Due to the contract agreement between COMOTI and the funding agency, the research activities and data presented in this paper will be presented in an Executive Summary Report and will be available for public use, after project closure.

Acknowledgments: The authors would like to thank Ron Szameitpreuss and his team from Extrude Hone GmbH, Germany, for their support during this research and S.C. Aeroteh S.A, Romania, for their support with balancing the AM closed impeller.

Conflicts of Interest: The authors declare no conflict of interest.

References

1. National Research Council. *3D Printing in Space. Washington*; The National Academies Press: Danvers, MA, USA, 2014. [CrossRef]
2. Campbell, I.; Diegel, O.; Kowen, J.; Mostow, N.; Wohlers, T. *Wohlers Report 2021, Additive Manufacturing and 3D Printing State of the Industry*; Wohlers Associates: Fort Collins, CO, USA, 2021.
3. Camañes, C.; Champel, B.; Mariotto, M.; Callegari, R.; te Nijenhuis, A.; Jan van Gerner, H.; van Es, J.; van den Berg, R. State of the Art of Space Thermal Control Systems, (IMPACTA—Innovative Mechanically Pumped Loop for ACTive Antennae). 2019. Available online: https://ec.europa.eu/research/participants/documents/downloadPublic?documentIds=080166e5c802fe77&appId=PPGMS (accessed on 17 June 2021).
4. Allison, T.C.; Rimpel, A.M.; Moore, J.J.; Wilkes, J.C.; Pelton, R.; Wygant, K. Manufacturing and Testing Experience with Direct Metal Laser Sintering for Closed Centrifugal Compressor Impellers. In Proceedings of the 43rd Turbomachinery & 30th Pump Users Symposia (Pump & Turbo 2014), Houston, TX, USA, 23–25 September 2014. [CrossRef]
5. Fernández, S.; Jiménez, M.; Porras, J.; Romero, L.; Espinosa, M.M.; Domínguez, M. Additive Manufacturing and Performance of Functional Hydraulic Pump Impellers in Fused Deposition Modeling Technology. *J. Mech. Des.* **2015**, *138*, 024501. [CrossRef]

6. Meboldt, M.; Klahn, C. (Eds.) Industrializing Additive Manufacturing. In *Proceedings of AMPA 2020*; Springer: New York, NY, USA, 2021.

7. Huber, M.; Hartmann, M.; Ess, J.; Loeffel, K.; Kränzler, T.; Rettberg, R. Process method for manufacturing impellers by Selective Laser Melting (SLM). In Proceedings of the International Conference on Additive Manufacturing in Products and Applications (AMPA) ETH, Zürich, Switzerland, 13–15 September 2017.

8. Yaghi, A.; Ayvar-Soberanis, S.; Moturu, S.; Bilkhu, R.; Afazov, S. Design against distortion for additive manufacturing. *Addit. Manuf.* **2019**, *27*, 224–235. [CrossRef]

9. Huber, M.; Ess, J.; Hartmann, M.; Würms, A.; Rettberg, R.; Kränzler, T.; Löffel, K. Process Setup for Manufacturing of a Pump Impeller by Selective Laser Melting. Industrializing Additive Manufacturing. In *Proceedings of Additive Manufacturing in Products and Applications—AMPA2017*; Springer: New York, NY, USA, 2021.

10. Thomas, D. *The Development of Design Rules for Selective Laser Melting*; University of Wales: Cardiff, Wales, 2009.

11. Mumtaz, K.; Hopkinson, N. Top surface and side roughness of Inconel 625 parts processed using selective laser melting. *Rapid Prototyp. J.* **2009**, *15*, 96–103. [CrossRef]

12. Yang, T.; Liu, T.; Liao, W.; MacDonald, E.; Wei, H.; Chen, X.; Jiang, L. The influence of process parameters on vertical surface roughness of the AlSi10Mg parts fabricated by selective laser melting. *J. Mater. Process. Technol.* **2018**, *266*, 26–36. [CrossRef]

13. Zhang, L.; Zhang, S.; Zhu, H.; Hu, Z.; Wang, G.; Zeng, X. Horizontal dimensional accuracy prediction of selective laser melting. *Mater. Des.* **2018**, *160*, 9–20. [CrossRef]

14. Kamarudin, K.; Wahab, S.; Shayfull, Z.; Ahmed, A.; Raus, A. Dimensional Accuracy and Surface Roughness Analysis for AlSi10Mg Produced by Selective Laser Melting (SLM). *MATEC Web Conf.* **2016**, *78*, 01077. [CrossRef]

15. Wang, D.; Wu, S.; Bai, Y.; Lin, H.; Yang, Y.; Song, C. Characteristics of typical geometrical features shaped by selective laser melting. *J. Laser Appl.* **2017**, *29*, 022007. [CrossRef]

16. Simonelli, M.; McCartney, D.G.; Barriobero-Vila, P.; Aboulkhair, N.T.; Tse, Y.Y.; Clare, A.; Hague, R. The Influence of Iron in Minimizing the Microstructural Anisotropy of Ti-6Al-4V Produced by Laser Powder-Bed Fusion. *Met. Mater. Trans. A* **2020**, *51*, 2444–2459. [CrossRef]

17. Gussone, J.; Bugelnig, K.; Barriobero-Vila, P.; da Silva, J.C.; Hecht, U.; Dresbach, C.; Sket, F.; Cloetens, P.; Stark, A.; Schell, N.; et al. Ultrafine eutectic Ti-Fe-based alloys processed by additive manufacturing—A new candidate for high temperature applications. *Appl. Mater. Today* **2020**, *20*, 100767. [CrossRef]

18. Grabowski, J.; Kozmel, T. New High-Performance Materials Tailored Specifically for Additive Manufacturing Processes. In Proceedings of the AHS International 74th Annual Forum & Technology Display, Phoenix, AZ, USA, 14–17 May 2018.

19. Liu, G.; Zhang, X.; Chen, X.; He, Y.; Cheng, L.; Huo, M.; Yin, J.; Hao, F.; Chen, S.; Wang, P.; et al. Additive manufacturing of structural materials. *Mater. Sci. Eng. R Rep.* **2021**, 100596. [CrossRef]

20. Shi, R.; Khairallah, S.A.; Roehling, T.T.; Heo, T.W.; McKeown, J.; Matthews, M.J. Microstructural control in metal laser powder bed fusion additive manufacturing using laser beam shaping strategy. *Acta Mater.* **2019**, *184*, 284–305. [CrossRef]

21. Roehling, T.T.; Shi, R.; Khairallah, S.A.; Roehling, J.D.; Guss, G.M.; McKeown, J.T.; Matthews, M.J. Controlling grain nucleation and morphology by laser beam shaping in metal additive manufacturing. *Mater. Des.* **2020**, *195*, 109071. [CrossRef]

22. Adiaconitei, A.; Vintila, I.; Mihalache, R.; Paraschiv, A.; Frigioescu, T.; Vladut, M.; Pambaguian, L. A Study on Using the Additive Manufacturing Process for the Development of a Closed Pump Impeller for Mechanically Pumped Fluid Loop Systems. *Materials* **2021**, *14*, 967. [CrossRef]

23. Valjak, F.; Bojčetić, N.; Nordin, A.; Godec, D. *Conceptual Design for Additive Manufacturing: An Explorative Study*; Cambridge University Press: Cambridge, UK, 2020.

24. Valjak, F.; Bojčetić, N. Conception of Design Principles for Additive Manufacturing. *Proc. Des. Soc. Int. Conf. Eng. Des.* **2019**, *1*, 689–698. [CrossRef]

25. Schumacher, F.; Watschke, H.; Kuschmitz, S.; Vietor, T. Goal Oriented Provision of Design Principles for Additive Manufacturing to Support Conceptual Design. *Proc. Des. Soc. Int. Conf. Eng. Des.* **2019**, *1*, 749–758. [CrossRef]

26. Mani, M.; Witherell, P.; Jee, H. Design Rules for Additive Manufacturing: A Categorization. In Proceedings of the ASME 2017 International Design Engineering Technical Conferences and Computers and Information in Engineering Conference, Cleveland, OH, USA, 6–9 August 2017.

27. Rosen, D.W. Research supporting principles for design for additive manufacturing. *Virtual Phys. Prototyp.* **2014**, *9*, 225–232. [CrossRef]

28. Vayre, B.; Vignat, F.; Villeneuve, F. Designing for Additive Manufacturing. *Procedia CIRP* **2012**, *3*, 632–637. [CrossRef]

29. Redwood, B.; Schöffer, B.; Garret, B. *The 3D Printing Handbook: Technologies, Design and Applications, 3D Hubs, B.V*; Coers & Roest: Amsterdam, The Netherlands, 2017.

30. Condruz, M.R.; Matache, G.; Paraschiv, A.; Frigioescu, I.F.P.; Badea, T. Microstructural and Tensile Properties Anisotropy of Selective Laser Melting Manufactured IN 625. *Materials* **2020**, *13*, 4829. [CrossRef]

31. Condruz, M.R.; Matache, G.; Paraschiv, A.; Badea, T.; Badilita, V. High Temperature Oxidation Behavior of Selective Laser Melting Manufactured IN 625. *Metals* **2020**, *10*, 668. [CrossRef]

32. Condruz, M.R.; Matache, G.; Paraschiv, A. Characterization of IN 625 recycled metal powder used for selective laser melting. *Manuf. Rev.* **2020**, *7*, 5. [CrossRef]

33. Vintila, I.S.; Adiaconitei, A.; Mihalache., R.; Paraschiv, A.; Frigioescu, T.; Condurachi, F.; Datcu, D. A study on reducing the adherent dross on additively manufactured closed impeller. *Sci. J. TURBO* **2021**, *VIII*, 33–41.
34. *Mechanical Vibration—Balance Quality Requirements for Rotors in a Constant (Rigid) State—Part 1: Specification and Verifi-Cation of Balance Tolerances, ISO 1940-1:2003*; International Organization for Standardization: Geneva, Switzerland, 2003.
35. *Standard Terminology for Additive Manufacturing—Coordinate Systems and Test Methodologies; ISO 52921:2013*; International Organization for Standardization: Geneva, Switzerland, 2013.
36. *Metallic Materials—Tensile Testing—Part 1: Method of Test at Room Temperature; ISO 6892-1:2009*; International Organization for Standardization: Geneva, Switzerland, 2009.
37. *Standard Specification for Additive Manufacturing Nickel Alloy (UNS N06625) with Powder Bed Fusion*; ASTM F3056-14e1; ASTM International: West Conshohocken, PA, USA, 2014.

Article

Double-Level Energy Absorption of 3D Printed TPMS Cellular Structures via Wall Thickness Gradient Design

Minting Zhong [1], Wei Zhou [1], Huifeng Xi [2], Yingjing Liang [1] and Zhigang Wu [1,*]

[1] School of Civil Engineering, Guangzhou University, Guangzhou 510006, China; 2111916048@e.gzhu.edu.cn (M.Z.); z18826737435@163.com (W.Z.); yjliang@gzhu.edu.cn (Y.L.)

[2] MOE Key Laboratory of Disaster Forecast and Control in Engineering, School of Mechanics and Construction Engineering, Jinan University, Guangzhou 510632, China; xihuifeng@jnu.edu.cn

* Correspondence: zgwu@gzhu.edu.cn

Abstract: This paper investigates the deformation mechanism and energy absorption behaviour of 316 L triply periodic minimal surface (TPMS) structures with uniform and graded wall thicknesses fabricated by the selective laser melting technique. The uniform P-surface TPMS structure presents a single-level stress plateau for energy absorption and a localized diagonal shear cell failure. A graded strategy was employed to break such localized geometrical deformation to improve the overall energy absorption and to provide a double-level function. Two segments with different wall thicknesses separated by a barrier layer were designed along the compression direction while keeping the same relative density as the uniform structure. The results show that the crushing of the cells of the graded P-surface TPMS structure occurs first within the thin segment and then propagates to the thick segment. The stress–strain response shows apparent double stress plateaus. The stress level and length of each plateau can be adjusted by changing the wall thickness and position of the barrier layer between the two segments. The total energy absorption of the gradient TPMS structure was also found slightly higher than that of the uniform TPMS counterparts. The gradient design of TPMS structures may find applications where the energy absorption requires a double-level feature or a warning function.

Keywords: triply periodic minimal surface; selective laser melting; 316 L stainless steel; energy absorption; deformation mechanism

Citation: Zhong, M.; Zhou, W.; Xi, H.; Liang, Y.; Wu, Z. Double-Level Energy Absorption of 3D Printed TPMS Cellular Structures via Wall Thickness Gradient Design. *Materials* **2021**, *14*, 6262. https://doi.org/10.3390/ma14216262

Academic Editor: Rubén Paz

Received: 17 September 2021
Accepted: 18 October 2021
Published: 21 October 2021

Publisher's Note: MDPI stays neutral with regard to jurisdictional claims in published maps and institutional affiliations.

1. Introduction

Porous metallic materials have been extensively researched, which can be attributed to their extraordinary properties, such as their high strength to weight ratio, excellent thermal insulation, and damping properties [1–3], which may find useful applications in bone implants [4], heat exchangers [5], and energy absorption [6]. Some popular porous structures are largely inspired by topologies that found in nature, such as body-centered cubic (BCC) [7], rhombic dodecahedron [8], honeycomb structures [9], and truss–lattice structures [10,11], which show geometry periodicity and topological homogeneity. However, these structures commonly suffer nodal stress concentration and may be subject to instable mechanical performance and premature failure [12–14].

Recently, the nature-inspired triply periodic minimal surface (TPMS) structures that have been found in urchins [15], butterfly wing scales [16], and the exoskeletons of beetles [17] have attracted a great deal of attention. TPMSs are mathematically rigorous, with a zero–mean curvature, and are composed of infinite and non-self-intersecting surfaces that are associated with a specific crystallographic space group symmetry [18]. A TPMS has a long-range ordered structure with a high specific surface area, high porosity, and stable mechanical properties. Its internal structure is interconnected, and its surface is smooth. These appealing properties of TPMS structures promotes uniform stress distribution on the surface of cell walls and avoids the stress concentration under loading. Hence, it is an ideal porous structure for structural energy absorption design.

Investigations on the structure design and mechanical performance of metallic TPMS structures have been conducted extensively since the emergence of the additive manufacturing (AM) technique. The compressive deformation characteristics of TPMS structures are similar to those of conventional metallic foams, showing three distinctive deformation stages, including an initial linear elastic stage, a middle stress plateau stage, and a final densification stage [19,20]. The energy absorption capability is mainly contributed from the plateau stage, and this normally requires a stable stress level and long plateau strain. Zhang et al. reported that the TPMS structures fabricated by selective laser melting (SLM) possess superior stiffness, plateau stress, and energy absorption ability relative to the BCC lattice [21]. Liang et al. also fabricated 316 L TPMS structures with different relative densities by SLM. They found that deformation of the cells concentrates much along the diagonal band regions, and when the relative density was lower than 0.35, the yield strength and energy absorption capacity of the P-surface were better than those of the G-surface [22]. Yang et al. studied the fatigue properties of G-surface TPMS structures and found the failure samples have nearly 45° fracture bands along the diagonal surface [23]. It can be seen that the stress concentration and the dominant diagonal shear failure of the uniform TPMS structures limits their energy absorption performance.

As a type of important porous structure, functionally graded structures are the optimal choice to tackle the issue of stress concentration and are promising for creating light weight and high energy absorption parts. The properties of functionally graded structures include the gradient wall thickness, pore size, and unit cell type along the loading direction. Fan et al. proposed that a graded wall thickness design can effectively improve the mechanical properties of the structure. Compared with the uniform thickness structure, the elastic modulus of the graded sample is increased by more than 94 MPa, and the cumulative energy absorption is increased by 15.4% [24]. Min et al. simulated the in-plane impact properties and energy absorption properties of concave hexagonal honeycomb materials with a gradient cell thickness [25]. They found that the concave hexagonal honeycomb materials with a positive thickness gradient have better energy absorption and impact resistance than the uniform structure. Al-Ketan et al. conducted mechanical tests on gradient samples in parallel and perpendicular to the grading direction to study the effect of the loading direction on the displayed deformation mechanism [26]. When the load is applied parallel to the grading direction, the collapse starts layer by layer, and the deformed layer densifies at the top of the next one, resulting in a hardening behavior. The test perpendicular to the grading direction showed a deformation mode along the diagonal shear band that was similar to the uniform structure. Yu et al. showed that the stress in the wall thickness-graded Schwarz P-surface structure gradually exceeds that in the uniform P-surface structure and that the total absorbed energy per unit volume W_{vD} is 1.5 times higher than that of the uniform P-surface structure [27]. Panesar et al. evaluated the mechanical properties of the proposed gradient thickness design and showed that the intersecting/grading/scaling lattice strategy is 40–50% better than the uniform structure in terms of specific stiffness [28].

To summarize the above literature findings, one can see that the gradient TPMS structure has great potential to achieve high mechanical performance and to improve the energy absorption ability of the structure. The gradient designs in the previous reports normally follow a linear grading of properties (wall thickness, lattice type, and cell size gradient) that is parallel to the loading direction. The middle stress plateau evolves into to a gradual strain hardening stage following the implementation of gradient design. In this study, we propose a different gradient design in the wall thickness. The distribution of different wall thickness is segmental and discrete in order to break the original localized deformation within the diagonal shear bands. Such a design provides a characteristic double-stage energy absorption characteristic during the plastic deformation of the structure. This may shed a light on novel gradient design for applications requiring multi-level energy absorption.

2. Materials and Methods

2.1. TPMS Models

The primitive P-surface TPMS structure was selected to study the gradient design and its effect on the deformation mechanism. The approximate value of the P-surface is defined by the below formula [29]:

$$\Phi_P(x,y,z) = cos(\omega x) + cos(\omega y) + cos(\omega z) = c \tag{1}$$

where Φ_p is a surface calculated at the isovalue c, $\omega = 2\pi/L$, and L is the length of the unit cell. The expansion of the surface in three-dimensional space can be controlled by modifying c. When $c = 0$, the resulting surfaces divide the space into sub-domains of equal volumes. Then, the structure of TPMS is realized by surface thickening into the form of a continuous shell with limited thickness [30]. The P-surface TPMS model has a unit cell length of 4 mm and an array of 8 × 8 × 8 tessellating cells. The TPMS structure with different wall thicknesses is obtained by using the plug-in Grasshopper in Rhino, and then the STL file is generated.

2.2. Gradient Design

Uniform and gradient P-surface TPMS structures were investigated. Figure 1 shows the comparison between the uniform (Uni-1) and gradient (Gra-1) models, which had a similar relative density of ~0.22. Figure 1a,b show the three-dimensional view of the uniform and gradient models, respectively. Figure 1c,d present the front face (*x-z* plane) of the models and shows the details of the wall thickness arrangement. Uni-1 apparently has a constant wall thickness of t = 0.36 mm throughout the model. In contrast, the wall thickness of Gra-1 distributes differently along the z-axis (loading direction). The top segment, which includes 1–3.5 layers, is relatively thinner, with a wall thickness of $t_1 = 0.27$ mm, while the bottom segment, which includes 4.5–8 layers, is thicker, with a wall thickness of $t_2 = 0.4$ mm. A transition or barrier layer is introduced between these two segments, covering the 3.5–4.5 layer. The gradient thickness of this barrier layer follows $t_1 \rightarrow 2t_2 \rightarrow t_2$, which is 0.27→1.08→0.54 mm in the case of Gra-1. The purpose of having $2t_2$ in the middle of the barrier layer is to introduce a dense layer to better separate the thin and thick segments under loading, i.e., to make a sharp stepping feature of the double stress plateaus. The sample ID, wall thickness, mass, relative density, and wall thickness distribution of all of the samples are detailed in Table 1.

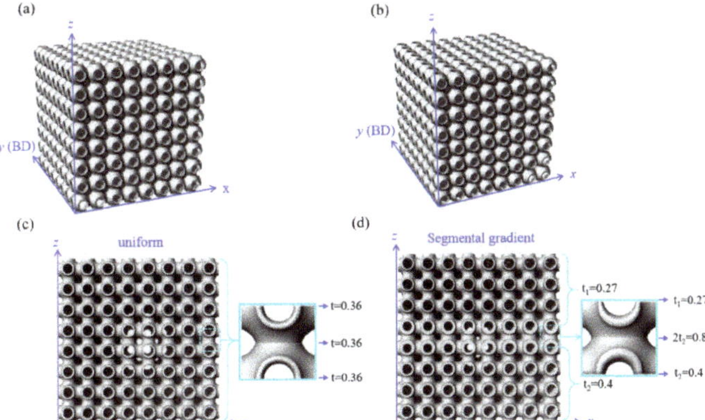

Figure 1. Structure models of the uniform and gradient TPMS structures. (**a**) Model of Uni-1, (**b**) model of Gra-1, (**c**) the front view of Uni-1, and (**d**) the front view of Gra-1 showing the gradient thickness distribution along z-axis. BD indicates building direction. BD ⊥ Gradient direction. The values labelled on the models are in mm.

Table 1. Sample ID, sample mass, relative density, and wall thickness distribution of the designed uniform and gradient TPMS structures.

Wall Thickness Type	Sample ID	Designed Mass (g)	Actual Mass (g)	Designed Relative Density (%)	Actual Relative Density (%)	Wall Thickness Distribution along z-axis (mm)
Uniform	Uni-1	55.87	58.45	21.37	22.36	1–8 layers: t = 0.36
	Uni-2	68.63	71.08	26.24	27.18	1–8 layers: t = 0.45
	Uni-3	80.47	80.99	30.77	30.97	1–8 layers: t = 0.52
Gradient	Gra-1	55.87	59.19	21.37	22.64	1–3.5 layers: t_1: 0.27 3.5–4.5 layer: t_1-$2t_2$-t_2: 0.27–0.8–0.4 4.5–8 layers: t_2: 0.4
	Gra-2	68.63	70.71	26.24	27.04	1–3.5 layers: t_1: 0.27 3.5–4.5 layers: t_1-$2t_2$-t_2: 0.27–1.08–0.54 4.5–8 layers: t_2: 0.54
	Gra-3	80.47	81.86	30.77	31.31	1–3.5 layers: t_1: 0.27 3.5–4.5 layer: t_1-$2t_2$-t_2: 0.27–1.34–0.67 4.5–8 layers: t_2: 0.67
	Gra-4	76.39	77.32	29.22	29.58	1–2 layers: t_1: 0.27 2–3 layer: t_1-$2t_2$-t_2: 0.27–1.08–0.54 3–8 layers: t_2: 0.54
	Gra-5	60.86	63.96	23.27	24.46	1–5 layers: t_1: 0.27 5–6 layer: t_1-$2t_2$-t_2: 0.27–1.08–0.54 6–8 layers: t_2: 0.54

2.3. Finite Element Analysis

The single cell model of the P-surface was first generated by the mathematical formula (Equation (1)) in Rhino software (version 6.23) and was then saved as an igs format file and imported to the Hypermesh software (version 14) for mesh distribution. Since the wall thickness of the TPMS model is much smaller than the size of the model, shell element (S4R) was used for the construction of the model. Abaqus (version 6.14-1 under academic license) was used to simulate the compressive behavior of the TPMS models. The meshed unit cell of the P-surface was imported into Abaqus, and the array function was used to form an $8 \times 8 \times 8$ model. The base material, 316 L stainless steel, was modelled using an elastic model and a simple rate-independent J2 plastic model with isotropic hardening in Abaqus. The material properties of 316 L stainless steel defined in Abaqus are given in Table 2. The wall thicknesses of the FE models were set to keep their relative densities identical to those of the tested samples. Two steel plates with infinite stiffness were established using a discrete rigid body. The displacement control was implemented to the top steel plate, with a strain rate that was the same as the compression experiments, while all the degrees of freedom of the bottom steel plate were fixed.

Table 2. Material parameters of the base material 316 L used in FE simulation.

Material Density (g/cm³)	Elasticity Type	Yield Strength (MPa)	Young's Modulus (MPa)	Poisson's Ratio	Plasticity Type	Hardening Rule
7.98	Isotropic	460	169,000	0.3	Von-Mises yield criterion	Isotropic

2.4. SLM Fabrication

Stainless steel 316 L gas-atomized powder was provided by AMC Powder with an average particle size of 45 µm (D50). Table 3 shows the chemical composition of SS316 L powder. The TPMS samples were fabricated in 99.99% argon using an HBD-100 machine (Guangdong Hanbang 3D Technology Co.,Ltd, Zhongshan, China). A laser power (P) of 160 W, scanning speed (v) of 1000 mm/s, hatch spacing (h) of 80 μm, and layer thickness (t) of 30 μm were used in this study. According to $E = P/(v \times h \times t)$, the delivered energy density (E) during the laser melting process can be calculated as 66.7 J/mm^3. A bidirectional orthogonal scanning strategy was adopted for each layer.

Table 3. Chemical composition of 316 L powder.

Element	Cr	Ni	Mo	Mn	Si	Cu	Fe
wt.%	17.35	12.02	2.74	1.36	0.33	0.23	bal.

2.5. Characterization Techniques

The surface morphology of the TPMS samples was examined using a scanning electron microscope (SEM, Japan Electronics Co., Ltd, Tokyo, Japan) and a sec JSM-7001F instrument (Japan Electronics Co., Ltd, Tokyo, Japan). The internal defects were checked using a nanoVoxel-3000 micro-CT instrument (Sanying Precision Instrument Co., Ltd., Tianjin, China). The quasi-static compression test of the TPMS samples was conducted using an ETM series electronic universal testing machine (Shenzhen SUNS Technology Stock Co., Ltd., Shenzhen, China) with a load capacity of 300 kN. The crosshead speed of 2 mm/min was used for testing. The compression direction was perpendicular to the building direction.

3. Results and Discussion

Figure 2 shows the surface quality of the as-fabricated Uni-1. Figure 2a is a photograph of Uni-1, and Figure 2b is a representative CT scan image of the x-z plane at the location indicated by the dashed lines. It can be seen that the surface contours show minimal geometrical defects. The internal cross-section of the sample presents a continuous connection of the shell walls without sharp horns or irregularities. However, there are some locations showing particle adhesion on the inner surface of the TPMS structure (arrowed). Figure 2c,d are the SEM micrographs at different locations of the sample, as circled in (a). It should be noted that the surface is covered with numerous partially melted 316 L particles, which is a common surface imperfection in powder-bed AM techniques. In general, no gross defects such as holes or cracks were found within the as-fabricated samples.

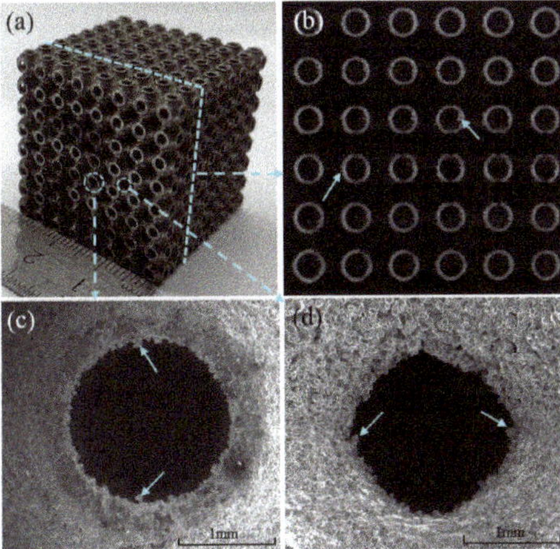

Figure 2. Printing geometry and surface quality of Uni-1. (**a**) A photograph of the overall appearance, (**b**) one of the CT scan images showing internal surface contour, and (**c**,**d**) SEM micrographs of selected locations (circled in (**a**)) on the x-z plane showing the presence of excessive residual particles.

Figure 3 shows the compressive properties and deforming characteristics of the uniform TPMS samples. Figure 3a displays the compressive stress–strain curves (solid lines) and the corresponding FE stress–strain curves (dashed lines) of Uni-1, Uni-2, and Uni-3. The stress–strain curves can be divided into three distinct deformation stages: (1) 0–3%: the linear elastic stage, (2) 3–65%: the stress plateau stage, and (3) >65%: the densification stage. Stress fluctuations can be observed on the stress plateau when the applied strain is greater than 40%. Such mechanical instability of the P-surface TPMS structure has also been reported by other authors, which is due to the localized bulking of the thin and curved shell walls [21]. The experimental results show a relatively shorter stress plateau and an earlier onset of the densification compared to the FE results for all the three samples. This is because the actual average wall thicknesses in the as-fabricated samples are slightly greater than the designed values, as seen in Table 1, which caused the early self-contacting of the shell walls.

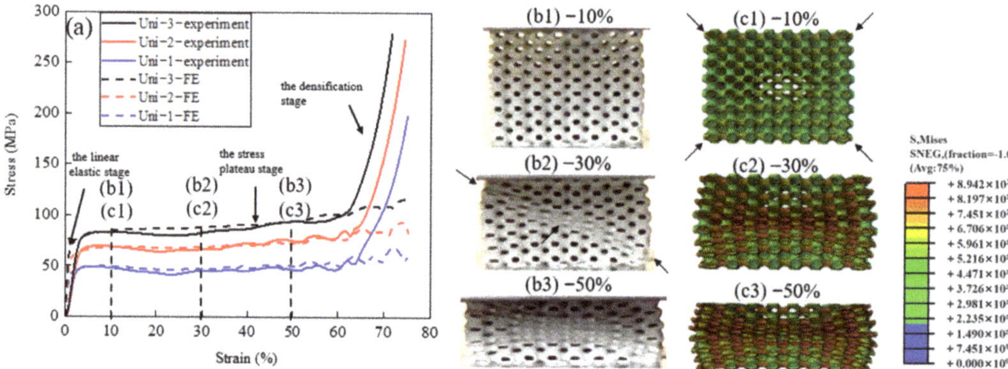

Figure 3. Deformation behaviour of uniform TPMS structures. (**a**) Stress–strain curves of Uni−1, Uni−2, and Uni−3, (**b1–b3**) snapshots of the side surface during compression experiment, and (**c1–c3**) Mises stress cloud diagrams of the side surface during FE simulation of Uni−1.

As the wall thickness increases (Uni−1→Uni−2→Uni−3), the yield strength increases progressively from 47 MPa to 83 Mpa. In the meantime, the length of stress plateau shortens slightly, and the stress fluctuation phenomenon weakens. This implies better mechanical stability in structures with greater wall thicknesses. Figure 3(b1–b3,c1–c3) are the snapshots of the side surface (x-z plane) of Uni-1 at different applied strains during the experiment and the corresponding FE Mises stress cloud diagrams during the simulation. The snapshots and Mises stress cloud diagrams are selected at 10%, 30% and 50% of deformation, as labeled in (a). Generally speaking, the cells in uniform P-surface collapse in layers and show a localized double shear band geometry on the side surface during the experiment, which is consistent with the deformation in the FE simulation. When the strain is 10%, the cell openings of the four corners show the largest stress magnitude (arrowed) and thus the largest cell deformation. At 30%, the local deformation pattern shows severe cell crushing along only one diagonal shear band in the experiment, while the model in the FE simulation shows a symmetric double shear band stress concentration and cell deformation. The asymmetrical deformation pattern is likely due to the uneven defect distribution in this sample. When the strain reached 50%, the cells were squeezed into symmetrical double shear bands in the experiment, and only the cells outside of the shear band geometry were not completely closed, as shown in both the experimental and simulated results.

Figure 4 shows the mechanical properties of Gra−1, Gra−2, and Gra−3 gradient structures. In this group, the wall thickness keeps constant at $t_1 = 0.27$ mm from 1–3.5 layers along the z-direction for all three samples, while the wall thickness t_2 of 4.5–8 layers is thicker, being 0.4 mm, 0.54 mm, and 0.67 mm for Gra−1, Gra−2, and Gra−3, respectively. This gives rise to a thinner upper segment and thicker lower segment in these samples.

The middle section of 3.5–4.5 layer is a wall thickness transition section and also acts as a barrier layer to separate the deformation of the thick segment from the thin one since the thickness gradient is designed as $t_1 \rightarrow 2t_2 \rightarrow t_2$.

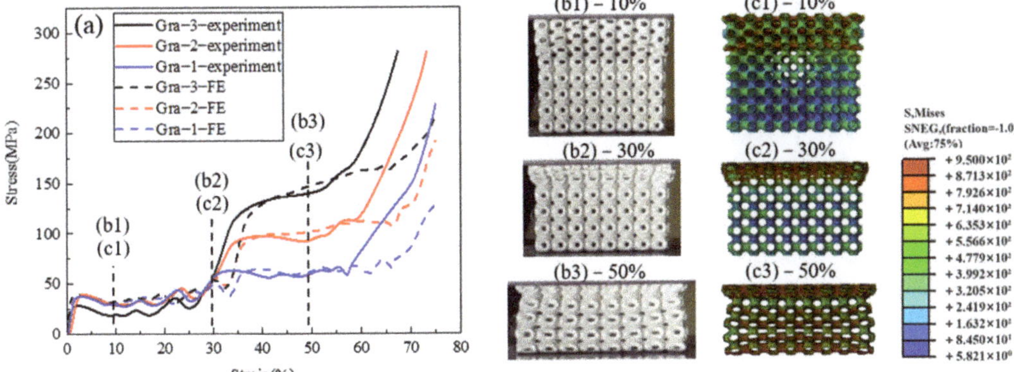

Figure 4. Deformation behaviour of the gradient TPMS structures with different heights of the second stress plateau. (**a**) stress–strain curves of Gra−1, Gra−2, and Gra−3 and (**b1–b3**) snapshots of Gra−3 at 10%, 30%, and 50% strain during the compression experiment. (**c1–c3**) Mises stress cloud diagrams of Gra−3 at 10%, 30%, and 50% strain during the FE simulation.

Figure 4a shows the stress–strain curves of Gra−1, Gra−2, and Gra−3, in which the solid lines are the experimental results and where the dashed lines are the FE simulation results. It can be seen that the gradient structures clearly present two distinct stress plateaus before and after 30% of the deformation strain. The first stress plateau continues to be almost unchanged at the lower stress level of ~40 MPa with the presence of some minor stress fluctuations, while the second stress plateau is higher and rises progressively from 60–140 MPa with the increase of t_2. In general, the FE simulation stress–strain responses show good agreement with the experimental results. Small discrepancies exist in the onset of the second stress plateau and in the densification between the experimental and simulation results. This is because the real wall thicknesses of the printed samples are slightly greater than the designed values, thus showing earlier self-contacting of the collapsed cell walls. Compared to the uniform structures (Figure 3a), the densification stage of the gradient structures appears to be less clearly presented, as gradual hardening is shown first and is then followed by the rapidly rising stress–strain slope. This is due to the gradual deformation of the barrier layer at the end of the deformation plateau and occurs simultaneously with the densification process of the 1–3.5 and 4.5–8 layers.

Figure 4(b1–b3,c1–c3) show the experimental snapshots and FE Mises stress diagrams at different applied strains on the side surface of Gra-3. When the structure is deformed to 10%, heavy deformation of the open cells can be seen within the upper thin segment, i.e., 1–3.5 layers. Close inspection on the side surface reveals a double diagonal deformation within the thin section (Figure 4(b1)). The corresponding FE Mises stress map (Figure 4(c1)) also shows the localized stress distribution along the double shear band. When the applied strain reaches 30%, the thin section is almost fully collapsed, while the cells in the lower thick section show minimal changes in shape (Figure 4(b2)). At the same time, the FE Mises stress map shows large stress values within the squeezed upper segment, and small values within the lower segment (Figure 4(c2)). When the structure is deformed to 50%, the thick segment shows clear crushed cells as deformation enters the middle of the second stress plateau (Figure 4(b3)). The corresponding FE Mises stress cloud map (Figure 4(c3)) shows a similar cell deformation appearance, and a second pair of double shear bands with high stress magnitudes also appear.

Figure 5 shows the deformation behaviour of another group of gradient samples, Gra−2, Gra−4, and Gra−5. In this group, t_1 and t_2 are kept constant at 0.27 and 0.54 mm, while the position of the barrier layer (0.27–1.08–0.54 mm) varies. To be specific, the barrier layer is located at the 3.5–4.5, 2–3, and 5–6 layer for Gra−2, Gra−4, and Gra−5, respectively. The simulation results are well in line with the experimental results. The main difference among the three samples is the change of plateau length. Figure 5(b1–b3,c1–c3) show the experimental snapshots and FE simulated Mises stress diagrams at different applied strains on the side surface of Gra−4 during compression. It can be seen that the localized deformation occurs sequentially from the thin segment to the thick segment separated by the barrier layer.

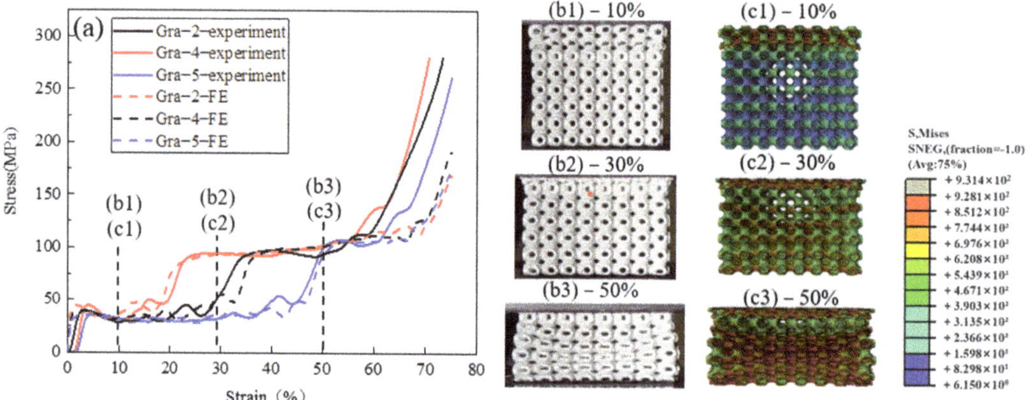

Figure 5. Deformation behaviour of the gradient TPMS structures with different lengths of the second stress plateau. (**a**) stress–strain curves of Gra−2, Gra−4, and Gra−5 samples, (**b1–b3**) snapshots of Gra−4 at 10%, 30%, and 50% strain during the compression experiment, and (**c1–c3**) Mises stress cloud diagrams of Gra−4 at 10%, 30%, and 50% strain during the FE simulation.

The specific energy absorption (SEA) is a meaningful engineering parameter to evaluate the energy absorption characteristics of a structure. It is defined as $SEA = \frac{W}{\rho}$, where W is the energy absorbed under compression at a given strain and where ρ is the mass density of the cellular structure. In this study, W was obtained by integrating the stress–strain curves up to a deformation strain of 65%, which is the onset of densification of the uniform structures. In general, the energy absorption values of the P-surface with gradient design slightly outperformed the uniform counterparts by 3–9%, as seen in Figure 6. This is because the segmental gradient design of the wall thickness breaks the localized deformation within the original double shear bands of the uniform structure and distributes the plastic deformation of the entire structure more sufficiently and evenly. The separation of the two segments with different wall thickness (relative density) improves the overall effectiveness of the cell deformation in the structure, thus improving the energy absorption performance.

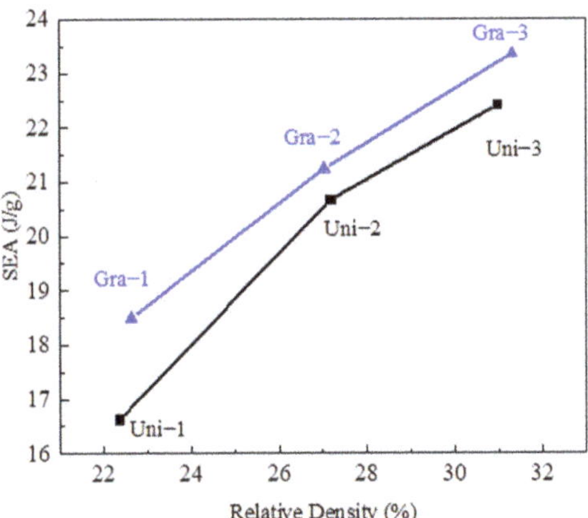

Figure 6. SEA comparison of the uniform and gradient TPMS structures with similar relative densities.

4. Conclusions

In this study, uniform and gradient P-surface TPMS structures with different relative densities were designed and fabricated using the SLM technique. The deformation mechanism and energy absorption behaviour of the structures were investigated, which led to the following conclusions:

1. The as-fabricated TPMS samples show overall satisfactory geometrical and surface quality with freedom from any gross internal or external defects. The presence of residual 316 L particles on the curved surfaces contributed to a rough surface outline as well as a slightly greater sample mass relative to the designed values.
2. The uniform TPMS structures show a typical localized deformation pattern within a double shear band geometry. The yield stress and plateau stress increase with increasing the relative density.
3. The gradient TPMS samples show a double-leveled deformation manner with a two stress plateaus on the stress–strain responses. The height and length ratio between the stress plateaus can be adjusted by changing the wall thickness in each segment and by adjusting the position of the barrier layer in the structure.
4. The specific energy absorption SEA of the gradient TPMS structures are slightly higher than the uniform counterparts with similar relative densities.

Author Contributions: Conceptualization, Z.W.; methodology, Z.W. and Y.L.; validation, M.Z.; formal analysis, M.Z. and W.Z.; investigation, M.Z., W.Z. and Z.W.; resources, H.X. and Y.L.; data curation, M.Z. and Z.W.; writing—original draft preparation, M.Z. and Z.W.; writing—review and editing, Z.W.; visualization, M.Z., W.Z. and Z.W.; supervision, Z.W.; project administration, Z.W.; funding acquisition, Y.L., H.X. and Z.W. All authors have read and agreed to the published version of the manuscript.

Funding: This research was funded by the National Natural Science Foundation of China (Grant Numbers: 52178193 and 12172151), the Natural Science Foundation of Guangdong Province (Grant Numbers: 2018A030313742 and 2020A1515011064), the Guangzhou University Postgraduate Innovation Ability Training Funding Program (Grant Numbers: 2019GDJC-M43 and 2020GDJC-M43), and the State Key Laboratory for Strength and Vibration of Mechanical Structures of Xi'an Jiaotong University (Grant Number: SV2018-KF-32). Z.W. would like to acknowledge the financial support from the Hundred Talents Program of Guangzhou University (Grant Number: RQ2021016).

Institutional Review Board Statement: Not applicable.

Informed Consent Statement: Not applicable.

Data Availability Statement: The raw data required to reproduce these findings are available from the corresponding author by request.

Conflicts of Interest: The authors declare no conflict of interest.

References

1. Bobbert, F.; Lietaert, K.; Eftekhari, A.A.; Pouran, B.; Ahmadi, S.; Weinans, H.; Zadpoor, A.A. Additively manufactured metallic porous biomaterials based on minimal surfaces: A unique combination of topological, mechanical, and mass transport properties. *Acta Biomater.* **2017**, *53*, 572–584. [CrossRef]
2. Rashed, M.G.; Ashraf, M.; Mines, R.A.W.; Hazell, P.J. Metallic microlattice materials: A current state of the art on manufactur-ing, mechanical properties and applications. *Mater. Des.* **2016**, *95*, 518–533. [CrossRef]
3. Wadley, H.N.G.; Fleck, N.A.; Evans, A.G. Fabrication and structural performance of periodic cellular metal sandwich struc-tures. *Compos. Sci. Technol.* **2003**, *63*, 2331–2343. [CrossRef]
4. Wauthle, R.; van der Stok, J.; Yavari, S.A.; Van Humbeeck, J.; Kruth, J.-P.; Zadpoor, A.A.; Weinans, H.; Mulier, M.; Schrooten, J. Additively manufactured porous tantalum implants. *Acta Biomater.* **2015**, *14*, 217–225. [CrossRef] [PubMed]
5. Wang, Y.; Huang, L.; Mei, D.; Feng, Y.; Qian, M. Numerical modeling of microchannel reactor with porous surface microstruc-ture based on fractal geometry. *Int. J. Hydrogen Energy* **2018**, *43*, 22447–22457. [CrossRef]
6. Liu, Y.; Li, S.; Zhang, L.; Hao, Y.; Sercombe, T. Early plastic deformation behaviour and energy absorption in porous β-type biomedical titanium produced by selective laser melting. *Scr. Mater.* **2018**, *153*, 99–103. [CrossRef]
7. Feng, Q.; Tang, Q.; Liu, Y.; Setchi, R.; Soe, S.; Ma, S.; Bai, L. Quasi-static analysis of mechanical properties of Ti6Al4V lattice structures manufactured using selective laser melting. *Int. J. Adv. Manuf. Technol.* **2017**, *94*, 2301–2313. [CrossRef]
8. Xiao, L.; Song, W.; Wang, C.; Liu, H.; Tang, H.; Wang, J. Mechanical behavior of open-cell rhombic dodecahedron Ti–6Al–4V lattice structure. *Mater. Sci. Eng A* **2015**, *640*, 375–384. [CrossRef]
9. Zhang, L.; Song, B.; Zhao, A.; Liu, R.; Yang, L.; Shi, Y. Study on mechanical properties of honeycomb pentamode structures fab-ricated by laser additive manufacturing: Numerical simulation and experimental verification. *Compos. Struct.* **2019**, *226*, 111199. [CrossRef]
10. Brandt, M. *Laser Additive Manufacturing: Materials, Design, Technologies, and Applications*; Woodhead Publishing: Cambridge, UK, 2016.
11. Zheng, X.; Lee, H.; Weisgraber, T.H.; Shusteff, M.; DeOtte, J.; Duoss, E.B.; Kuntz, J.D.; Biener, M.M.; Ge, Q.; Jackson, J.A.; et al. Ultralight, ultrastiff mechanical metamaterials. *Science* **2014**, *344*, 1373–1377. [CrossRef]
12. Deshpande, V.S.; Fleck, N.A.; Ashby, M.F. Effective properties of the octet-truss lattice material. *J. Mech. Phys. Solids* **2001**, *49*, 1747–1769. [CrossRef]
13. Dong, L.; Deshpande, V.; Wadley, H. Mechanical response of Ti–6Al–4V octet-truss lattice structures. *Int. J. Solids Struct.* **2015**, *60–61*, 107–124. [CrossRef]
14. Meza, L.R.; Das, S.; Greer, J.R. Strong, lightweight, and recoverable three-dimensional ceramic nanolattices. *Science* **2014**, *345*, 1322–1326. [CrossRef]
15. Nissen, H.-U. Crystal Orientation and Plate Structure in Echinoid Skeletal Units. *Science* **1969**, *166*, 1150–1152. [CrossRef]
16. Michielsen, K.; Stavenga, D.G. Gyroid cuticular structures in butterfly wing scales: Biological photonic crystals. *J. R. Soc. Interface* **2007**, *5*, 85–94. [CrossRef]
17. Galusha, J.W.; Richey, L.R.; Gardner, J.S.; Cha, J.N.; Bartl, M.H. Discovery of a diamond-based photonic crystal structure in beetle scales. *Phys. Rev. E* **2008**, *77*, 050904. [CrossRef]
18. Hyde, S.; Blum, Z.; Landh, T.; Lidin, S.; Ninham, B.; Andersson, S.; Larsson, K. *The Language of Shape: The Role of Curvature in Condensed Matter: Physics, Chemistry and Biology*; Elsevier: Amsterdam, The Netherlands, 1996.
19. Ramamurty, U.; Paul, A. Variability in mechanical properties of a metal foam. *Acta Mater.* **2004**, *52*, 869–876. [CrossRef]
20. Gibson, L.J. Mechanical Behavior of Metallic Foams. *Annu. Rev. Mater. Sci.* **2000**, *30*, 191–227. [CrossRef]
21. Zhang, L.; Feih, S.; Daynes, S.; Chang, S.; Wang, M.Y.; Wei, J.; Lu, W.F. Energy absorption characteristics of metallic triply peri-odic minimal surface sheet structures under compressive loading. *Addit. Manuf.* **2018**, *23*, 505–515.
22. Liang, Y.; Zhou, W.; Liu, Y.; Yang, Y.; Xi, H.; Wu, Z. Energy Absorption and Deformation Behaviour of 3D Printed TPMS Stain-less Steel Cellular Structures Under Compression. *Steel Res. Int.* **2021**, *92*, 2000411. [CrossRef]
23. Yang, L.; Yan, C.; Cao, W.; Liu, Z.; Song, B.; Wen, S.; Zhang, C.; Shi, Y.; Yang, S. Compression–compression fatigue behaviour of gyroid-type triply periodic minimal surface porous structures fabricated by selective laser melting. *Acta Mater.* **2019**, *181*, 49–66. [CrossRef]
24. Fan, X.; Tang, Q.; Feng, Q.; Ma, S.; Song, J.; Jin, M.; Guo, F.; Jin, P. Design, mechanical properties and energy absorption capabil-ity of graded-thickness triply periodic minimal surface structures fabricated by selective laser melting. *Int. J. Mech. Sci.* **2021**, *204*, 106586. [CrossRef]
25. Min, Y.; Fengxiang, X.; Mingyuan, G. In-plane impact performance of honeycomb material with gradient thickness and nega-tive Poisson's ratio. *J. Plast. Eng.* **2021**, *6*, 192–199.
26. Al-Ketan, O.; Lee, D.-W.; Rowshan, R.; Abu Al-Rub, R.K. Functionally graded and multi-morphology sheet TPMS lattices: De-sign, manufacturing, and mechanical properties. *J. Mech. Behav. Biomed. Mater.* **2020**, *102*, 103520. [CrossRef]
27. Yu, S.; Sun, J.; Bai, J. Investigation of functionally graded TPMS structures fabricated by additive manufacturing. *Mater. Des.* **2019**, *182*, 108021. [CrossRef]

28. Panesar, A.; Abdi, M.; Hickman, D.; Ashcroft, I. Strategies for functionally graded lattice structures derived using topology optimisation for Additive Manufacturing. *Addit. Manuf.* **2018**, *19*, 81–94. [CrossRef]
29. Von Schnering, H.G.; Nesper, R. Nodal surfaces of Fourier series: Fundamental invariants of structured matter. *Z. Für Phys. B Condens. Matter* **1991**, *83*, 407–412. [CrossRef]
30. Yan, C.; Hao, L.; Hussein, A.; Young, P.; Raymont, D. Advanced lightweight 316L stainless steel cellular lattice structures fabricated via selective laser melting. *Mater. Des.* **2014**, *55*, 533–541. [CrossRef]

Article

Flexible Pressure Sensor with Micro-Structure Arrays Based on PDMS and PEDOT:PSS/PUD&CNTs Composite Film with 3D Printing

Yiwei Shao, Qi Zhang *, Yulong Zhao *, Xing Pang, Mingjie Liu, Dongliang Zhang and Xiaoya Liang

State Key Laboratory for Manufacturing Systems Engineering, School of Mechanical Engineering, Xi'an Jiaotong University, Xi'an 710049, China; imporeed@stu.xjtu.edu.cn (Y.S.); px2014@stu.xjtu.edu.cn (X.P.); liumingjie@stu.xjtu.edu.cn (M.L.); zhangdl666@stu.xjtu.edu.cn (D.Z.); liang_xiaoya@stu.xjtu.edu.cn (X.L.)
* Correspondence: zhq0919@xjtu.edu.cn (Q.Z.); zhaoyulong@xjtu.edu.cn (Y.Z.)

Abstract: Flexible pressure sensors are widely used in different fields, especially in human motion, robot monitoring and medical treatment. Herein, a flexible pressure sensor consists of the flat top plate, and the microstructured bottom plate is developed. Both plates are made of polydimethylsiloxane (PDMS) by molding from the 3D printed template. The contact surfaces of the top and bottom plates are coated with a mixture of poly (3,4-ethylenedioxythiophene) poly (styrene sulfonate) (PEDOT:PSS) and polyurethane dispersion (PUD) as stretchable film electrodes with carbon nanotubes on the electrode surface. By employing 3D printing technology, using digital light processing (DLP), the fabrication of the sensor is low-cost and fast. The sensor models with different microstructures are first analyzed by the Finite Element Method (FEM), and then the models are fabricated and tested. The sensor with 5×5 hemispheres has a sensitivity of 3.54×10^{-3} S/kPa in the range of 0–22.2 kPa. The zero-temperature coefficient is -0.0064%FS/$^\circ$C. The durability test is carried out for 2000 cycles, and it remains stable during the whole test. This work represents progress in flexible pressure sensing and demonstrates the advantages of 3D printing technology in sensor processing.

Keywords: flexible pressure sensor; microstructure; 3D printing; composite film

Citation: Shao, Y.; Zhang, Q.; Zhao, Y.; Pang, X.; Liu, M.; Zhang, D.; Liang, X. Flexible Pressure Sensor with Micro-Structure Arrays Based on PDMS and PEDOT:PSS/PUD&CNTs Composite Film with 3D Printing. *Materials* **2021**, *14*, 6499. https://doi.org/10.3390/ma14216499

Academic Editor: Rubén Paz

Received: 23 September 2021
Accepted: 28 October 2021
Published: 29 October 2021

Publisher's Note: MDPI stays neutral with regard to jurisdictional claims in published maps and institutional affiliations.

1. Introduction

Due to the wide application of flexible pressure sensors in human motion [1–3], robot monitoring [4,5] and medical treatment [6–9], research on flexible pressure sensors has become popular in the past decade. Such sensors with high flexibility can convert pressure information input into an electrical signal output. Among them, capacitive and piezoresistive sensors are mostly reported. Capacitive sensors usually have two electrodes, sandwiching a dielectric layer. When pressure is applied, the distance between the electrodes changes, the dielectric constant of the dielectric layer changes or the electrode area changes [1,9–25]. By using different dielectric materials, changing the electrode size, etc., capacitive sensors meeting different requirements can be designed. However, due to the limitation of the size of electrodes, the capacitance is small, and the stray capacitance is relatively large, which is susceptible to external interference and affects the measurement accuracy.

The piezoresistive sensor has a simple structure, and the circuit design is easy [6]. This kind of sensor usually deforms under pressure and detects the pressure by the resistance change in three different ways: (1) The piezoresistive characteristics of the material itself [2,4,26]; (2) Doped carbon nanotubes, graphene, silver nanowires and other conductive fillers form more electrical contacts or a tunneling effect [3,5,27–32]; (3) The contact resistance of the contact area between the two electrodes changes [6,8,33–37]. For the third type of piezoresistive sensors, the common way to improve performance is to add fine microstructures onto the electrode contact surface. Compared with a device without

microstructures [32], when an external pressure is applied within a certain range, the microstructure increases the change in the contact area between the two electrodes, improving the sensitivity of the sensor. By adjusting the shape and size of the microstructure, different performance sensors can be easily obtained. Micro pyramids are the most common microstructure in pressure sensors [6,33,34,36]. As reported by Choong's work in 2014, a piezoresistive pressure sensor was achieved by coating a compressible substrate containing an array of micro pyramids, which was obtained by replication of silicon process templates, and had a sensitivity of 4.88 kPa^{-1} over a wide range of pressures from 0.37 to 5.9 kPa [6]. Moreover, microcylindrical and other structures were also reported and fabricated in a similar method [22,35,37].

Although the performance of these sensors is excellent and the silicon process is mature and precise, the fabrication process has its disadvantages. The silicon process is not only time-consuming and costly but can also generally only process planar structures. Complex surface structures, such as curved surfaces, are hard to process. It is necessary to study other fabrication processes, such as three-dimensional (3D) printing technologies [38]. Three-dimensional printing can directly generate parts of any shape from computer graphics data without machining, thereby greatly shortening the product development cycle and reducing costs. Moreover, 3D printing can also print some complicated appearances, such as curved surfaces and inclined surfaces, which are hard to fabricate by traditional silicon processes, expanding more ideas to design novel sensors.

Here, by 3D printing, the reasonable design and manufacturing method of a flexible pressure sensor is proposed. FEM is applied to simulate the stress distribution and resistance of sensors with different microstructures and density distributions. The results show that the sensor performance is dependent on the shape of microstructures, and the hemisphere is found to be a proper geometric structure that achieves the best sensor performance in terms of sensitivity and measurement range compared to pyramid and cone, while sensors with different hemisphere density distributions have similar responses but different measurement ranges. To simplify the manufacturing process, the templates with microstructures are fabricated by 3D printing. By molding the templates with microstructures, PDMS bottom plates with microstructures are created, and flat top plates without microstructures are obtained in the same way. Then, a thin layer of PEDOT:PSS and PUD mixture is evenly applied to the contact surface of the top and bottom plates and CNTs are distributed on the top of this layer of film electrode. Finally, the flat top plate and microstructured bottom plate are assembled to form the pressure before testing.

2. Materials and Methods

2.1. Sensor Design

The schematic diagram of the proposed flexible pressure sensors is shown in Figure 1a. The sensor was composed of upper and lower parts. The main body of the two parts was a flat square plate with a thickness of 1.3 mm and a side length of 15.0 mm. The surface of the lower part was evenly distributed with micro-structure protrusions, while the upper part was a completely flat plate.

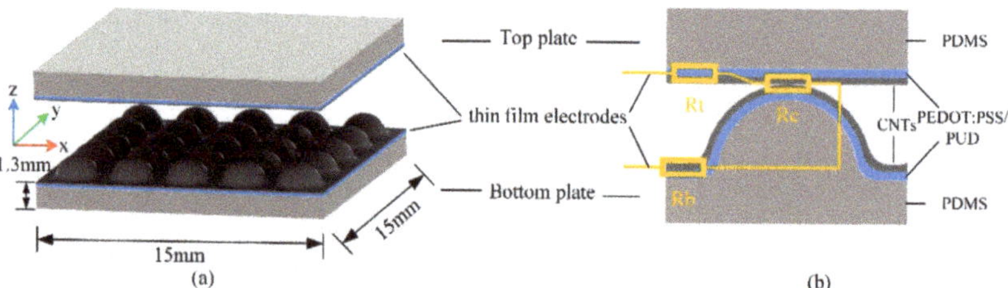

Figure 1. The proposed pressure sensor: (**a**) basic structure; (**b**) simplified sensor model.

Here we designed 6 different structures for the lower part: 2 × 2 hemispheres, 3 × 3 hemispheres, 4 × 4 hemispheres, 5 × 5 hemispheres with a diameter of 2.0 mm, 5 × 5 cones with a diameter of 2.0 mm on the bottom and 5 × 5 pyramids with a side length of 2.0 mm on the bottom, as shown in Figure 2.

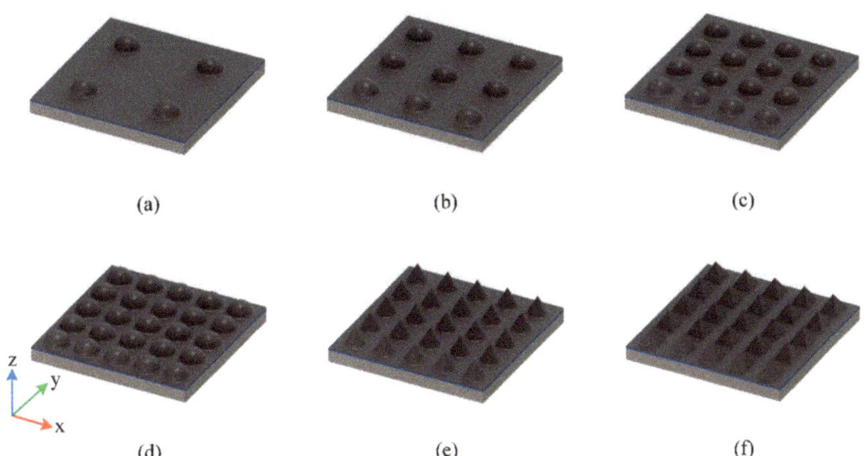

Figure 2. Schematic diagram of the designed sensors with 6 different structures: (**a**) 2 × 2 hemispheres; (**b**) 3 × 3 hemispheres; (**c**) 4 × 4 hemispheres; (**d**) 5 × 5 hemispheres; (**e**) 5 × 5 cones; (**f**) 5 × 5 pyramids.

To analyze the change trend of resistance under pressure, the sensor model was simplified [6], as shown in Figure 1b. The total resistance (R) consists of top plate resistance (R_t), contact resistance (R_c) and bottom plate resistance (R_b):

$$R = R_t + R_c + R_b \tag{1}$$

When the pressure is applied, the change of resistance could be described:

$$\Delta R = R - R_0 = (R_t + R_c + R_b) - (R_{t0} + R_{c0} + R_{b0}) \tag{2}$$

Since the top plate resistance was almost invariable, Equation (2) could be simplified to Equation (3):

$$\Delta R \approx (R_c - R_{c0}) + (R_b - R_{b0}) \tag{3}$$

According to Ohm's law,

$$R = \frac{\rho L}{A} \tag{4}$$

where L was the length of the resistor, A was the cross-sectional area and ρ was the resistivity.

Equation (3) was rewritten to Equation (5):

$$\Delta R = \rho_c \left(\frac{L_c}{A_c} - \frac{L_{c0}}{A_{c0}} \right) + \rho_b \left(\frac{L_b}{D_b C_b} - \frac{L_{b0}}{D_{b0} C_{b0}} \right) \tag{5}$$

where A_c was the contact area, L_c was the film electrodes thickness at the contact area, D_b was the film thickness at the side of micro-structure, L_b was the film length at the side of micro-structure and C_b was the perimeter of the contact area.

When the pressure was applied, L_c and L_b decreased, and A_c, D_b and C_b increased, resulting in the resistance of the sensor decreasing.

2.2. Sensor Simulation

To analyze the influence of microstructure protrusions of different shapes and sizes on the performance of the sensor, the finite element method was introduced. The software used for simulation was COMSOL 5.6. Six sensor models with the different microstructures mentioned above were imported into the software, both top and bottom plates were modeled by hyperelasticity and the material model was the Mooney–Rivlin model, two parameters, with C_{10} 2.4×10^5 Pa, C_{01} 6.6×10^4 Pa and bulk modulus κ 1.25×10^7 Pa. Then external displacement was applied, pushing the top plate downward to make contacts with the microstructures on the bottom plate. The pressure distance increased from 0 to 0.6 mm, and the change curve between displacement and force and resistance was obtained. The specific results were as follows.

In the simulation results of the pressure change with the displacement of the depression, the protrusions of different structures were compared. As shown in Figure 3a, when the number of protrusions was 5×5, different structures had different pressure curves. In the same depression displacement of 0.6 mm, the sensor with hemispherical structures had the largest range of 44.09 kPa, while the sensor with pyramidal structures and the conical structures had a similar range, which is less than a quarter of that of the sensor with hemispherical structures.

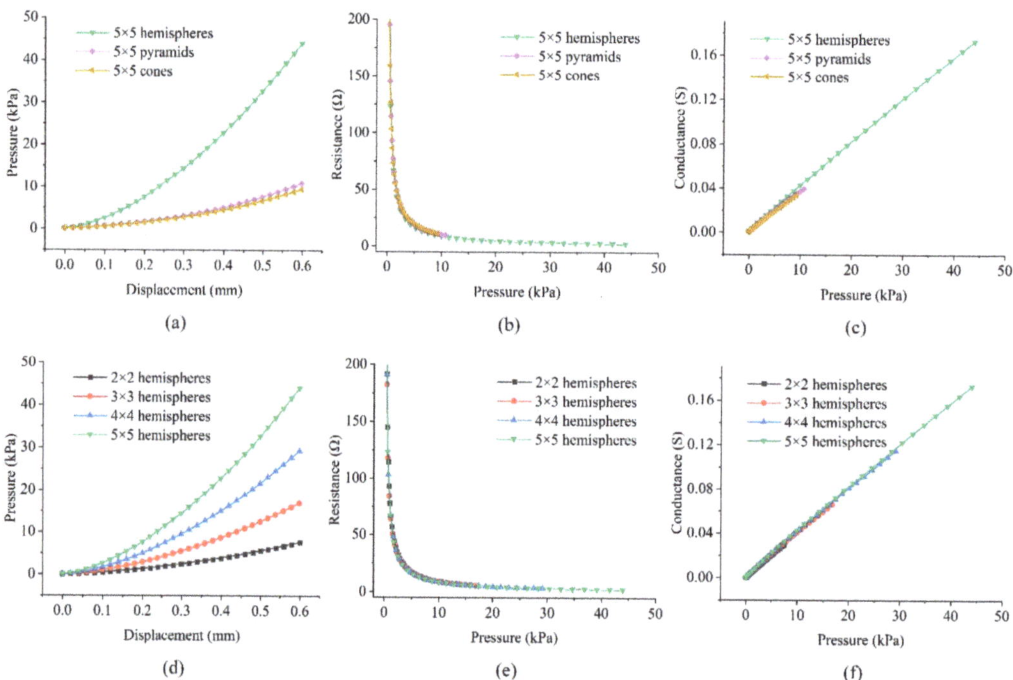

Figure 3. FEM results of the sensors with different microstructures: (**a–c**) Sensors with 5×5 hemispheres, pyramids and cones; (**d–f**) Sensors with different density distribution of hemispheres.

Next, we studied the hemispheric protrusions with different distribution densities on the bottom plate surface of the same size. As shown in Figure 3d, it was obvious that as the number of hemispheres increases, the pressure also increases. Analysis of the data showed that the pressure was proportional to the number of hemispheres, and each hemisphere provided a pressure of approximately 1.76 kPa when the pressure displacement was 0.6 mm. This was also in line with common sense, and the sensor can be regarded as a parallel connection of different numbers of single hemispherical structure sensors. In this way, the higher the number of hemispheres, the larger the measurement range of the sensor.

Then the change of the sensor's resistance with pressure was studied. As shown in Figure 3b,e, 6 types of sensors with different types and numbers of protrusions were compared. Although the structure and number of protrusions cause a huge difference in range, the resistance-pressure curves of all sensors in the same range were very similar. In the range of 0–5 kPa, the resistance rapidly decreased from the maximum value to 20 Ω. When the pressure was higher than 5 kPa, the resistance decreased very slowly. The inverse of the resistance was used to get the conductance. As shown in Figure 3c,f, the conductance changed almost linearly with pressure. The conductance-pressure curves of 5 × 5 hemispheres, pyramids and cones were relatively similar, while the conductance-pressure curves of sensors with different numbers of hemisphere protrusions almost overlap. No matter how the number of hemispheres changed, the output resistance of the sensor remained unchanged under pressure. Therefore, we could design sensors with different hemispherical density distributions according to the different ranges required by the measurement requirements, without the need for major modifications to the subsequent circuit.

Meanwhile, the stress of the sensors with pyramid and cone structures was concentrated on the top of the microstructure, where the top plate and the microstructures on the bottom plate were in contact. The stress of the sensor with a hemispherical structure was relatively dispersed, and the maximum stress was much smaller than that of the sensor with pyramid and cone structures. When the applied displacement was 0.6 mm, the maximum von Mises stress of pyramid, cone and hemisphere structures were 6.40×10^5, 7.24×10^5 and 2.15×10^5 Pa, respectively, as shown in Figure 4. Larger stress would cause the conductive film to be easily damaged.

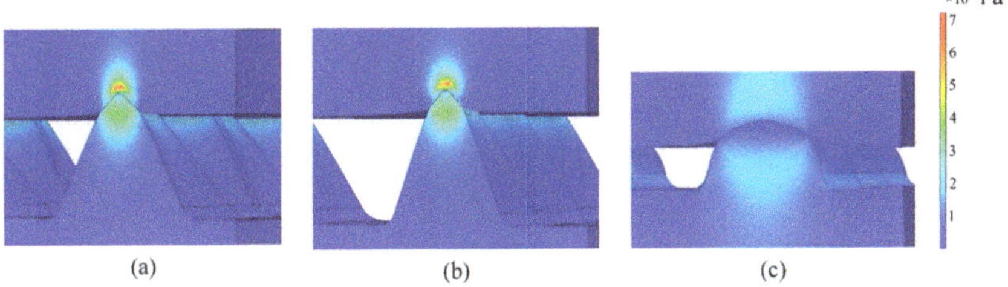

(a)   (b)   (c)

Figure 4. Von Mises stress of (**a**) pyramid, (**b**) cone and (**c**) hemisphere structures under 0.6 mm of displacement.

2.3. Sensor Fabrication

The templates with micro-structure were designed, as shown in Figure 5 below.

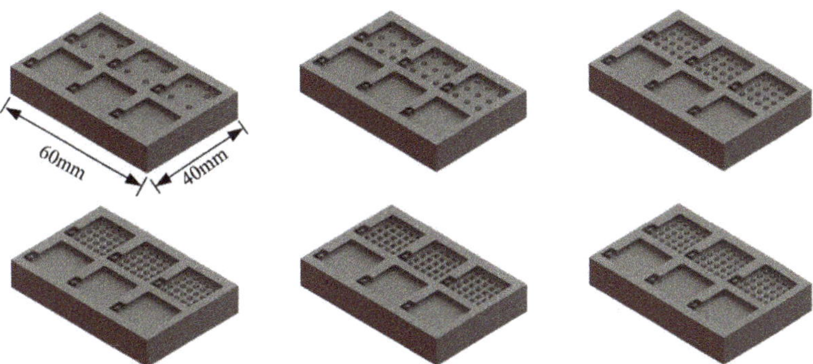

Figure 5. Designed templates.

The templates were divided into two parts, one part was the top plate, and the other was the bottom plate with different structure depressions. High-temperature-resistant resin (High Temp, Formlabs, Somerville, MA, USA) was used for processing the template with a 3D Digital Light Processing (DLP) printer (M-Jewelry U50, MAKEX, Ningbo, China). This printer is a desktop 3D printer with a lateral resolution of 50 μm and a thickness resolution of 5 μm. The high-temperature resin offers a heat deflection temperature (HDT) of 238 °C at 0.45 MPa. During printing, the template is printed from bottom to top in a vertical orientation.

After printing, we rinsed the template thoroughly with isopropanol and ethanol to remove resin residue. Then, the template was exposed to ultraviolet (UV) light in a UV curing machine (Cure3D, MAKEX, Ningbo, China) for 1 h to ensure complete curing. After that, the template was placed in an air convection oven (PG-2J, ESPEC, Osaka, Japan) and heated to 120 °C for 3 h and cooled with the oven to eliminate the internal stress of the template and prevent deformation in the pouring process, as shown in Figure 6a.

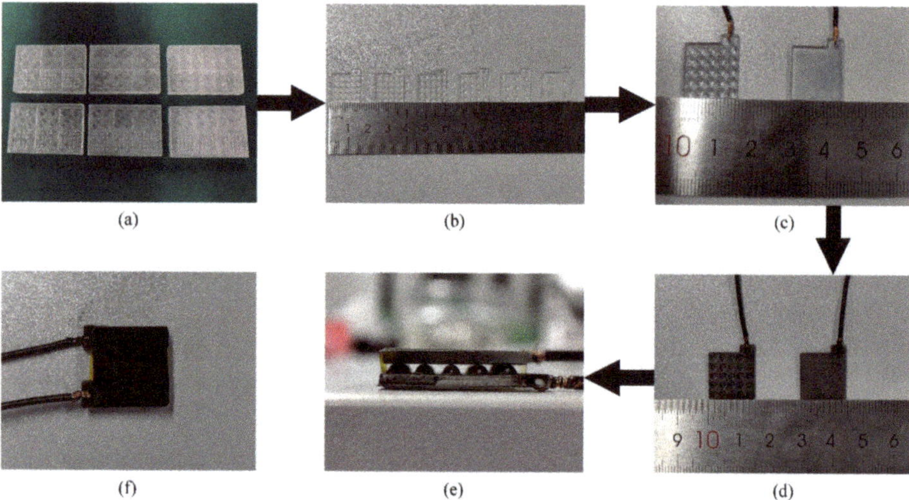

Figure 6. The fabrication process of sensors: (**a**) Template printing; (**b**) Template after cleaning and curing; (**c**) Top and bottom plates; (**d**) Plates with film electrodes; (**e,f**) Assembled sensor.

Next, the PDMS elastomer and curing agent (Sylgard 184, Dow Corning, Midland, MI, USA) were mixed evenly at a ratio of 10:1 and vacuumed to remove bubbles. The 3D printed template was heated to 100 °C in the oven for 4 h after the mixture was cast. After that, the top and bottom plates were obtained by peeling off the cured PDMS from the template, as shown in Figure 6b.

Then oxygen plasma was used to clean the top and bottom plates to enhance the hydrophilicity of the PDMS surface by the low-pressure plasma system (V6-G, Pink, Wertheim, Germany). Subsequently, a mixture of PEDOT:PSS (P Jet 700 N, Clevios, Hanau, Germany) and PUD (Tekspro 7360, WANHUA, Yantai, China), with a ratio of 9:1, was spread evenly by drop-casting to the top and bottom plate surfaces and cured completely at 70 °C for 30 min, as shown in Figure 6c. Then the carbon nanotube's water-based coating (XFEC01, XFnano, Nanjing, China) was diluted to 1%wt and applied on the PEDOT:PSS/PUD films in the same way to achieve a double-layer conductive film, as shown in Figure 6d.

Finally, the top and bottom plates were tied together by the Kapton tape, and the fabricated pressure sensor is shown in Figure 6e,f.

To measure the thickness of the film, the film step was prepared, and the film thickness was observed and measured with a 3D measuring laser microscope (LEXT OLS4100,

OLYMPUS, Tokyo, Japan). Five cross-sections of the film step were measured, as shown in Figure 7a. Due to the unevenness of the film near the step (120–140 μm in the *x*-axis position), the average film thickness was calculated from the film thickness between 350 and 450 μm in the *x*-axis position, which was 4.99 ± 0.50 μm, as shown in Figure 7b.

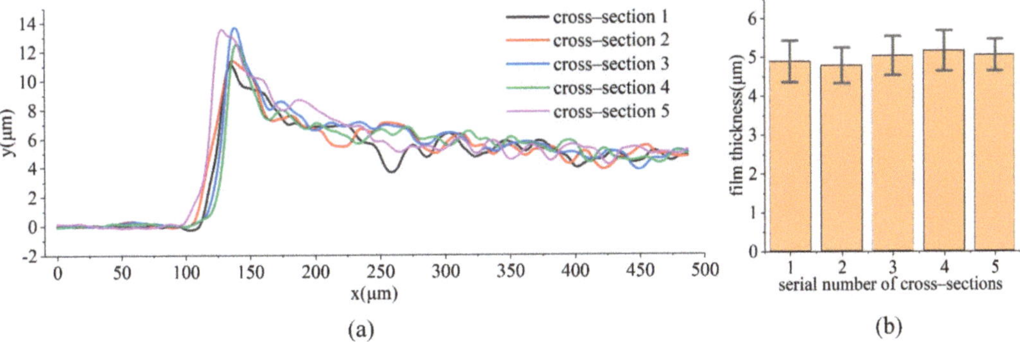

Figure 7. (**a**) The thickness of five cross-sections of the film; (**b**) Average thickness of the film between 350–450 μm in the *x*-axis position.

3. Results

3.1. Experimental Setup

The test bench mainly consisted of two parts, a compression test machine (PT-1198G, POOTAB, Dongguan, China) and a high-precision multimeter (8846A, Fluke, Everett, WA, USA) [35], as shown in Figure 8. The sensor was placed on the test machine plate, and the force was applied to the sensor by adjusting the displacement of the squeeze head. The resistance change of the sensor was read through the high-precision multimeter. The experimental setup was put in a 25 °C clean room, where the environmental impact can be minimized.

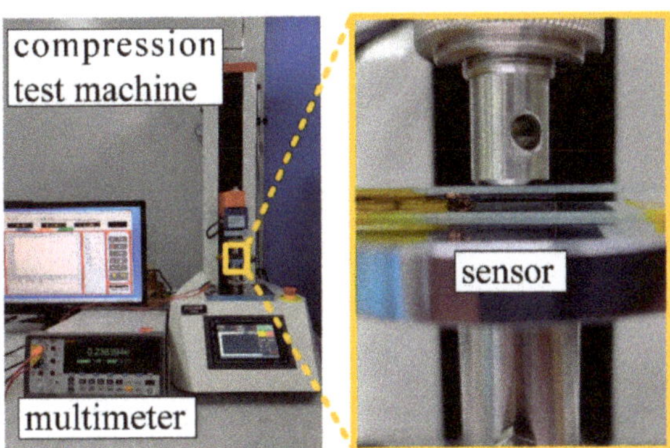

Figure 8. Experimental Setup.

3.2. Static Test

In order to evaluate the static performance of sensors with different microstructure arrays, we tested the six types of sensors processed above. The test range of the sensor with 2 × 2 hemisphere structure was 0–13.3 kPa, and the range of the rest of the five sensors was 0–22.2 kPa. The force load started from 0 kPa, and increased by 2.22 kPa each time, with each step holding for 20 s. After reaching the maximum range, it decreased by 2.22 kPa each time. The pressure was also held for 20 s, and the resistance change of the whole period was read through the high-precision multimeter. As shown in Figure 9a,b, at low pressure ranges (0–4 kPa), the sensor resistance decreased sharply, but at high pressure (>4 kPa), since the microstructure deformation of the sensor saturated with the increase in pressure, the resistance decrease speed is significantly reduced. This result was consistent with the conclusion of the previous FEM result.

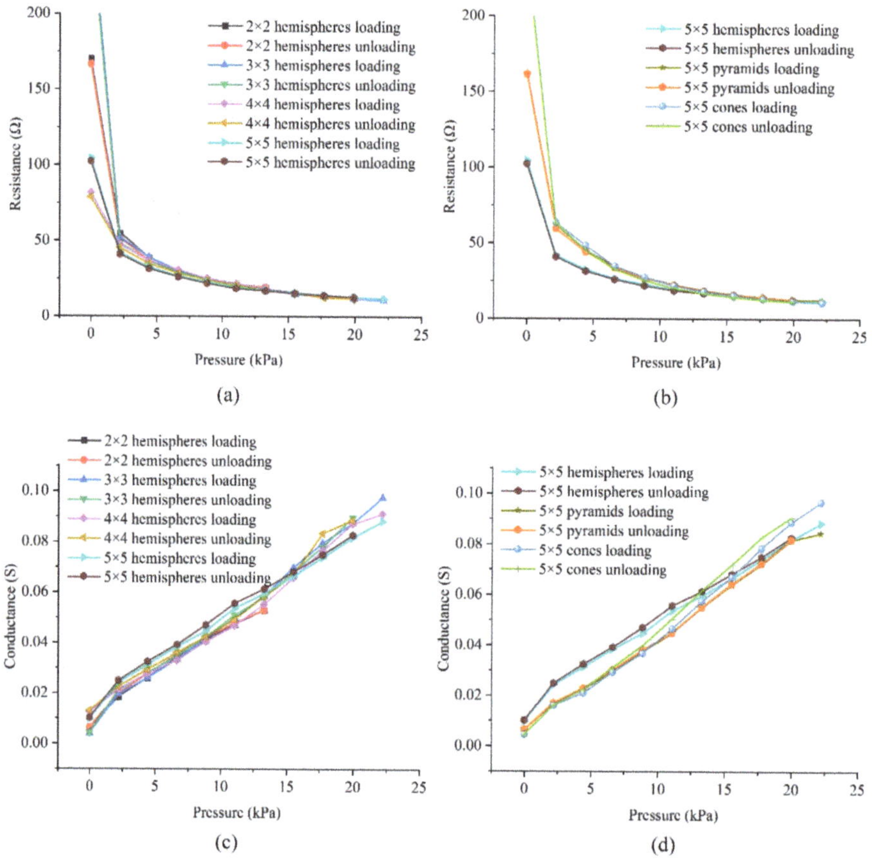

Figure 9. Sensors static test results: (**a**,**c**) Sensors with different hemispheres distribution densities; (**b**,**d**) Sensors with different structures.

By converting resistance to conductance, as shown in Figure 9c,d, it can be seen that the conductance-pressure curves of the six sensors are almost linear and almost overlap, which is similar to the simulation conclusion. Taking a 5×5 hemisphere structure sensor as an example, it can be obtained by the calculation that the sensitivity of the sensor is 3.54×10^{-3} S/kPa in the range of 0–22.2 kPa. The maximum conductance difference between loading and unloading was 0.0023 S at 8.89 kPa, and the hysteresis was 1.41%FS. This was due to the viscoelasticity of the flexible film and the substrate, which could only be minimized but not completely eliminated. The largest deviation of the resistance appeared at 0 kPa; it was considered that the contact between the top and bottom plates was not sufficient, so the deviation was relatively large.

3.3. Temperature Test

The temperature test of the pressure sensor with 5×5 hemispheres was conducted within a temperature range from 0 to 80 °C in the air convection oven. The output data is shown in Figure 10, and the maximum change of the output conductance was −0.302 mS at 80 °C. The zero-temperature coefficient was calculated, which was −0.0064%FS/°C. It might be caused by the thermal expansion of the PDMS substrate and the PEDOT:PSS/PUD film electrodes.

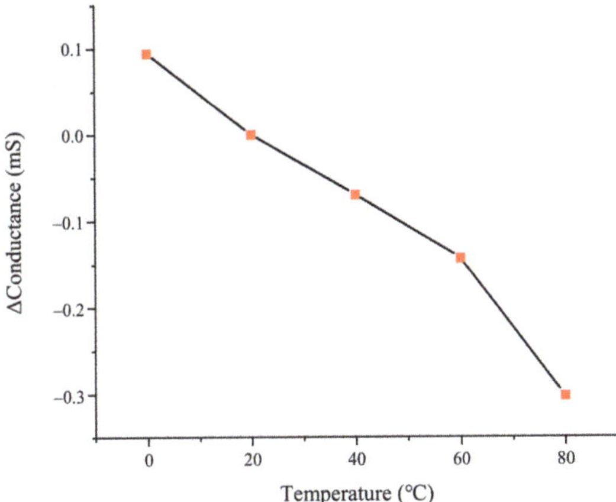

Figure 10. Thermal zero drift of the sensor.

3.4. Durability Test

By controlling the movement of the compression test machine, the sensor with 5×5 hemispheres had undergone 2000 loading/unloading tests, and each loading/unloading process took 2 s. The data was read by the high-precision multimeter, and the results are shown in Figure 11.

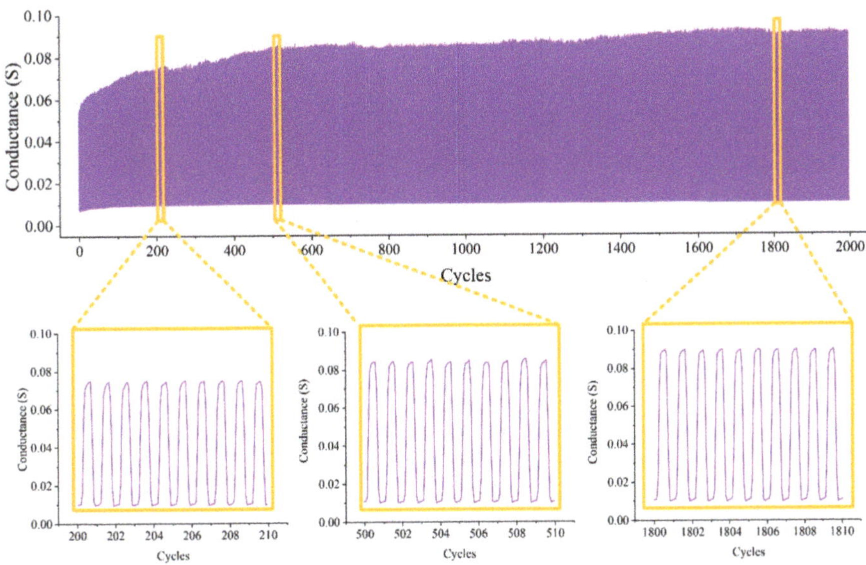

Figure 11. Durability test results.

Before 500 cycles, there was an obvious increase in conductance from 0.06 to 0.082 S during loading. The conductance of unloading remained stable at about 0.01 S during the whole test. Judging from the magnified images of 500–510 cycles and 1800–1810 cycles, the conductance change curve of each load/unload cycle was almost the same, which also reflected the good repeatability of the sensor. It can be seen from the rise and fall speed of the resistance that the response speed of the sensor was very fast.

3.5. Effect of Test on Electrode Films

The surfaces of the sensor before and after the test were observed using the scanning electron microscope (SEM) (Hitachi, SU8010, Japan), as shown in Figure 12. Figure 12a–c was the sensor image before the test. Under low magnification, stripes could be seen on the sensor surface, which was due to the 3D printing mold formed layer by layer, and the layer thickness was set to 50 μm during printing. Under high magnification, the distribution of carbon tubes on the surface of the sensor could be clearly seen. It can be seen in the electron microscope image of the sensor after the durability test that although the electrode film seemed intact at low magnification (Figure 12d), slight cracks on the electrode film can be found at high magnification (Figure 12e). At higher magnification, the loose CNTs have been compacted and embedded in the lower PEDOT:PSS/PUD hybrid film (Figure 12f).

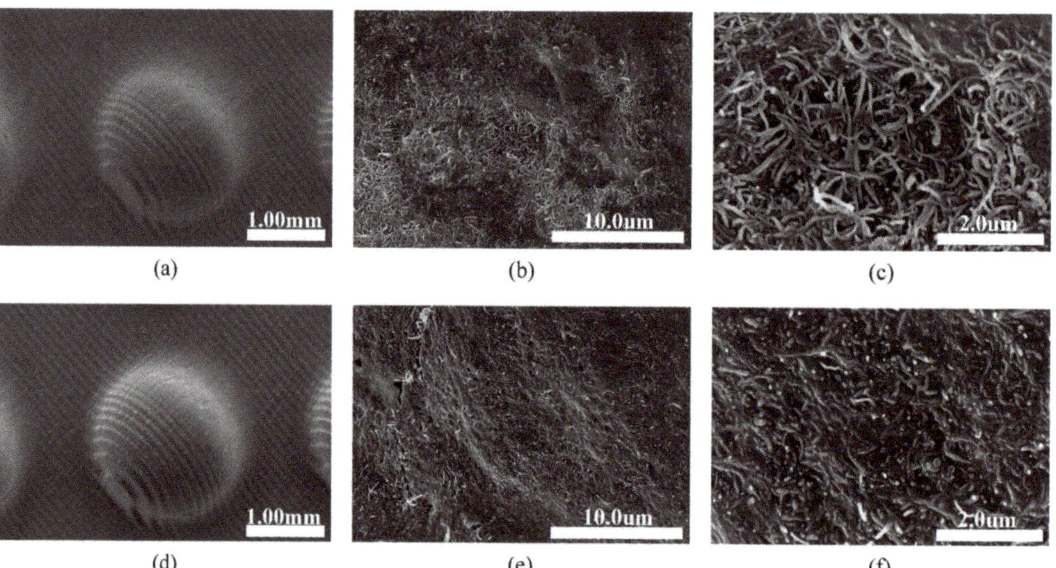

Figure 12. SEM images of sensors before (**a–c**) and after (**d–f**) the test: (**a,d**) Surface of the structure; (**b,c,e,f**) Top of the hemisphere under higher magnification.

4. Conclusions

In this work, we discussed the influence of protrusions with different morphologies and different density distributions on flexible pressure sensors performance and developed a new method of manufacturing sensors based on 3D printing technology. The elastic top and bottom plates of the sensor with micro-structure protrusions were made of PDMS, and the surfaces of the plates were coated with a conductive film, which was made from a mixture of PEDOT:PSS and PUD, and a layer of CNTs covered it. The top and bottom plates were obtained by replicating a 3D printed template. Compared with the previous templates obtained by the silicon process, 3D printing has the advantages of time-saving, low cost and being able to process unique shapes. By using FEM simulation, three sensors with different types of microstructures, including pyramid, cone and hemisphere, were analyzed, and the results show that the sensor with the hemispherical structure had the best performance. Furthermore, sensors with different hemisphere density distributions had similar responses. The difference was that the range of the sensor increased as the hemisphere density increased. All sensors were fabricated and tested. The sensor with 5×5 hemispheres had a sensitivity of 3.54×10^{-3} S/kPa and hysteresis of 1.41%FS in the range of 0–22.2 kPa. The zero-temperature coefficient was -0.0064%FS/°C. The durability test was carried out for 2000 cycles, and the sensor remained stable during the test. These results prove that 3D printing has great potential in the field of sensor manufacturing. In future research, higher precision 3D printers and suiTable 3D printing materials will optimize the processing and improve the performance of the sensor.

Author Contributions: Conceptualization, Y.Z.; methodology, Y.Z. and Y.S.; investigation, Y.S., X.P. and M.L.; writing—original draft preparation, Y.S.; writing—review and editing, Q.Z., D.Z. and X.L.; funding acquisition, Y.Z. and Q.Z. All authors have read and agreed to the published version of the manuscript.

Funding: This research was funded by the National Natural Science Foundation of China (Grant No. 51805425 and U20A20296) and the Ningbo major special project of the Plan "Science and Technology Innovation 2025" (Grant No. 2020Z023).

Institutional Review Board Statement: Not applicable.

Informed Consent Statement: Not applicable.

Data Availability Statement: Data sharing not applicable.

Conflicts of Interest: The authors declare no conflict of interest.

References

1. Laszczak, P.; Jiang, L.; Bader, D.L.; Moser, D.; Zahedi, S. Development and validation of a 3D-printed interfacial stress sensor for prosthetic applications. *Med. Eng. Phys.* **2015**, *37*, 132–137. [CrossRef]
2. Lu, N.; Lu, C.; Yang, S.; Rogers, J. Highly Sensitive Skin-Mountable Strain Gauges Based Entirely on Elastomers. *Adv. Funct. Mater.* **2012**, *22*, 4044–4050. [CrossRef]
3. Amjadi, M.; Pichitpajongkit, A.; Lee, S.; Ryu, S.; Park, I. Highly Stretchable and Sensitive Strain Sensor Based on Silver Nanowire–Elastomer Nanocomposite. *ACS Nano* **2014**, *8*, 5154–5163. [CrossRef]
4. Park, Y.L.; Majidi, C.; Kramer, R.; Berard, P.; Wood, R.J. Hyperelastic pressure sensing with a liquid-embedded elastomer. *J. Micromech. Microeng.* **2010**, *20*, 6. [CrossRef]
5. Xiao, X.; Yuan, L.; Zhong, J.; Ding, T.; Liu, Y.; Cai, Z.; Rong, Y.; Han, H.; Zhou, J.; Wang, Z.L. High-Strain Sensors Based on ZnO Nanowire/Polystyrene Hybridized Flexible Films. *Adv. Mater.* **2011**, *23*, 5440–5444. [CrossRef]
6. Choong, C.L.; Shim, M.B.; Lee, B.S.; Jeon, S.; Ko, D.S.; Kang, T.H.; Bae, J.; Lee, S.H.; Byun, K.E.; Im, J.; et al. Highly Stretchable Resistive Pressure Sensors Using a Conductive Elastomeric Composite on a Micropyramid Array. *Adv. Mater.* **2014**, *26*, 3451–3458. [CrossRef] [PubMed]
7. Liu, Z.; Ma, Y.; Ouyang, H.; Shi, B.; Li, N.; Jiang, D.; Xie, F.; Qu, D.; Zou, Y.; Huang, Y.; et al. Transcatheter Self-Powered Ultrasensitive Endocardial Pressure Sensor. *Adv. Funct. Mater.* **2019**, *29*, 1807560. [CrossRef]
8. He, K.; Hou, Y.; Yi, C.; Li, N.; Sui, F.; Yang, B.; Gu, G.; Li, W.; Wang, Z.; Li, Y.; et al. High-performance zero-standby-power-consumption-under-bending pressure sensors for artificial reflex arc. *Nano Energy* **2020**, *73*, 104743. [CrossRef]
9. Tee, B.C.K.; Chortos, A.; Dunn, R.R.; Schwartz, G.; Eason, E.; Bao, Z. Tunable Flexible Pressure Sensors using Microstructured Elastomer Geometries for Intuitive Electronics. *Adv. Funct. Mater.* **2014**, *24*, 5427–5434. [CrossRef]
10. Mannsfeld, S.C.B.; Tee, B.C.K.; Stoltenberg, R.M.; Chen, C.; Barman, S.; Muir, B.V.O.; Sokolov, A.N.; Reese, C.; Bao, Z.N. Highly sensitive flexible pressure sensors with microstructured rubber dielectric layers. *Nat. Mater.* **2010**, *9*, 859–864. [CrossRef]
11. Fan, F.R.; Lin, L.; Zhu, G.; Wu, W.Z.; Zhang, R.; Wang, Z.L. Transparent Triboelectric Nanogenerators and Self-Powered Pressure Sensors Based on Micropatterned Plastic Films. *Nano Lett.* **2012**, *12*, 3109–3114. [CrossRef]
12. Hu, W.; Niu, X.; Zhao, R.; Pei, Q. Elastomeric transparent capacitive sensors based on an interpenetrating composite of silver nanowires and polyurethane. *Appl. Phys. Lett.* **2013**, *102*, 083303. [CrossRef]
13. Schwartz, G.; Tee, B.C.K.; Mei, J.; Appleton, A.L.; Kim, D.H.; Wang, H.; Bao, Z. Flexible polymer transistors with high pressure sensitivity for application in electronic skin and health monitoring. *Nat. Commun.* **2013**, *4*, 1859. [CrossRef] [PubMed]
14. Boutry, C.M.; Nguyen, A.; Lawal, Q.O.; Chortos, A.; Rondeau-Gagné, S.; Bao, Z. A Sensitive and Biodegradable Pressure Sensor Array for Cardiovascular Monitoring. *Adv. Mater.* **2015**, *27*, 6954–6961. [CrossRef]

15. Kwon, D.; Lee, T.-I.; Shin, J.; Ryu, S.; Kim, M.S.; Kim, S.; Kim, T.-S.; Park, I. Highly Sensitive, Flexible, and Wearable Pressure Sensor Based on a Giant Piezocapacitive Effect of Three-Dimensional Microporous Elastomeric Dielectric Layer. *ACS Appl. Mater. Interfaces* **2016**, *8*, 16922–16931. [CrossRef] [PubMed]

16. Li, T.; Luo, H.; Qin, L.; Wang, X.W.; Xiong, Z.P.; Ding, H.Y.; Gu, Y.; Liu, Z.; Zhang, T. Flexible Capacitive Tactile Sensor Based on Micropatterned Dielectric Layer. *Small* **2016**, *12*, 5042–5048. [CrossRef] [PubMed]

17. Cho, S.H.; Lee, S.W.; Yu, S.; Kim, H.; Chang, S.; Kang, D.; Hwang, I.; Kang, H.S.; Jeong, B.; Kim, E.H.; et al. Micropatterned Pyramidal Ionic Gels for Sensing Broad-Range Pressures with High Sensitivity. *ACS Appl. Mater. Interfaces* **2017**, *9*, 10128–10135. [CrossRef] [PubMed]

18. Lee, K.; Lee, J.; Kim, G.; Kim, Y.; Kang, S.; Cho, S.; Kim, S.; Kim, J.-K.; Lee, W.; Kim, D.-E.; et al. Rough-Surface-Enabled Capacitive Pressure Sensors with 3D Touch Capability. *Small* **2017**, *13*, 1700368. [CrossRef]

19. Bae, G.Y.; Han, J.T.; Lee, G.; Lee, S.; Kim, S.W.; Park, S.; Kwon, J.; Jung, S.; Cho, K. Pressure/Temperature Sensing Bimodal Electronic Skin with Stimulus Discriminability and Linear Sensitivity. *Adv. Mater.* **2018**, *30*, 1803388. [CrossRef]

20. Boutry, C.M.; Kaizawa, Y.; Schroeder, B.C.; Chortos, A.; Legrand, A.; Wang, Z.; Chang, J.; Fox, P.; Bao, Z. A stretchable and biodegradable strain and pressure sensor for orthopaedic application. *Nat. Electron.* **2018**, *1*, 314–321. [CrossRef]

21. Boutry, C.M.; Beker, L.; Kaizawa, Y.; Vassos, C.; Tran, H.; Hinckley, A.C.; Pfattner, R.; Niu, S.; Li, J.; Claverie, J.; et al. Biodegradable and flexible arterial-pulse sensor for the wireless monitoring of blood flow. *Nat. Biomed. Eng.* **2019**, *3*, 47–57. [CrossRef] [PubMed]

22. Yang, J.; Luo, S.; Zhou, X.; Li, J.; Fu, J.; Yang, W.; Wei, D. Flexible, Tunable, and Ultrasensitive Capacitive Pressure Sensor with Microconformal Graphene Electrodes. *ACS Appl. Mater. Interfaces* **2019**, *11*, 14997–15006. [CrossRef]

23. Yang, J.C.; Kim, J.-O.; Oh, J.; Kwon, S.Y.; Sim, J.Y.; Kim, D.W.; Choi, H.B.; Park, S. Microstructured Porous Pyramid-Based Ultrahigh Sensitive Pressure Sensor Insensitive to Strain and Temperature. *ACS Appl. Mater. Interfaces* **2019**, *11*, 19472–19480. [CrossRef]

24. Asghar, W.; Li, F.; Zhou, Y.; Wu, Y.; Yu, Z.; Li, S.; Tang, D.; Han, X.; Shang, J.; Liu, Y.; et al. Piezocapacitive Flexible E–Skin Pressure Sensors Having Magnetically Grown Microstructures. *Adv. Mater. Technol.* **2020**, *5*, 1900934. [CrossRef]

25. Ruth, S.R.A.; Beker, L.; Tran, H.; Feig, V.R.; Matsuhisa, N.; Bao, Z. Rational Design of Capacitive Pressure Sensors Based on Pyramidal Microstructures for Specialized Monitoring of Biosignals. *Adv. Funct. Mater.* **2020**, *30*, 1903100. [CrossRef]

26. Ma, X.; Zhang, Q.; Guo, P.; Tong, X.; Zhao, Y.; Wang, A. Residual Compressive Stress Enabled 2D-to-3D Junction Transformation in Amorphous Carbon Films for Stretchable Strain Sensors. *ACS Appl. Mater. Interfaces* **2020**, *12*, 45549. [CrossRef] [PubMed]

27. Cai, J.-H.; Li, J.; Chen, X.-D.; Wang, M. Multifunctional polydimethylsiloxane foam with multi-walled carbon nanotube and thermo-expandable microsphere for temperature sensing, microwave shielding and piezoresistive sensor. *Chem. Eng. J.* **2020**, *393*, 124805. [CrossRef]

28. Gong, S.; Schwalb, W.; Wang, Y.; Chen, Y.; Tang, Y.; Si, J.; Shirinzadeh, B.; Cheng, W. A wearable and highly sensitive pressure sensor with ultrathin gold nanowires. *Nat. Commun.* **2014**, *5*, 3132. [CrossRef]

29. Amjadi, M.; Yoon, Y.J.; Park, I. Ultra-stretchable and skin-mountable strain sensors using carbon nanotubes-Ecoflex nanocomposites. *Nanotechnology* **2015**, *26*, 11. [CrossRef] [PubMed]

30. Zheng, Y.; Li, Y.; Dai, K.; Liu, M.; Zhou, K.; Zheng, G.; Liu, C.; Shen, C. Conductive thermoplastic polyurethane composites with tunable piezoresistivity by modulating the filler dimensionality for flexible strain sensors. *Compos. Part A Appl. Sci. Manuf.* **2017**, *101*, 41–49. [CrossRef]

31. Tallman, T.N.; Gungor, S.; Wang, K.W.; Bakis, C.E. Tactile imaging and distributed strain sensing in highly flexible carbon nanofiber/polyurethane nanocomposites. *Carbon* **2015**, *95*, 485–493. [CrossRef]

32. Wang, Y.; Wang, L.; Yang, T.T.; Li, X.; Zang, X.B.; Zhu, M.; Wang, K.L.; Wu, D.H.; Zhu, H.W. Wearable and Highly Sensitive Graphene Strain Sensors for Human Motion Monitoring. *Adv. Funct. Mater.* **2014**, *24*, 4666–4670. [CrossRef]

33. Li, H.; Wu, K.; Xu, Z.; Wang, Z.; Meng, Y.; Li, L. Ultrahigh-Sensitivity Piezoresistive Pressure Sensors for Detection of Tiny Pressure. *ACS Appl. Mater. Interfaces* **2018**, *10*, 20826–20834. [CrossRef]

34. Chou, H.-H.; Nguyen, A.; Chortos, A.; To, J.W.F.; Lu, C.; Mei, J.; Kurosawa, T.; Bae, W.-G.; Tok, J.B.H.; Bao, Z. A chameleon-inspired stretchable electronic skin with interactive colour changing controlled by tactile sensing. *Nat. Commun.* **2015**, *6*, 8011. [CrossRef]

35. Park, H.; Jeong, Y.R.; Yun, J.; Hong, S.Y.; Jin, S.; Lee, S.J.; Zi, G.; Ha, J.S. Stretchable Array of Highly Sensitive Pressure Sensors Consisting of Polyaniline Nanofibers and Au-Coated Polydimethylsiloxane Micropillars. *ACS Nano* **2015**, *9*, 9974–9985. [CrossRef] [PubMed]

36. Kim, K.; Jung, M.; Kim, B.; Kim, J.; Shin, K.; Kwon, O.S.; Jeon, S. Low-voltage, high-sensitivity and high-reliability bimodal sensor array with fully inkjet-printed flexible conducting electrode for low power consumption electronic skin. *Nano Energy* **2017**, *41*, 301–307. [CrossRef]

37. Jeong, C.; Ko, H.; Kim, H.-T.; Sun, K.; Kwon, T.-H.; Jeong, H.E.; Park, Y.-B. Bioinspired, High-Sensitivity Mechanical Sensors Realized with Hexagonal Microcolumnar Arrays Coated with Ultrasonic-Sprayed Single-Walled Carbon Nanotubes. *ACS Appl. Mater. Interfaces* **2020**, *12*, 18813–18822. [CrossRef] [PubMed]

38. Liu, M.; Zhang, Q.; Zhao, Y.; Liu, C.; Shao, Y.; Tong, X. A high sensitivity micro-pressure sensor based on 3D printing and screen-printing technologies. *Meas. Sci. Technol.* **2020**, *31*, 035106. [CrossRef]

Article

Topology Optimisation for Compliant Hip Implant Design and Reduced Strain Shielding

Nathanael Tan and Richard J. van Arkel *

Department of Mechanical Engineering, Imperial College London, London SW7 2AZ, UK
* Correspondence: r.vanarkel@imperial.ac.uk

Abstract: Stiff total hip arthroplasty implants can lead to strain shielding, bone loss and complex revision surgery. The aim of this study was to develop topology optimisation techniques for more compliant hip implant design. The Solid Isotropic Material with Penalisation (SIMP) method was adapted, and two hip stems were designed and additive manufactured: (1) a stem based on a stochastic porous structure, and (2) a selectively hollowed approach. Finite element analyses and experimental measurements were conducted to measure stem stiffness and predict the reduction in stress shielding. The selectively hollowed implant increased peri-implanted femur surface strains by up to 25 percentage points compared to a solid implant without compromising predicted strength. Despite the stark differences in design, the experimentally measured stiffness results were near identical for the two optimised stems, with 39% and 40% reductions in the equivalent stiffness for the porous and selectively hollowed implants, respectively, compared to the solid implant. The selectively hollowed implant's internal structure had a striking resemblance to the trabecular bone structures found in the femur, hinting at intrinsic congruency between nature's design process and topology optimisation. The developed topology optimisation process enables compliant hip implant design for more natural load transfer, reduced strain shielding and improved implant survivorship.

Keywords: stress shielding; total hip replacement; femoral component; lattice; 3d printing; aseptic loosening; bone remodelling; internal structures; biomimicry

Citation: Tan, N.; van Arkel, R.J. Topology Optimisation for Compliant Hip Implant Design and Reduced Strain Shielding. *Materials* **2021**, *14*, 7184. https://doi.org/10.3390/ma14237184

Academic Editor: Rubén Paz

Received: 31 August 2021
Accepted: 22 November 2021
Published: 25 November 2021

Publisher's Note: MDPI stays neutral with regard to jurisdictional claims in published maps and institutional affiliations.

1. Introduction

Total hip arthroplasty (THA) is one of the most successful surgical procedures in modern medicine, providing a treatment option for debilitating diseases such as end-stage osteoarthritis and osteonecrosis of the femoral head. The majority of hip implants last for 25 years [1]. However, low failure rates still translate to many patients requiring revision surgery, as ~2 million procedures are performed worldwide each year [2]. For example, in the UK, 35,000 patients have required revision surgery [3]. These revision procedures cost more and have worse outcomes [4]. Hence, researchers and industry continually strive to advance technology to improve outcomes and survivorship.

Aseptic loosening is the most common reason for revision surgery [3]. One of the causes of aseptic loosening is strain shielding, which results from typical hip implant materials having moduli one to two orders of magnitude greater than the bone they replace/are implanted into [5–7]. Less load is transferred through the proximal femoral bone, resulting in a loss of the mechanical stimulus that drives bone formation. Over time, this leads to bone loss and increases the risk of loosening and/or periprosthetic fractures. It also increases the complexity of any revision surgeries as there is less bone available in which to implant a revision prosthesis.

The advent of commercial metal additive manufacturing provides the opportunity to manufacture hip stems with intricate cellular geometries, resulting in stems that are less stiff than their solid counterparts [7–17]. Most have adopted a lattice size optimisation algorithm, which is inherently limited by a minimum feature size. These designs are therefore subject to practical limiting factors, such as fatigue resistance and manufacturability of miniscule structures [10–20]. Indeed, researchers studying size-optimising lattices for femoral stems have commented that maintaining fatigue strength with small features is the most challenging criterion to fulfil when developing an optimal solution [9].

An alternative approach could be to borrow topology optimisation techniques from the field of structural optimisation [21,22] and directly use the results for hip implant design, without using lattice structures. As topology optimisation does not presume a priori material distribution, the algorithm has the liberty to leave specific locations void of material, whereas the corresponding size optimisation algorithm would be unable to remove unit elements from the lattice, merely reducing their sizes to a specified minimum. Such an approach has been demonstrated to reduced strain shielding in fracture plates [23], and has shown success in modelling the growth of internal bone structures when implemented with geometric constraints [24]. Thus, it holds great potential in producing implants that more closely match bone's natural characteristics, reducing any strain shielding effect.

Topology optimisation is most used to minimise compliance (i.e., maximise stiffness) for a given volume fraction to provide mass reduction [22]. Thus, its application here might seem counterintuitive: Femoral stems are too stiff, hence there is need to increase compliance. The technique is useful, however, as imposing a volume fraction of less than one will always result in a concomitant increase in the minimum compliance when compared to a volume fraction equal to one. Thus, this approach will inherently produce more compliant stems despite maximizing the stiffness and strength at the set volume fraction. Indeed, the approach could enable a design tool for research across the risk/benefit spectrum by aiming for high/low global compliance objectives: A lower compliance target might enable more natural load transfer in the femur and reduce strain shielding but comes with risk that the design may be less robust against adverse loading scenarios (e.g., falls) and suboptimal implantations (e.g., stem undersizing [25,26]). Approaches to risk management vary between designers/manufacturers. Therefore, this research focuses on the development of the topology optimisation process rather than what the desired global compliance should be.

A factor that would aid translation would be preserving the outside shape of the hip stem. Implant manufacturers have spent decades optimising surgical instrumentation, interference fits, implant finishes and coatings. A design that maintains the outer shape of existing clinical implants could be advantageous as it could be implanted with minimal deviation from designs that have decades of clinical data supporting their use.

Therefore, this paper aimed to develop a topology optimisation process for designing the internal geometries of femoral hip stems for reduced strain shielding. Two proof-of-concept designs were manufactured and tested: (1) a porous implant, similar to those previously described but utilising a stochastic trabecular-like structure rather than cellular lattices, and (2) a novel selectively hollowed implant that maintains the outer shape of the implant such that only the stiffness of the stem is reduced, with no other design changes.

2. Materials and Methods

2.1. Topology Optimisation

2.1.1. Solid Isotropic Material with Penalisation Method

Femoral component topology was optimised via the Solid Isotropic Material with Penalisation (SIMP) method [21,22,27] (Figure 1). All elements were initially set to the desired volume fraction. Then, for each element in each iteration, the sensitivity of its compliance to change in volume fraction was calculated. The volume fraction of elements with sensitivity higher/lower than the optimality criteria was then increased/decreased, respectively, while ensuring the global volume fraction was maintained. The algorithm effectively removes material from the least sensitive element first, retaining maximum stiffness per unit of material deducted thus minimising the compliance of the model.

A MATLAB function, TOP3D [27], was adopted and modified to develop the optimised hip implants. It performed SIMP topology optimisation on voxel models to find the design variables x of n total elements to minimise the structure's compliance $c(\tilde{x})$ while maintaining a material volume $v(\tilde{x})$ less than or equal to a volume fraction $\bar{v}$. By default, the material distribution $\tilde{x}$ was the density-filtered design variables.

Optimisation Problem Statement	
Find:	$x = [x_1\ x_2\ x_3 \dots\ x_n]^T$
Minimize:	$c(\tilde{x}) = F^T U(\tilde{x}) = U(\tilde{x})^T K(\tilde{x}) U(\tilde{x})$
	$v(\tilde{x}) = \tilde{x}^T v - \bar{v} \leq 0$
Subject to:	$x \in \mathbb{R}^n$
	$0 \leq x \leq 1$

As the SIMP method changed each element's stiffness based on design variables per iteration, the derivation of each element's stiffness matrix was altered from conventional finite element methods:

$$k^0 = \int_{-1}^{+1} \int_{-1}^{+1} \int_{-1}^{+1} B^T C^0 B d\xi_1 d\xi_2 d\xi_3 \tag{1}$$

Equation (1): Element stiffness matrix per unit Young's modulus.

$$E_i(x_i) = E_{min} + x_i^p (E_0 - E_{min}) \tag{2}$$

Equation (2): Penalised element Young's modulus.

$$k_i(x_i) = E_i(x_i) k^0 \tag{3}$$

Equation (3): Element's Young's modulus as a function of design variable.

Equation (1) was used to derive the element stiffness matrix per unit Young's modulus using the deformation matrix for a unit cube B and constitutive matrix per-unit Young's modulus C^0. Equation (2) was used to calculate each element's Young's modulus as a penalised interpolation between a defined minimum E_{min} and material's Young's Modulus E_0 based on its design variable x_i and the chosen penalisation factor p. Equation (3) was then used to compute each element's stiffness matrix by multiplying the per-unit Young's modulus stiffness by the element's Young's modulus. The sum of element stiffness matrices created the global stiffness matrix for that iteration. Standard finite element method steps were then taken to assemble the global force matrix F and calculate nodal displacements U.

During each iteration, each element's design variable was adjusted incrementally based on the sensitivity of that element's compliance with respect to its design variable. For elements with sensitivity greater than an optimality criterion, the design variable was increased, and those less than it decreased. The bisection method was used to determine the optimality criterion which yielded the minimum global compliance per iteration of the optimisation process.

The SIMP method often employs a density filter to mitigate numerical instabilities, such as the checkerboard problem, ensuring feasible solutions are obtained. Whereas the density filter ensures material continuity, it often results in greyscale solutions which require a non-uniform material Young's Modulus to actualise. It has been proposed that filtering sensitivities, instead of densities, could provide a more black-and-white solution with minimal compromise on the optimality of solutions [28]. Being able to manufacture optimal solutions with minimal alterations and simplifications would both preserve the theoretical performance of the design and could enable patient-specific implants that could not only account for bone geometry [29], but also daily load conditions. Both the density and sensitivity filter were used to develop a greyscale and black-and-white solution of equivalent global stiffness. These solutions were developed into porous and selectively hollowed implants, respectively.

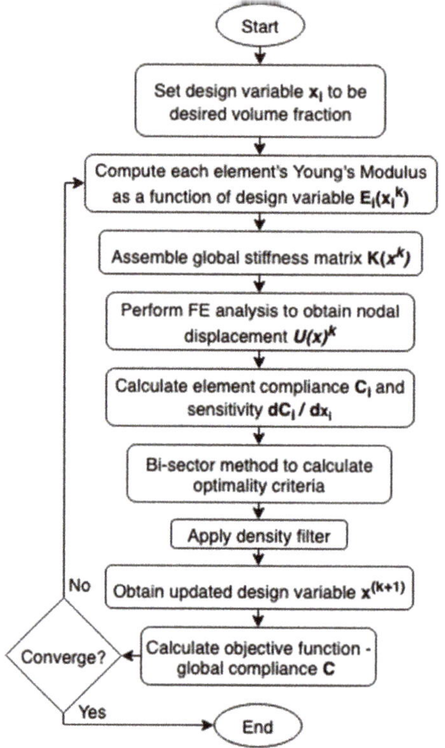

Figure 1. Overview of the Solid Isotropic Material with Penalisation optimisation method.

2.1.2. Model Preparation

A Sawbones femur model (Model #3403, Sawbones) and a representative implant CAD model [30] were downloaded. A standard femoral neck resection was performed on the proximal femur, and the femur prepared for implantation by creating a cut out of the implant's shape in the femur CAD model with Boolean subtract. The STL file was converted to voxel models in MATLAB using a ray intersection method [31]. As bone is an anisotropic material that comprises 2 distinctly different regions, a scanning function was also added to assign the outer cortical bone and inner trabecular bone with a Young's modulus of 15 GPa and 0.8 GPa. The outermost surface of the implant was excluded from the optimisation to ensure continuity and contact between the implant and the femur.

2.1.3. Loading Conditions

Previous work applying topology optimisation to the growth of a femur [24,32] highlighted the importance of considering the relative frequency of different load cases which simulate different daily activities. TOP3D was further modified to sequentially apply unique load cases with varying frequencies, simulating the peak hip joint reaction force during common activities of daily living [33] (Table 1). Muscles forces from the gluteal muscles and iliopsoas were added based on typical loads during gait [34–36] to demonstrate that muscle loading can be incorporated into the topology optimisation process. A load case of 1-1-1-2-3-4 (numbered in Table 1) was originally modelled, as prescribed by previous work [24]. However, this was later modified to a 1-2-1-3-1-4 sequence, as it was found that sequential gait loading led to premature convergence.

Table 1. Magnitudes and directions of load cases 1–4.

Load Case	Action	Load (N)	Superior-Medial Angle (°)	Superior-Anterior Angle (°)
1	Walk	1925	17	11
2	Jog	3065	15	15
3	Sit down	1360	20	11
4	Stand up	1600	24	8

2.2. Implant Development

Implants were designed to achieve the same global compliance as a solid cortical bone implant. Whereas implants predominantly replace trabecular bone, the solid cortical implant reference was used as a proof of concept, exemplifying how the optimised implants can be tuned to a specific objective function. The reference objective function was obtained by calculating the compliance of a solid cortical bone implant subject to identical boundary conditions and one complete loading sequence. A parametric study was conducted to guide a trial-and-error process to produce optimal greyscale and black-and-white solutions that matched this global compliance objective function within 3%.

2.2.1. Porous Implant

The femoral stem was filled with a stochastic porous structure with infinitely thin stuts generated in Rhinoceros 3D (McNeel Europe, Barcelona, Spain). A charge field was then constructed and super-imposed over the strut array, where the magnitude of each point charge was determined by the design variable of the corresponding voxel from greyscale solution. Thicknesses were then prescribed to the struts depending on the charge field's magnitude at each strut's centre. The thicknesses were based on previous research which related laser parameters to strut thicknesses to mechanical properties to enable stiffness matched porous structures to be manufactured [19,37], leading to an implant design with 65% porosity and pore size range of 0.1–2 mm. The solid proximal and distal ends to the implant were then combined with the porous structure to finalise the implant (Figure 2). A tapering interface enabled a gradual transition between the porous and solid material regions.

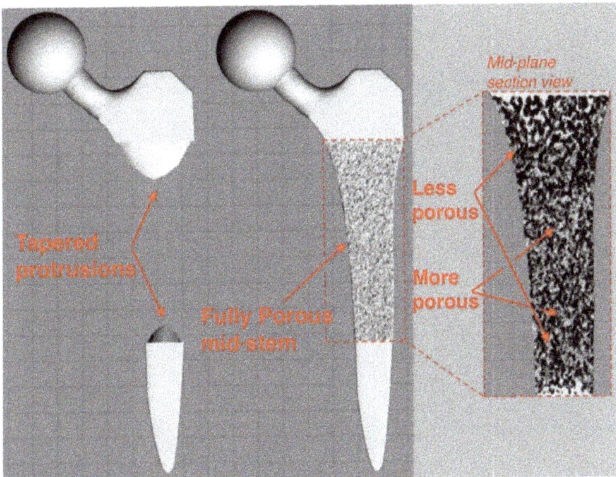

Figure 2. Tapered protrusions (left) enabled a gradual transition between the porous and solid regions of the porous implant (right).

2.2.2. Selectively Hollowed Implant

The black-and-white solution was first converted to a STL file (Figure 3, left) using 3D Slicer [38] before the internal geometry was extracted and positioned within the standard implant model. Minor post processing was performed using Meshmixer (Autodesk Ltd, Birmingham, UK) to smooth kinks in the internal geometry and add distal drainage holes to enable non-fused powder removal post additive manufacture. The volume fraction of the of the hollowed stem was 40%.

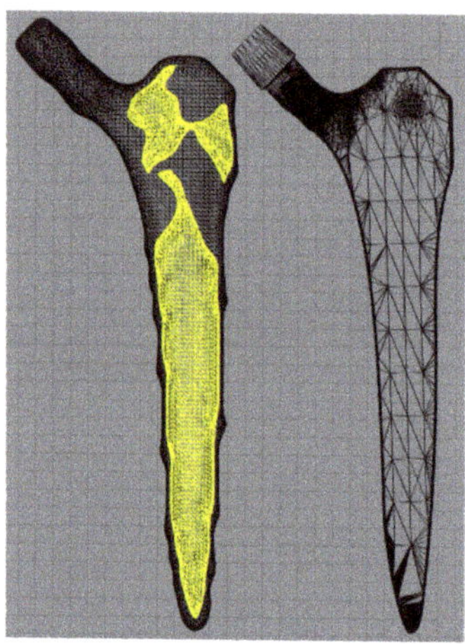

Figure 3. Internal geometry (yellow) separated from the rough external geometry (left) before being placed inside a standard implant (right).

2.3. Finite Element Analysis

Finite element analyses were performed in Fusion360 (Autodesk Ltd, Birmingham, UK). For all tests, the software's automatic mesh generation and refinement feature was used with parabolic order elements. The initial element size was set to 5% of the model-based size, and the adaptive refinement control was set to high, with 10 iterations to achieve convergence within a tolerance of 5%. Von Mises stress was chosen as the convergence criteria (as opposed to displacement), as subsequent tests required stresses or strains to be investigated, and stresses tend to converge more slowly than displacements.

2.3.1. Implant Strength

The strength of the novel selectively hollowed implant was compared to that of the solid stem through simulating BS ISO 7206-4-2010. The stem was angled 10°/9° in the coronal/sagittal planes, respectively, and a uniformly distributed vertical load of 2300 N was applied on the flat face of the femoral neck, producing a statically equivalent bending moment as a load applied through a spherical femoral head centre. First, stresses throughout the stem were observed visually to ensure no locations exceeded the yield stress of titanium. Second, maximum stress values were extracted and tabulated from two locations of interest: (1) at the femoral neck, as this is historically the weakest part of the hip implant; and (2) a transverse plane 10 mm proximal of the distal fixed face, away from the fixed boundary, at a location where the selectively hollowed implant had thin wall sections.

2.3.2. Strain Shielding

The strain shielding stimulus was evaluated by quantifying peri-implanted strains following implantation of the solid and the novel selectively hollowed implants compared to the native femur. The native femur model was prepared by first transecting the complete sawbones femur model such that the distal tip of the femoral stem would be at least 10 mm from the fixed boundary condition applied to the distal femur. The femoral head was sliced at 45° in the superior-lateral direction, ensuring that the bending moment produced from forces applied on this surface would be equivalent to that produced by a point load in the centre of the femoral head. The trabecular bone and cortical bone regions of the sawbones model were again assigned a Young's modulus of 0.8 GPa and 15 GPa, respectively, and the implants were inserted via Boolean subtract.

In total, 2300 N of load was applied to either the sliced femoral head (for the native femur) or the flat end of the implant (once implanted) in six loading directions (Table 2). The first load case represented the direction of gait loading most frequently imposed during optimisation, but with higher loading magnitude. Five other loading directions were chosen to investigate the correlation between reduction in strain shielding and loading direction, varying angles in each plane independently. These angles were based on the minimum, average and maximum angles of peak joint reaction force (JRF) [39] measured with instrumented hip implants during a range of daily activities [40], thus spanning the range of loads that may be expected during typical use. Importantly, these additional load cases meant that the implant's performance was evaluated under load cases not used during the optimisation process.

Table 2. Summary of the loading directions simulated strain shielding tests. The top row is equivalent to the peak gait loading angle used during the optimisation process. The other rows represent typical variations in the direction of the joint reaction force during activities of daily living.

Description	Superior-Medial Angle (°)	Superior-Anterior Angle (°)
Gait angles from optimisation	17	11
Min angle of JRF in coronal plane	2	5
Max angle of JRF in coronal plane	21	5
Average angle of JRF in both planes	12	5
Min angle of JRF in sagittal plane	12	−5
Max angle of JRF in sagittal plane	12	15

Maximum principal strains were measured on the bone surface along the perimeter of the proximal-most plane for each of Gruen zones 4–7 [41]. Percentage strain was calculated by dividing peri-implant strain values with the native femoral strain values. It has been suggested that a reduction in strain of more than 50% (percentage strain < 50%) predicts bone resorption in that region [7]. By comparing percentage strain of the solid and selectively hollowed implants, improvements to strain shielding were quantified, and changes to trend along the length of the femur were observed.

2.4. Experimental Tests

Femoral heads were fitted to the stems in CAD, and the solid, selectively hollowed and porous designs were manufactured from 316L-0407 stainless steel powder in a powder bed fusion machine (AM250, Renishaw PLC, Gloucestershire, UK) in 50 μm layers. Different laser parameters were used for different regions of the stem: Porous structures were manufactured in line with previous research [19], while solid material was manufactured with standard laser parameters provided by the machine manufacturer and applied to the parts through their build preparation software (QuantAM, Renishaw PLC, UK).

For mechanical testing, the distal 40 mm of the implants was inserted into custom-made steel cylinders and held at 10° in the coronal plane using laboratory clamps. Poly methyl methyl acrylic (PMMA, Simplex, Kemdent) was poured into the cylinder, which, when cured, fixed the implant in this position (Figure 4). The steel cylinder had machined features that enabled it to attach to the base of a screw-driven materials testing machine equipped with a 5 kN load cell (model 5565, Instron, High Wycombe, UK). The fixture was positioned such that the femoral head was directly in the centre of a compression platen (accessory T1223-1021, Instron). Each sample was loaded up to 1500 N at a compression rate of 1 mm/min before being unloaded. Each sample was tested consecutively 3 times, and the average final deflection at 1500 N was calculated. The relative stiffness of each implant was calculated as the force applied per unit deflection of the femoral head.

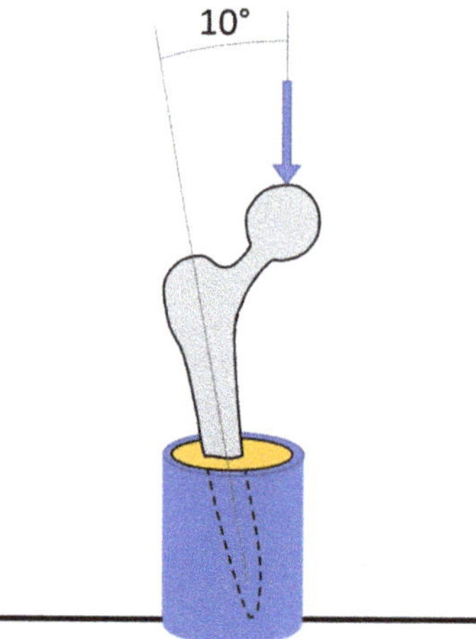

Figure 4. Diagram of experimental tests showing a hip implant (grey) secured in a metallic pot (blue) with PMMA (yellow).

3. Results

3.1. Stem Development

Porous and selectively hollowed implants were successful designed and manufactured with the topology optimisation process (Figure 5). It was found that the 1-1-1-2-3-4 loading pattern led to premature convergence. Therefore, finite element analysis and experimental testing was only performed on the 1-2-1-3-1-4 loading pattern designs. Under this loading condition, the selectively hollowed implant had material distribution similar to the principal directions of the trabecular structure in the native femur (Figure 6).

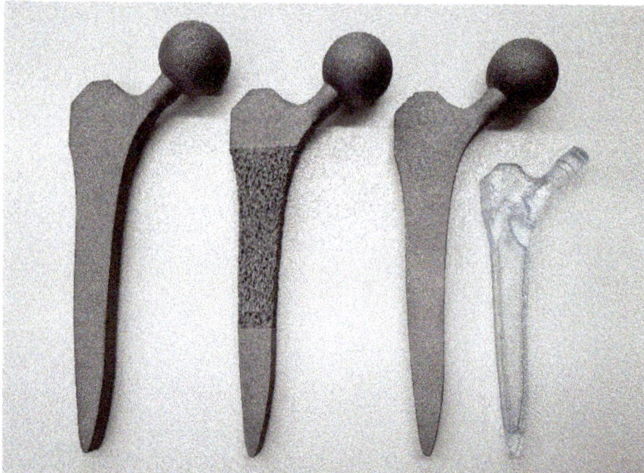

Figure 5. The successfully manufactured solid (left), porous (middle) and selectively hollowed (right) hip implants. The solid and selectively hollowed implants look identical. Thus, a clear resin plastic version of the selectively hollowed implant is also shown to indicate the internal structure of the selectively hollowed implant (right).

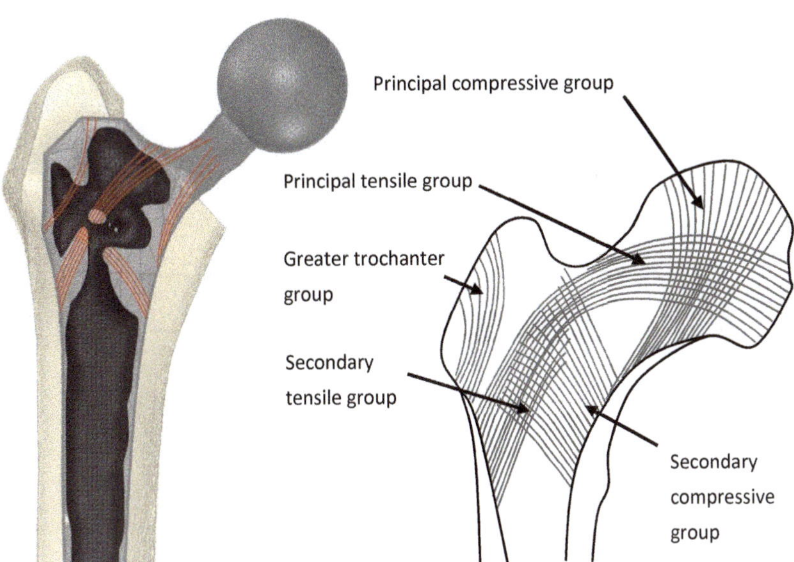

Figure 6. (Left) An example selectively hollowed stem implanted in a femur. Material remaining in the implant is indicative of the principal remodelling direction for trabecular bone in the femur (red lines). (Right) Representation of trabecular groups in the native femur.

3.2. Finite Element Analyses

3.2.1. Strength

For both the solid and selectively hollowed implants, the maximum stress was located at the femoral neck (Figure 7), with the maximum stress marginally lower in the selectively hollowed implant. However, internal femoral stem stresses were approximately 50% greater in the selectively hollowed implant (Figures 7 and 8).

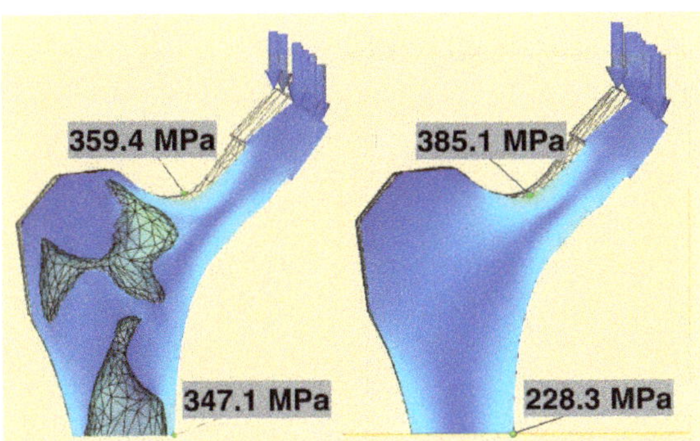

Figure 7. Maximum neck and stem stresses in the selectively hollowed (left) and solid (right) implants. The selectively hollowed implant had marginally lower neck stresses, but higher internal stem stresses compared to its solid equivalent. The image is cropped 10 mm from the distal fixed face.

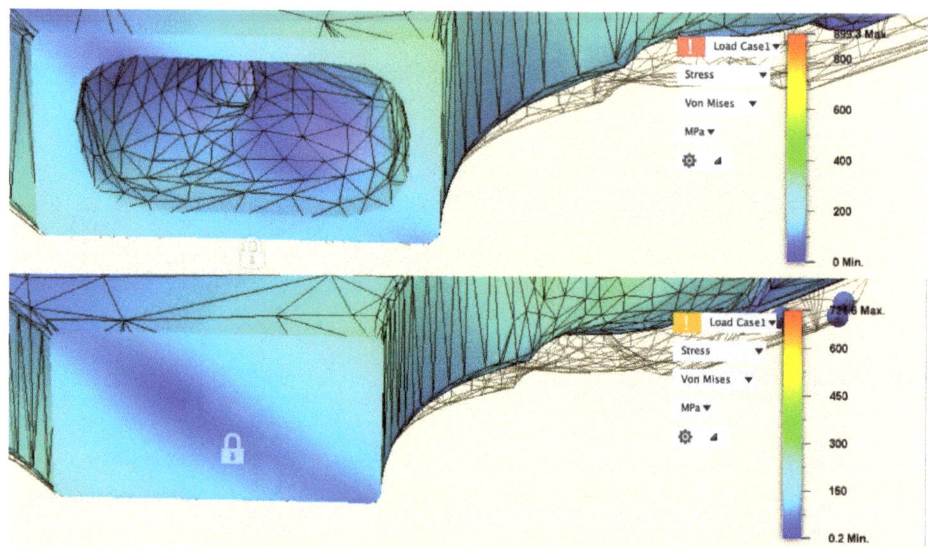

Figure 8. Stress distribution in the stem 10 mm proximal of the fixed face.

3.2.2. Strain Shielding

For the gait load case, both the solid and selectively hollowed implants were found to strain shield the femur, but the effect was less pronounced for the selectively hollowed implant (Table 3, Figure 9). The finite element analysis predicted that this exemplar selectively hollowed implant would lead to an 8-percentage-point (PP) reduction in strain shielding in Gruen zone 5, a 15 PP reduction in zone 6, and a 25 PP reduction in zone 7 compared to the solid implant.

Table 3. Maximum surface bone strain and percentage of native bone strain for the selectively hollowed, solid implant and native femur for peak gait loading.

Model	Gruen Zone 4		Gruen Zone 5		Gruen Zone 6		Gruen Zone 7	
	Max Strain ($\mu\varepsilon$)	% of Native	Max Strain ($\mu\varepsilon$)	% of Native	Max Strain ($\mu\varepsilon$)	% of Native	Max Strain ($\mu\varepsilon$)	% of Native
Native femur	3040	100	2670	100	3010	100	2630	100
Solid Implant	3210	106	1860	70	1460	49	520	20
Selectively Hollowed Implant	3210	106	2080	78	1940	64	1180	45

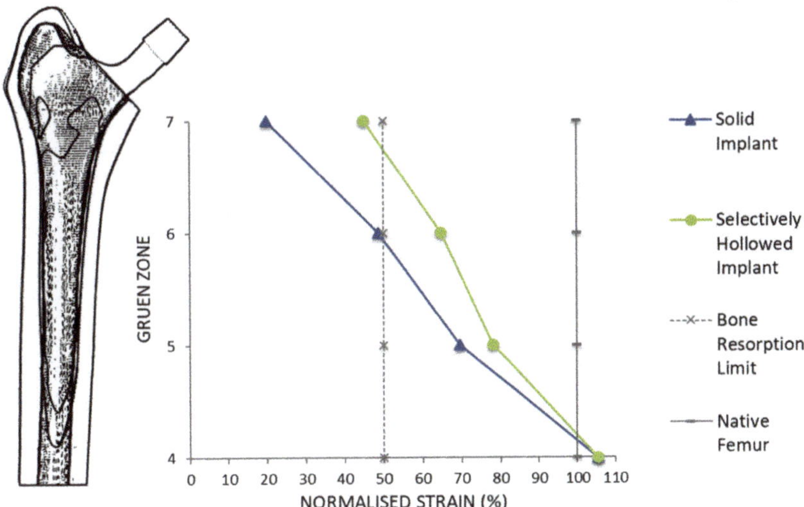

Figure 9. Peri-implant surface bone strain as a percentage of the strain measured for the native femur for different Gruen zones. A typical bone resorption limit of 50% was also indicated. The selectively hollowed implant strain shielded the femur less than the solid implant.

The angle of the force affected the amount of strain shielding but not the percentage point difference between the implants. Increasing the angle of the implant in the coronal implant led to increased strain shielding for both implants, with a near constant percentage point difference between the two implant designs (Figure 10 left). Changing the angle of load in the sagittal plane resulted in a minimum at 5°, again with a near-constant percentage point difference between the two implant designs (Figure 10 right). For all load cases, the selectively hollowed implant reduced the strain shielding effect compared to the solid equivalent.

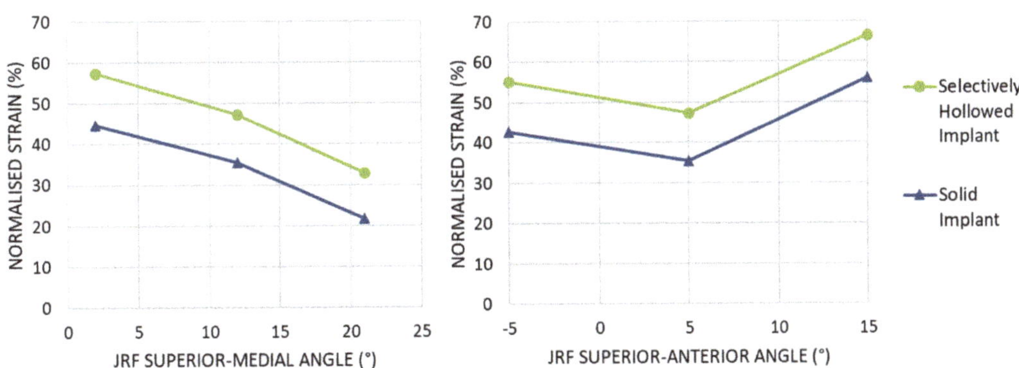

Figure 10. Peri-implant surface bone strain as a percentage of the strain measured for the native femur at Gruen zone 6 for different angles of joint reaction force (JRF). (Left) Variation in the coronal plane. (Right) Variation in the sagittal plane. For all load cases examined, the selectively hollowed implant resulted in less strain shielding than the solid reference implant.

3.3. Experimental Stiffness Test

The porous and selectively hollowed implants were both more compliant than the solid reference implant: There was a decrease in the equivalent stiffness of 39% and 40%, respectively (Figure 11). There was minimal difference between the stiffnesses of the two compliant implant designs, with the porous implant being only 3% stiffer than the selectively hollowed implant.

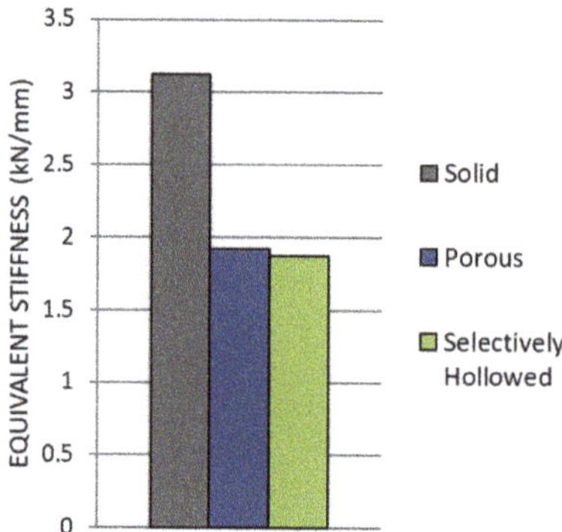

Figure 11. Comparison of experimental stiffness measurements for the solid reference, porous and selectively hollowed implant measured experimentally.

The equivalent stiffness of the novel selectively hollowed implant was experimentally measured to be 1.9 kN/mm, which was 77% of the value predicted by the finite element analysis (2.4 kN/mm).

4. Discussion

This most important finding of this study is that topology optimisation can be used to design porous and selectively hollowed femoral hip stems with increased compliance for reduced strain shielding. The resulting stems had only a 3% difference in stiffness when additive manufactured despite the stark differences in their design. This demonstrates the broad scope for application of the topology optimisation approach presented and how different filters can be used to design different implant variations from an otherwise identical optimisation process. The developed process includes global optimisation for multiple consecutive load cases and enables either conservative or radical stem design through optimising to different global compliance objective functions.

Other research groups have found that lattice-based stem designs are effective for reducing strain shielding [7–16], with innovations including the inclusion of fatigue life constraints during the optimisation process [7–9]. The use of a trabecular-like stochastic porous structure in this study is potentially advantageous compared to more regular lattice designs as stochastic structures enable control of anisotropy [42], which may result in an implant that is more tolerant to unanticipated load cases. Other researchers have optimised the outer shape of hip implants [5,43–48], or optimised the modulus/density distribution within implants [49,50], with similar proposed benefits for preventing strain shielding. However, to the authors' knowledge, this is the first paper to demonstrate that a strain shielding reduction can be gained by selectively hollowing the implant using topology optimisation with multiple load cases, and the first to additively manufacture and test a proof-of-concept device. The observed similarity between the selectively hollowed stem and trabecular structures (Figure 6) is likely explained by the inherent link between structural topology optimisation and the mechanisms that are believed to drive bone remodelling: The two concepts are mathematically matched [51]. Aside from its increased compliance and the need for a powder drainage hole, this novel selectively hollowed stem is indistinguishable from the solid counterpart (Figure 5), which may aid clinical translation, where each design change requires extensive testing and validation prior to clinical trial. For example, there is a complex interplay between broach design, technique and the stem to ensure suitable initial implant fixation [52,53]. Changing the outside shape of the stem would require the design of a new broach and re-validation to ensure that the desired fixation is achieved. Further, the selectively hollowed design inherently avoids transitions between porous and solid on the outer surface, which may lead to improved fatigue life compared to a porous stem by reducing the number of stress concentrations on the tensile face of the implant under bending loads.

The focus of this paper was to develop the multiple load case topology optimisation approach for hip implant design, rather than to produce a specific "best" design. Hence, simplifications were made to enable research to focus on the process rather than modelling complexity: (1) The bone was modelled with only a single property each for cortical/trabecular bone, whereas the bone was both inhomogeneous and anisotropic [54]. The process developed here could be applied to a calibrated CT scanned femoral bone model to capture regional variation in properties. (2) Loads based on only four daily activities were applied with loading from only two muscles. The activities simulated were based on previous works [24]. If a different set was chosen, or if activities had been included with greater temporal variation (in addition to just peak loading), then a different "optimal" stem would have resulted. Further, in vivo, there are 22 muscles spanning the hip joint, with load varying dynamically throughout activities [36]. The process developed is such that more physiological load cases can be added as desired, and future research could investigate which activities and phases of activity to include in the optimisation process. One might hypothesize that the inclusion of more physiological load cases would lead to a solution with greater similarity to the principal orientations and densities of trabeculae in the native femur. (3) The femoral stems were manufactured in 316L due to machine availability. Whereas this is a material used clinically, titanium alloy and cobalt-chrome alloy stems are more common. The topology optimisation process developed can inherently be used to find optimal stems in any material. (4) The FEA assumed that the implants and bone were fully bonded. This assumption represents an implant that has fully integrated into the body via bone in/on-growth [8] and has proven effective for predicting long-term bone remodelling [55,56]. Prior to this, implants are initially fixed via press-fit and friction [57], with bone growing into/onto the implant when the interfacial stress-strain state is appropriate and the relative micromotion between the implant and bone is around 110 μm [58]. Previous research into the effects of bone-prosthesis bonding has indicated that strain shielding is less during initial press-fit than when fully bonded [59]. As such, the fully bonded assumption made likely represents a worst-case scenario for strain shielding that is correlated with longer-term bone remodelling around the implant. (5) Machine compliance was not accounted for during the experimental stiffness tests. Further, stems were potted with PMMA (Figure 4) to fix them in the materials testing machine. PMMA has a modulus ~100 times less than steel leading to additional deformation under loading. Hence, the experimental result underestimated the stem stiffness, which helps explain the finding that the experimentally measured stiffness was 77% of the FEA predicted value. This does not affect the conclusions which focus on comparing the stems, rather than the absolute values achieved. Key factors, such as the orientation of the stem in the text fixtures, were controlled to ensure the validity of the relative comparisons made.

In conclusion, a topology optimisation process has been developed and applied to femoral hip stem design, resulting in two proof-of-concept designs that look radically different but achieve the same reduction in stiffness compared to a traditional solid implant. The approach can account for multiple load cases and enables design for different target global compliance. It provides an exciting new avenue for designing hip implants that enable more natural load transfer in the proximal femur for the reduced risk of strain shielding, bone loss and improved survivorship.

Author Contributions: Investigation, N.T.; Supervision, R.J.v.A. All authors have read and agreed to the published version of the manuscript.

Funding: The Medical Device Prototype & Manufacture Unit was funded by the Engineering and Physical Sciences Research Council (Grant number EP/R042721/1). RvA is funded by the Engineering and Physical Sciences Research Council (Grant number EP/S022546/1) to pre-clinically analyse new implant fixation technologies and optimise implants for in vivo performance.

Acknowledgments: The authors would like to thank Umar Hossain for his support with CAD packages, and Stephen Johnson and Renishaw PLC for their assistance manufacturing the prototypes.

Conflicts of Interest: The authors declare no conflict of interest.

Nomenclature

B	Standard deformation matrix for a unit cube
C^0	Standard constitutive matrix per unit Young's modulus
$c(\tilde{x})$	Total compliance as a function of filtered design variable
E_0	Material Young's Modulus
E_{min}	Non-zero minimum Element Young's Modulus
E_i	Element Young's Modulus
F	Global force matrix
H_{ij}	Weight factor ranging from 1 at the centre of element i to 0 at the centre of element j at a radius of R away from element i
$K(\tilde{x})$	Global stiffness matrix as a function of filtered design variable
k^0	Element stiffness matrix per unit Yong's modulus
k_i	ith Element stiffness
N_i	Neighbourhood of element i, defined as the volumetric space from centre of element i to centre of neighbouring element j
p	Penalisation factor
$U(\tilde{x})$	Global displacement matrix as a function of filtered design variable
v_j	Unit volume per neighbour element j
$v(\tilde{x})$	Total Volume Fraction as a function of filtered design variable
$\bar{v}$	Target volume fraction
x_i	ith Element design variable
x_j	Design variable of neighbour element j
x_i^p	Penalised ith element design variable
$\tilde{x}$	Filtered design variable

References

1. Evans, J.T.; Walker, R.; Blom, A.W.; Whitehouse, M.; Sayers, A. How long does a hip replacement last? A systematic review and meta-analysis of case series and national registry reports with more than 15 years of follow-up. *Lancet* **2019**, *393*, 647–654. [CrossRef]
2. Pabinger, C.; Lothaller, H.; Portner, N.; Geissler, A. Projections of hip arthroplasty in OECD countries up to 2050. *Hip Int.* **2018**, *28*, 498–506. [CrossRef] [PubMed]
3. NJR. *17th Annual Report*; National Joint Registry for England and Wales: Hemel Hempstead, UK, 2020; Available online: https://reports.njrcentre.org.uk/ (accessed on 31 August 2021).
4. Weber, M.; Renkawitz, T.; Voellner, F.; Craiovan, B.; Greimel, F.; Worlicek, M.; Grifka, J.; Benditz, A. Revision surgery in total joint replacement is cost-intensive. *BioMed Res. Int.* **2018**, *2018*, 8987104. [CrossRef] [PubMed]
5. Cilla, M.; Checa, S.; Duda, G.N.; Hernández, M.C. Strain shielding inspired re-design of proximal femoral stems for total hip arthroplasty. *J. Orthop. Res.* **2017**, *35*, 2534–2544. [CrossRef] [PubMed]
6. Sumner, D.R. Long-term implant fixation and stress-shielding in total hip replacement. *J. Biomech.* **2015**, *48*, 797–800. [CrossRef]
7. Arabnejad, S.; Johnston, B.; Tanzer, M.; Pasini, D. Fully porous 3D printed titanium femoral stem to reduce stress-shielding following total hip arthroplasty. *J. Orthop. Res.* **2017**, *35*, 1774–1783. [CrossRef]
8. Wang, Y.; Arabnejad, S.; Tanzer, M.; Pasini, D. Hip implant design with three-dimensional porous architecture of optimized graded density. *J. Mech. Des.* **2018**, *140*, 111406. [CrossRef]
9. He, Y.; Burkhalter, D.; Durocher, D.; Gilbert, J.M. Solid-lattice hip prosthesis design: Applying topology and lattice optimization to reduce stress shielding from hip implants. In *2018 Design of Medical Devices Conference*; American Society of Mechanical Engineers: New York, NY, USA, 2018.

10. Hanks, B.; Dinda, S.; Joshi, S. Redesign of the femoral stem for a total hip arthroplasty for additive manufacturing. In *International Design Engineering Technical Conferences and Computers and Information in Engineering Conference*; American Society of Mechanical Engineers: New York, NY, USA, 2018.

11. Abbas, R.S.; Al Ali, M.; Sahib, A.Y. Designing femoral implant using stress based topology optimization. In Proceedings of the World Congress on Engineering, London, UK, 4–6 June 2018.

12. Kladovasilakis, N.; Tsongas, K.; Tzetzis, D. Finite element analysis of orthopedic hip implant with functionally graded bioinspired lattice structures. *Biomimetics* **2020**, *5*, 44. [CrossRef]

13. Corona-Castuera, J.; Rodriguez-Delgado, D.; Henao, J.; Castro-Sandoval, J.C.; Poblano-Salas, C.A. Design and fabrication of a customized partial hip prosthesis employing CT-Scan data and lattice porous structures. *ACS Omega* **2021**, *6*, 6902–6913. [CrossRef]

14. Abate, K.M.; Nazir, A.; Jeng, J.-Y. Design, optimization, and selective laser melting of vin tiles cellular structure-based hip implant. *Int. J. Adv. Manuf. Technol.* **2021**, *112*, 2037–2050. [CrossRef]

15. Jetté, B.; Brailovski, V.; Dumas, M.; Simoneau, C.; Terriault, P. Femoral stem incorporating a diamond cubic lattice structure: Design, manufacture and testing. *J. Mech. Behav. Biomed. Mater.* **2018**, *77*, 58–72. [CrossRef] [PubMed]

16. Burton, H.E.; Eisenstein, N.M.; Lawless, B.M.; Jamshidi, P.; Segarra, M.A.; Addison, O.; Shepherd, D.E.; Attallah, M.; Grover, L.M.; Cox, S.C. The design of additively manufactured lattices to increase the functionality of medical implants. *Mater. Sci. Eng. C* **2018**, *94*, 901–908. [CrossRef] [PubMed]

17. Mehboob, H.; Tarlochan, F.; Mehboob, A.; Chang, S.-H. Finite element modelling and characterization of 3D cellular microstructures for the design of a cementless biomimetic porous hip stem. *Mater. Des.* **2018**, *149*, 101–112. [CrossRef]

18. Mower, T.M.; Long, M.J. Mechanical behavior of additive manufactured, powder-bed laser-fused materials. *Mater. Sci. Eng. A* **2016**, *651*, 198–213. [CrossRef]

19. Ghouse, S.; Babu, S.; Van Arkel, R.J.; Nai, K.; Hooper, P.; Jeffers, J.R. The influence of laser parameters and scanning strategies on the mechanical properties of a stochastic porous material. *Mater. Des.* **2017**, *131*, 498–508. [CrossRef]

20. Ghouse, S.; Babu, S.; Nai, K.; Hooper, P.; Jeffers, J.R. The influence of laser parameters, scanning strategies and material on the fatigue strength of a stochastic porous structure. *Addit. Manuf.* **2018**, *22*, 290–301. [CrossRef]

21. Rozvany, G.I.N.; Zhou, M.; Birker, T. Generalized shape optimization without homogenization. *Struct. Multidiscip. Optim.* **1992**, *4*, 250–252. [CrossRef]

22. Xia, L.; Xia, Q.; Huang, X.; Xie, Y.M. Bi-directional evolutionary structural optimization on advanced structures and materials: A comprehensive review. *Arch. Comput. Methods Eng.* **2016**, *25*, 437–478. [CrossRef]

23. Al-Tamimi, A.A.; Quental, C.; Folgado, J.; Peach, C.; Bartolo, P. Stress analysis in a bone fracture fixed with topology-optimised plates. *Biomech. Model. Mechanobiol.* **2019**, *19*, 693–699. [CrossRef]

24. Park, J.; Sutradhar, A.; Shah, J.J.; Paulino, G.H. Design of complex bone internal structure using topology optimization with perimeter control. *Comput. Biol. Med.* **2018**, *94*, 74–84. [CrossRef]

25. Bolland, B.J.; Wilson, M.J.; Howell, J.R.; Hubble, M.J.; Timperley, A.J.; Gie, G.A. An analysis of reported cases of fracture of the universal exeter femoral stem prosthesis. *J. Arthroplast.* **2017**, *32*, 1318–1322. [CrossRef]

26. Al-Dirini, R.M.A.; O'Rourke, D.; Huff, D.; Martelli, S.; Taylor, M. Biomechanical Robustness of a Contemporary Cementless Stem to Surgical Variation in Stem Size and Position. *J. Biomech. Eng.* **2018**, *140*, 091007. [CrossRef]

27. Liu, K.; Tovar, A. An efficient 3D topology optimization code written in Matlab. *Struct. Multidiscip. Optim.* **2014**, *50*, 1175–1196. [CrossRef]

28. Sigmund, O. Morphology-based black and white filters for topology optimization. *Struct. Multidiscip. Optim.* **2007**, *33*, 401–424. [CrossRef]

29. Jun, Y.; Choi, K. Design of patient-specific hip implants based on the 3D geometry of the human femur. *Adv. Eng. Softw.* **2010**, *41*, 537–547. [CrossRef]

30. Aldubaisi, A. Femoral Stem for Hip Prosthesis as a Part of Hip Implant. 2011. Available online: https://grabcad.com/library/femoral-stem (accessed on 10 January 2019).

31. Aitkenhead, A.H. Mesh Voxelisation. MATLAB Central File Exchange. 2013. Available online: https://uk.mathworks.com/matlabcentral/fileexchange/27390-mesh-voxelisation (accessed on 12 November 2018).

32. Geraldes, D.M.; Modenese, L.; Phillips, A.T.M. Consideration of multiple load cases is critical in modelling orthotropic bone adaptation in the femur. *Biomech. Model. Mechanobiol.* **2016**, *15*, 1029–1042. [CrossRef]

33. Bergmann, G.; Bender, A.; Dymke, J.; Duda, G.; Damm, P. Standardized loads acting in hip implants. *PLoS ONE* **2016**, *11*, e0155612. [CrossRef]

34. Cristofolini, L. A critical analysis of stress shielding evaluation of hip prostheses. *Crit. Rev. Biomed. Eng.* **1997**, *25*, 409–483. [CrossRef]

35. Delp, S.L.; Anderson, F.C.; Arnold, A.S.; Loan, P.; Habib, A.; John, C.T.; Guendelman, E.; Thelen, D.G. OpenSim: Open-source software to create and analyze dynamic simulations of movement. *IEEE Trans. Biomed. Eng.* **2007**, *54*, 1940–1950. [CrossRef]

36. Modenese, L.; Phillips, A.; Bull, A.M.J. An open source lower limb model: Hip joint validation. *J. Biomech.* **2011**, *44*, 2185–2193. [CrossRef]

37. Ghouse, S.; Reznikov, N.; Boughton, O.R.; Babu, S.; Ng, G.; Blunn, G.; Cobb, J.; Stevens, M.M.; Jeffers, J.R. The design and in vivo testing of a locally stiffness-matched porous scaffold. *Appl. Mater. Today* **2019**, *15*, 377–388. [CrossRef] [PubMed]

38. Kikinis, R.; Pieper, S.D.; Vosburgh, K.G. 3D Slicer: A platform for subject-specific image analysis, visualization, and clinical support. In *Intraoperative Imaging and Image-Guided Therapy*; Springer: Berlin/Heidelberg, Germany, 2014; pp. 277–289. Available online: https://www.slicer.org/ (accessed on 31 August 2021).
39. Van Arkel, R. On the biomechanics of ligaments and muscles throughout the range of hip motion. In *Mechanical Engineering*; Imperial College London: London, UK, 2015.
40. Bergmann, G.; Deuretzbacher, G.; Heller, M.; Graichen, F.; Rohlmann, A.; Strauss, J.; Duda, G.N. Hip contact forces and gait patterns from routine activities. *J. Biomech.* **2001**, *34*, 859–871. [CrossRef]
41. Gruen, T.A.; McNeice, G.M.; Amstutz, H.C. "Modes of failure" of cemented stem-type femoral components: A radiographic analysis of loosening. *Clin. Orthop. Relat. Res.* **1979**, *141*, 17–27. [CrossRef]
42. Hossain, U.; Ghouse, S.; Nai, K.; Jeffers, J.R. Controlling and testing anisotropy in additively manufactured stochastic structures. *Addit. Manuf.* **2021**, *39*, 101849. [CrossRef]
43. Kharmanda, G.; Antypas, I.; Dyachenko, A. Effective multiplicative formulation for shape optimization: Optimized Austin-Moore stem for hip prosthesis. *Int. J. Mech. Eng. Technol.* **2019**, *10*, 1–11.
44. Rahchamani, R.; Soheilifard, R. Three-dimensional structural optimization of a cementless hip stem using a bi-directional evolutionary method. *Comput. Methods Biomech. Biomed. Eng.* **2020**, *23*, 1–11. [CrossRef] [PubMed]
45. Tanino, H.; Ito, H.; Higa, M.; Omizu, N.; Nishimura, I.; Matsuda, K.; Mitamura, Y.; Matsuno, T. Three-dimensional computer-aided design based design sensitivity analysis and shape optimization of the stem using adaptive p-method. *J. Biomech.* **2006**, *39*, 1948–1953. [CrossRef]
46. Huiskes, R.; Boeklagen, R. Mathematical shape optimization of hip prosthesis design. *J. Biomech.* **1989**, *22*, 793–804. [CrossRef]
47. Nicolella, D.P.; Thacker, B.H.; Katoozian, H.; Davy, D.T. The effect of three-dimensional shape optimization on the probabilistic response of a cemented femoral hip prosthesis. *J. Biomech.* **2006**, *39*, 1265–1278. [CrossRef]
48. Latham, B.; Goswami, T. Effect of geometric parameters in the design of hip implants paper IV. *Mater. Des.* **2004**, *25*, 715–722. [CrossRef]
49. Sun, C.; Wang, L.; Kang, J.; Li, D.; Jin, Z. Biomechanical optimization of elastic modulus distribution in porous femoral stem for artificial hip joints. *J. Bionic Eng.* **2018**, *15*, 693–702. [CrossRef]
50. Fraldi, M.; Esposito, L.; Perrella, G.; Cutolo, A.; Cowin, S.C. Topological optimization in hip prosthesis design. *Biomech. Model. Mechanobiol.* **2009**, *9*, 389–402. [CrossRef]
51. Jang, I.G.; Kim, I.Y.; Kwak, B.M. Analogy of strain energy density based bone-remodeling algorithm and structural topology optimization. *J. Biomech. Eng.* **2008**, *131*, 011012. [CrossRef] [PubMed]
52. Warth, L.C.; Grant, T.W.; Naveen, N.B.; Deckard, E.R.; Ziemba-Davis, M.; Meneghini, R.M. Inadequate metadiaphyseal fill of a modern taper-wedge stem increases subsidence and risk of aseptic loosening: Technique and distal canal fill matter! *J. Arthroplast.* **2020**, *35*, 1868–1876. [CrossRef] [PubMed]
53. Bätz, J.; Messer-Hannemann, P.; Lampe, F.; Klein, A.; Püschel, K.; Morlock, M.M.; Campbell, G.M. Effect of cavity preparation and bone mineral density on bone-interface densification and bone-implant contact during press-fit implantation of hip stems. *J. Orthop. Res.* **2019**, *37*, 1580–1589. [CrossRef] [PubMed]
54. Munford, M.J.; Ng, K.C.G.; Jeffers, J.R.T. Mapping the multi-directional mechanical properties of bone in the proximal tibia. *Adv. Funct. Mater.* **2020**, *30*, 2004323. [CrossRef]
55. Weinans, H.; Huiskes, R.; Grootenboer, H.J. Effects of fit and bonding characteristics of femoral stems on adaptive bone remodeling. *J. Biomech. Eng.* **1994**, *116*, 393–400. [CrossRef]
56. Weinans, H.; Huiskes, R.; Van Rietbergen, B.; Sumner, D.R.; Turner, T.M.; Galante, J.O. Adaptive bone remodeling around bonded noncemented total hip arthroplasty: A comparison between animal experiments and computer simulation. *J. Orthop. Res.* **1993**, *11*, 500–513. [CrossRef]
57. Van Arkel, R.J.; Ghouse, S.; Milner, P.E.; Jeffers, J.R. Additive manufactured push-fit implant fixation with screw-strength pull out. *J. Orthop. Res.* **2018**, *36*, 1508–1518. [CrossRef]
58. Kohli, N.; Stoddart, J.C.; van Arkel, R.J. The limit of tolerable micromotion for implant osseointegration: A systematic review. *Sci. Rep.* **2021**, *11*, 1–11. [CrossRef]
59. McNamara, B.P.; Cristofolini, L.; Toni, A.; Taylor, D. Relationship between bone-prosthesis bonding and load transfer in total hip reconstruction. *J. Biomech.* **1997**, *30*, 621–630. [CrossRef]

Article

Transient Phase-Driven Cyclic Deformation in Additively Manufactured 15-5 PH Steel

Tu-Ngoc Lam [1,2], Yu-Hao Wu [1], Chia-Jou Liu [1], Hobyung Chae [3], Soo-Yeol Lee [3,*], Jayant Jain [4,*], Ke An [5] and E-Wen Huang [1,*]

1   Department of Materials Science and Engineering, National Yang Ming Chiao Tung University, Hsinchu 30010, Taiwan; lamtungoc1310@gmail.com (T.-N.L.); paradise7139@gmail.com (Y.-H.W.); jasmine861025@gmail.com (C.-J.L.)
2   Department of Physics, College of Education, Can Tho University, Can Tho 900000, Vietnam
3   Department of Materials Science and Engineering, Chungnam National University, Daejeon 34134, Korea; highteen5@cnu.ac.kr
4   Department of Materials Science and Engineering, Indian Institute of Technology, New Delhi 110016, India
5   Chemical and Engineering Materials Division, The Spallation Neutron Source, Oak Ridge National Laboratory, Oak Ridge, TN 37831, USA; kean@ornl.gov
*   Correspondence: sylee2012@cnu.ac.kr (S.-Y.L.); Jayant.Jain@mse.iitd.ac.in (J.J.); EwenHUANG@nctu.edu.tw (E.-W.H.)

Abstract: The present work extends the examination of selective laser melting (SLM)-fabricated 15-5 PH steel with the 8%-transient-austenite-phase towards fully-reversed strain-controlled low-cycle fatigue (LCF) test. The cyclic-deformation response and microstructural evolution were investigated via in-situ neutron-diffraction measurements. The transient-austenite-phase rapidly transformed into the martensite phase in the initial cyclic-hardening stage, followed by an almost complete martensitic transformation in the cyclic-softening and steady stage. The compressive stress was much greater than the tensile stress at the same strain amplitude. The enhanced martensitic transformation associated with lower dislocation densities under compression predominantly governed such a striking tension-compression asymmetry in the SLM-built 15-5 PH.

Keywords: 15-5 PH stainless steel; in-situ neutron diffraction; selective laser melting; low-cycle fatigue; martensite transformation

Citation: Lam, T.-N.; Wu, Y.-H.; Liu, C.-J.; Chae, H.; Lee, S.-Y.; Jain, J.; An, K.; Huang, E.-W. Transient Phase-Driven Cyclic Deformation in Additively Manufactured 15-5 PH Steel. *Materials* **2022**, *15*, 777. https://doi.org/10.3390/ma15030777

Academic Editor: Rubén Paz

Received: 23 December 2021
Accepted: 17 January 2022
Published: 20 January 2022

Publisher's Note: MDPI stays neutral with regard to jurisdictional claims in published maps and institutional affiliations.

1. Introduction

Selective laser melting (SLM), one of the most commonly used approaches in additive manufacturing (AM) technique, provides great abilities in fabricating layer by layer-built parts with complex geometry and customization, which significantly affects the anisotropy in mechanical properties with respect to the building directions [1–5]. The heterogeneous microstructure with the formation of defects in the SLM-manufactured alloys is inherently different from the homogeneous microstructure of the conventional cast and wrought alloys, causing unpredictable mechanical and fatigue properties of the SLM-fabricated components [2,6–8]. During the consecutive cycles of rapid heating and solidification under laser melting, the occurrence of defects, nucleation and growth, and phase transformation induces a metastable microstructure of the additive manufactured metals [2,9,10].

The imperfections such as pores are strongly governed by the welding environments, which significantly decreased the mechanical properties of the steel welded joints [11]. Garcias et al. compared the material properties fabricated via conventional manufacturing and AM methods in which porosity and lack of fusion in the additive manufactured steels were found to have a negative effect on their mechanical and fatigue responses [12]. Post processing may reduce the imperfections and improve the properties of additive manufactured alloys [12]. Tailoring the intricate AM process with plenty fabrication conditions associated with various post treatments is full of challenges but a requisite for great optimization of mechanical and fatigue performances of the additive manufactured alloys.

The AM process parameters enable the controllable microstructure of precipitation hardened stainless steels consisting of austenite and martensite phases and thus strongly govern their mechanical properties [13–16]. The retained austenite produced in the additive manufactured steels during AM process enabled the transformation-induced plasticity [17]. Investigations on the influence of non-equilibrium retained austenite γ-phase and the strain-induced martensitic transformation under loading in the additive manufactured 15-5 PH and 17-4 PH steels have been extensively reported [4,14,18–25]. Huang et al. examined the tensile properties of the 15-5 PH steel with a tunable fraction of metastable transient-austenite-phase, which can be successfully tailored via the SLM process. They demonstrated that a higher fraction of transient austenite γ-phase was beneficial to the tensile properties of the SLM-built 15-5 PH steel [4]. Specifically, the SLM-built 15-5 PH with the 18%-transient-austenite-phase revealed a more impressive strain hardening effect beyond yielding, while the specimen with the 8%-transient-austenite-phase disclosed similar deformation behavior as the as-quenched martensite steel [4]. Chae et al. demonstrated the principal strengthening mechanisms via various heat treatments in tailoring the mechanical properties of the additive manufactured 15-5 PH steels [26].

In the other alloy systems, deformation-induced phase development was found, such as in the cobalt-based superalloys during monotonic and cyclic deformation [27]. Meanwhile, although comprehensive studies have been devoted to deciphering the strain-induced phase transformation under monotonic tensile loading, the cyclic-induced phase transformation subjected to low-cycle-fatigue (LCF) deformation in the SLM-built 15-5 PH steel conducted by in-situ neutron diffraction is unclear and thus intriguing to examine.

We herein extend our earlier works [4,24,28] towards the microstructural evolution of α-matrix and transient γ-phase under cyclic continuous tension-compression loading. We explore the fully-reversed strain-controlled LCF deformation at a strain amplitude of $\pm 1.0\%$ in the SLM-built 15-5 PH with the 8%-transient-austenite-phase at room temperature. In-situ neutron-diffraction measurement was carried out to examine the cyclic-stress response during symmetric tension-compression cyclic loading by monitoring the microstructural evolutions [29], such as the dislocations [30], which are beneficial to elucidate the corresponding deformation mechanisms governing fatigue behavior of the SLM-built 15-5 PH. The possible factors governing the applied stress response upon cyclic loading were reported.

2. Materials and Methods

The additive manufactured 15-5 PH steel was fabricated from the PH1 powder via SLM. The chemical composition was composed of 14.35 wt% Cr, 4.03 wt% Ni, 2.71 wt% Cu, 0.01 wt% Mn, 0.99 wt% Si, 0.5 wt% Mo, 0.74 wt% Nb, <0.01 wt% C, and Fe balance. The laser power of 195 W, spot size of 70 μm, and scanning rate of 1200 mm/s were used. The vertically-built SLM 15-5 PH steel with the 8%-transient-austenite-phase used in this study was produced such that the building direction was parallel to the loading direction, shown in the inset of Figure 1. More details of the 8%-transient-austenite-phase specimen can be referred in our previous study [4].

In-situ neutron-diffraction measurements were carried out using the VULCAN engineering diffractometer at the Spallation Neutron Source (SNS) of the Oak Ridge National Laboratory (ORNL) [31–34]. The neutron-diffraction profiles were recorded simultaneously in both loading and transverse directions using two orthogonal detectors situated at $\pm 90°$ from the incident neutron beam. The axial and transverse detectors collect the diffraction patterns from crystallographic orientations parallel and perpendicular to the applied loading direction, respectively. The LCF test was conducted with a maximum tensile strain of 1.0% and a maximum compressive strain -1.0% at a cyclic frequency of 1 Hz. The cylindrical dog-bone-shaped specimen used for LCF loading had a total length of 70 mm, a gauge length of 10 mm, and a diameter of 6 mm, shown in the inset of Figure 1.

Neutron-diffraction profiles were recorded during LCF test in a strain-controlled mode at maximum tensile and compressive strains of 1.0% and −1.0% of each fatigue cycle until the specimen was fractured. The procedures for in-situ neutron-diffraction experiments and data analysis followed our earlier protocol [4]. The neutron-diffraction patterns were fitted using the general structure analysis system II (GSAS-II) [35] for phase characterization and profile analysis. Moreover, the development of dislocation density during LCF response for both α-matrix and transient γ-phase was determined from the peak-width evolution following Tomota's method [36]. The evolution of phase fraction was obtained from the refined peak intensity as a function of fatigue cycles.

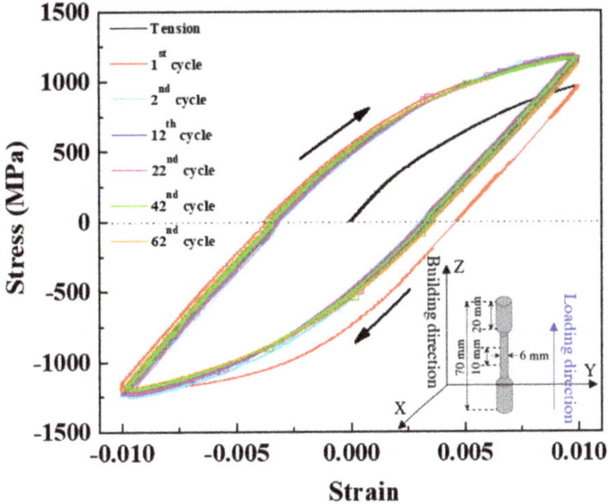

Figure 1. Stress-strain curves under LCF test in the SLM-built 15-5 PH steel. Schematic illustration of the vertically-built SLM 15-5 PH specimen is shown in the inset.

3. Results

3.1. Cyclic Stress-Strain Curves

Figure 1 presents the hysteresis loops at several selected fatigue cycles under LCF test at a strain amplitude of ±1.0% in the SLM-built 15-5 PH steel. The initial loading was in tension. In the 1st cycle, the stress value was 955 MPa at 1.0% strain. During unloading and compression, the plastic deformation started, and the stress value was 1158 MPa at −1.0% strain, respectively. In reloading to tension, the stress value increased to 1118 MPa at 1.0% strain, originating from the significant strain hardening from the 1st to 2nd cycles. In the 2nd cycle, during unloading to zero, plastic strain decreased but the stress value at −1.0% strain was higher compared to the 1st cycle. The softening and saturation stages then occurred up to the 62nd cycle at which the specimen was fractured.

3.2. Cyclic-Induced Martensitic Transformation

Figure 2 depicts the evolution of in-situ neutron-diffraction patterns at 1.0% strain at the 1st and 62nd cycles in both axial loading and transverse directions. The diffraction profiles of body-centered cubic (BCC) α-matrix and face-centered cubic (FCC) γ-phase austenite assigned for each (h k l) were illustrated as the blue and red arrows, respectively. The intensity of (1 1 0) peak of α-matrix in the transverse direction (Figure 2c) was greater by 30% than that in the axial loading direction in the 1st cycle (Figure 2a). Upon cycling, the intensity of α-matrix peaks (ferrite and martensite) significantly increased while that of γ-phase peaks dramatically decreased and almost disappeared at the 62nd cycle in both axial loading (Figure 2b) and transverse directions (Figure 2d). The neutron results indicated the cyclic-induced martensitic transformation in the SLM-built 15-5 PH during continuous tension-compression loading.

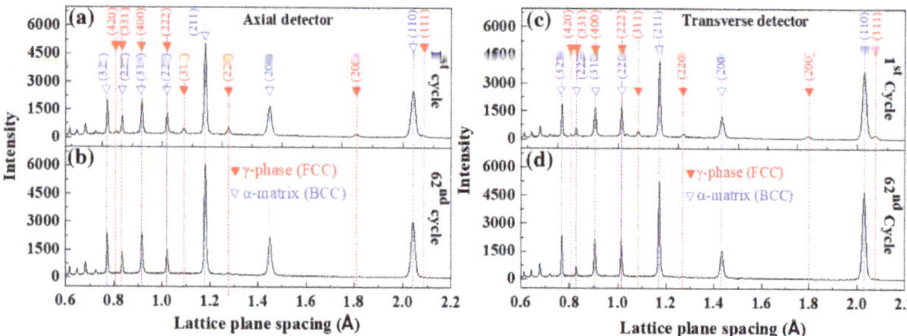

Figure 2. In-situ neutron-diffraction profiles at the maximum tensile strain of 1.0% in the axial loading direction at the (**a**) 1st cycle and (**b**) 62nd cycle. Those in the transverse direction at the (**c**) 1st cycle and (**d**) 62nd cycle in the SLM-built 15-5 PH steel.

3.3. Tension-Compression Asymmetry Behavior

Figure 3a describes the cyclic-stress amplitudes at the tensile strain of 1.0% (red circles ○) and compressive strain of −1.0% (red squares □) as a function of fatigue cycles from 1 to 62 under LCF response in the SLM-built 15-5 PH steel. Under tensile deformation, the specimen experienced an initial cyclic hardening in which the applied stress significantly increased from 955 MPa at the 1st cycle to 1177 MPa at the 22nd cycle, followed by a slightly cyclic-softening and steady behavior up to the 62nd cycle. Meanwhile, under compressive deformation, there was an increase of cyclic stress from 1158 MPa to 1241 MPa at the 2nd cycle where the cyclic-hardening/softening transition occurred, followed by a more obvious cyclic-softening and steady behavior until the 62nd cycle. Moreover, the compressive stress was much higher than the tensile stress at the same strain amplitude, suggesting an asymmetric response in tensile-compressive deformation in the SLM-built 15-5 PH steel. Such a remarkable tension-compression asymmetry behavior is one of the typical characteristics of the additive manufactured alloys [37,38].

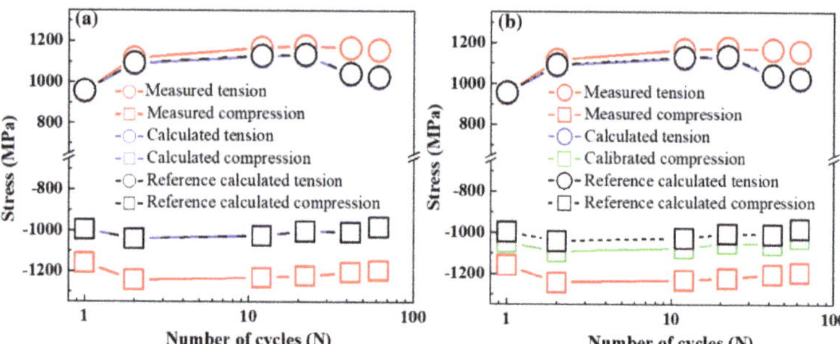

Figure 3. (**a**) The measured stress, calculated stress, and reference calculated stress at the maximum tensile and compressive deformations as a function of fatigue cycles. (**b**) The calculated compressive stress in (**a**) was replaced by the calibrated compressive stress.

Since the applied stresses vary differently under tension and compression at the same strain amplitude, clarifying the relationship between the macroscopic applied stress and microscopic lattice strain is necessary. We followed Wang et al.'s [39] and Young et al.'s [40] methods to estimate the effective phase stresses in understanding the role of α-matrix and γ-phase under applied stresses. We calculated the average von Mises equivalent stress for each phase [41] on the assumption that two orthogonal transverse strain components

($\sigma_{11} = \sigma_{22}$) are equal for cylindrical dog-bone sample. We applied the generalized Hooke's law shown as below.

$$\sigma_{11}^{\alpha} = \sigma_{22}^{\alpha} = \frac{E}{(1+v)(1-2v)}\{(1-v)\varepsilon_{11}^{\alpha} + v(\varepsilon_{22}^{\alpha} + \varepsilon_{33}^{\alpha})\}, \tag{1}$$

$$\sigma_{33}^{\alpha} = \frac{E}{(1+v)(1-2v)}\{(1-v)\varepsilon_{33}^{\alpha} + v(\varepsilon_{22}^{\alpha} + \varepsilon_{11}^{\alpha})\}, \tag{2}$$

where E is Young's modulus, v is Poisson's ratio, σ_{33} and ε_{33} are the stress and strain components along the axial loading direction, respectively, σ_{11} and σ_{22} are the stress components in the transverse direction, ε_{11} and ε_{22} are the strain components in the transverse direction, and α is the BCC α-matrix. The same calculation is applied for the FCC γ-phase by substituting γ for α.

The calculated stress ($\sigma_{ij}^{calculated}$) can be estimated as follows

$$\sigma_{ij}^{calculated} = \sigma_{ij}^{\alpha}(1-f) + \sigma_{ij}^{\gamma}f, \tag{3}$$

where f is the volume fraction of the FCC γ-phase, σ_{ij}^{α} and σ_{ij}^{γ} are the stress components of the BCC α-matrix and FCC γ-phase, respectively.

Figure 3a presents the calculated tensile and compressive stresses obtained from the generalized Hooke's law shown in the blue circles ($\circ$) and squares ($\square$), respectively, while the reference calculated tensile and compressive stresses using the reference Young's modulus [42] were depicted in the black circles ($\circ$) and squares ($\square$), respectively. It can be seen from Figure 3a that the calculated stresses were almost the same as the reference calculated stresses under both tensile and compressive deformations. Furthermore, the measured stress was much closer to the reference calculated stress under tension than under compression.

3.4. Effect of Porous Structure on the Tension-Compression Asymmetry Behavior

The outstanding tension-compression asymmetry behavior may be governed by various reasonable factors such as porous structure, phase transformation, and dislocation density. The pores, microvoids, or inclusions commonly exist in the additive manufactured steels due to thermal gradients or unmelted powder during the additive manufacturing process [37,43–45]. Thus, we first assumed that porous structure is one of the possible causes resulting in the tension-compression asymmetry behavior. To correlate the mechanical properties of porous materials with their relative densities, we recalled the Ashby and Gibson model (1997) as follows [46]:

$$\sigma_y = \sigma_{y0} \times \rho_{rel}^{3/2}, \tag{4}$$

$$E_{eff} = E_0 \times \rho_{rel}^{2}, \tag{5}$$

where σ_Y and E_{eff} are the yield strength and effective modulus of the porous material, respectively; σ_{y0} and E_0 are the yield strength and elastic modulus of the bulk material, respectively; and ρ_{rel} is the relative density of the porous material.

We herein assumed that when the specimen experiences tensile deformation, the internal pores are stretched and the volume becomes larger. However, the internal pores are squeezed or even disappeared as the specimen undergoes compressive deformation. The deformation is presumably highly influenced by large internal porosity in tension. Therefore, the parameters of the porous material were expressed as the case of tension while the parameters of the bulk material were represented as the case of compression. Based on the measured Young's moduli obtained from the slopes of lattice strain versus engineering stress curves under tension and compression, the relative densities of α-matrix and γ-phase calculated from Equation (5) were 0.970 and 0.794, respectively. The stress components of α-matrix and γ-phase of the porous material under compression were accordingly adjusted by

Equation (4) instead of Equation (2). The calibrated compressive stress was then calculated by Equation (3), which was shown in the green squares (□) in Figure 3b. Compared with the calculated compressive stress, the calibrated compressive stress was closer to the measured compressive stress. Despite taking the porous structure into consideration, the calibrated compressive stress was still inconsistent with the measured compressive stress, suggesting other factors governing the tension-compression asymmetry behavior besides the porous structure.

3.5. Effect of Microscopic Change on the Tension-Compression Asymmetry Behavior

Since the two phases of α-matrix and γ-phase co-existed and there was martensitic transformation in the SLM-built 15-5 PH steel during cyclic loading, exploring the contribution of each phase to the macroscopically measured stresses under both tension and compression is necessary. Figure 4a,b present the microscopically individual stresses of α-matrix and γ-phase as a function of fatigue cycles, respectively. The individual stress of α-matrix disclosed an obvious symmetry under tension and compression; however, the individual stress of γ-phase revealed an evident tension-compression asymmetry behavior upon cycling. The macroscopic tension-compression asymmetry was probably originated from the major contribution of γ-phase instead of α-matrix. It should be noted that the phase fraction should be taken into account when calculating the macroscopic stress of the specimen by Equation (3). Particularly, the phase fraction of γ-phase austenite (8 wt%) was much lower than that of α-phase ferrite (92 wt%) before the fatigue test. Moreover, the γ-phase austenite rapidly transformed to the martensite phase up to the 2nd cycle, followed by a nearly complete martensitic transformation upon further cycling, as seen in the phase fraction evolution of α-matrix and γ-phase in Figure 4c,d, respectively. Since there is a very small fraction of γ-phase austenite upon cycling, its effect on the macroscopic stress should be trivial. Therefore, the tension-compression asymmetry of γ-phase had a negligible influence on the macroscopic tension-compression asymmetry behavior in the SLM-built 15-5 PH.

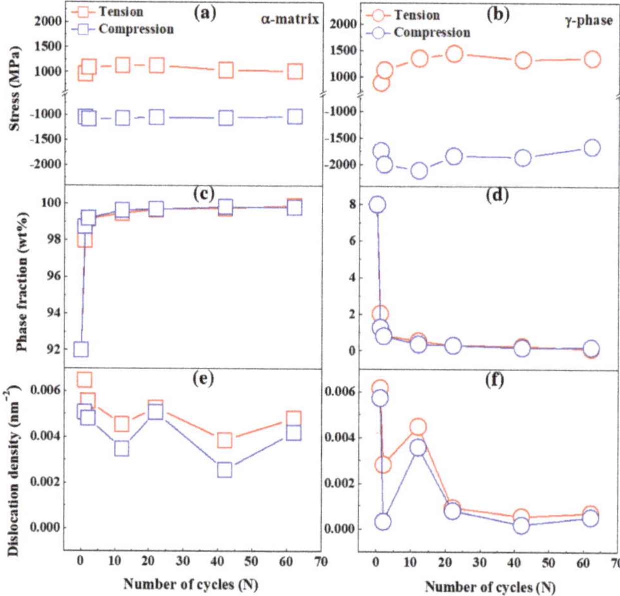

Figure 4. The microscopically individual stresses of the (**a**) α-matrix and (**b**) γ-phase; the phase fraction of the (**c**) α-matrix and (**d**) γ-phase; the dislocation density of the (**e**) α-matrix and (**f**) γ-phase as a function of fatigue cycles.

Accompanying with the martensitic transformation, examination of microstructural evolution of the individual phases during cyclic loading is indispensable to understand the tension-compression asymmetry in the SLM-built 15-5 PH steel. Figure 4e,f show the dynamic evolution of dislocation densities of α-matrix and γ-phase as a function of fatigue cycles, respectively. The evolution of dislocations under tension and compression disclosed similar trends in both α-matrix and γ-phase, however, the dislocations exhibited lower densities in compression than in tension for both α-matrix and γ-phase, implying the different contribution of dislocation activities to the tension-compression asymmetry in the SLM-built 15-5 PH specimen.

4. Discussion

Since the pores are usually formed in the fabricated alloys during SLM process and they strongly affect their mechanical properties, we further calculated the porosity of the SLM-built 15-5 PH by the Equation (6) [47].

$$P(\%) = \left(1 - \frac{\rho_{measured}}{\rho_{theoretical}}\right) * 100, \tag{6}$$

where $\rho_{measured}$ is the measured density of the SLM-built 15-5 PH specimen and $\rho_{theoretical}$ is the theoretical density of the material.

The measured density of the SLM-built 15-5 PH measured by the Archimedes method was 7.759 g/cm^3, which was similar to the density of the SLM-built 17-4PH steel with similar manufacturing parameters measured by Hengfeng Gu et al. [48]. The porosity of the SLM-built 15-5 PH specimen was determined to be 0.53%, which is insufficient to have a major impact on the tension-compression asymmetry behavior [49]. The results inferred that the porosity had relatively minor contributions to the tension-compression asymmetry behavior in the SLM-built 15-5 PH specimen.

The phase fraction evolution of both α-matrix and γ-phase was similar under tension and compression in Figure 4c,d, however, there was a noticeable difference in the amount of martensitic transformation between tension and compression. A slightly higher fraction of the cyclic-induced martensite phase was visible under compression than under tension in the cyclic-hardening region, while no evident discrepancy of martensitic transformation was seen in the cyclic-softening and steady stage. Since the martensite phase is harder than the other phases, the more pronounced fraction of cyclic-induced martensite phase under compression significantly contributes to the higher cyclic compressive stress at the same strain amplitude and thus the more hardening behavior under compression.

In Figure 4e,f, although the dislocation densities of both α-matrix and γ-phase decreased with increasing fatigue cycles, the dislocation densities of γ-phase dramatically reduced upon cycling. Such a drastic degradation of dislocation density of γ-phase leads to the cyclic-softening region [50]. Furthermore, the dislocation densities of both α-matrix and γ-phase were lower under compression rather than under tension. The decreased dislocation densities under the fatigue test were also observed under the monotonic tensile test in the SLM-built 15-5 PH, which was assigned to the coalescence of dislocations during martensitic transformation [4]. Lower dislocation densities under compression promote the nucleation and growth of the martensite phase [50], which is beneficial to the higher cyclic stress under compression.

5. Conclusions

We examined the cyclic-stress response under strain-controlled LCF test in the SLM-built 15-5 PH steel with 8%-transient-austenite-phase and identified the possible underlying factors. The SLM-built 15-5 PH specimen was fractured at the 62nd cycle upon cycling loading at a strain amplitude of ±1.0%. The compressive stresses of 1158 MPa and 1241 MPa were higher than the tensile stresses of 955 MPa and 1118 MPa at the 1st and 2nd cycles, respectively. The effect of inherent pores and porosity on the SLM-built 15-5 PH plays a secondary role in higher cyclic compressive stress. Such a hardening response under compression was mainly ascribed to the primary martensitic transformation coupled with lower dislocation densities. Understanding the principal microstructural changes governing the applied stress amplitude upon cycling is helpful for the design of fatigue-resistant additive manufactured alloys.

Author Contributions: Conceptualization, E.-W.H.; investigation, C.-J.L. and Y.-H.W.; resources, H.C., S.-Y.L., J.J. and K.A.; writing—original draft preparation, T.-N.L.; writing—review and editing, S.-Y.L. and E.-W.H.; supervision, E.-W.H.; project administration, E.-W.H. All authors have read and agreed to the published version of the manuscript.

Funding: Research conducted at ORNL's Spallation Neutron Source was sponsored by the Scientific User Facilities Division, Office of Basic Energy Sciences, US Department of Energy. The authors are grateful to the support of the Ministry of Science and Technology (MOST) Programs MOST 110-2224-E-007-001 and MOST 108-2221-E-009-131-MY4. This work was financially supported by the "Center for the Semiconductor Technology Research" from The Featured Areas Research Center Program within the framework of the Higher Education Sprout Project by the Ministry of Education (MOE) in Taiwan. Also supported in part by the Ministry of Science and Technology, Taiwan, under Grant MOST 110-2634-F-009-027. This work was supported by the Higher Education Sprout Project of the National Yang Ming Chiao Tung University and the MOE, Taiwan. SYL was supported by the National Research Foundation (NRF) grant funded by the Korean Government (2021R1A4A1031494, 2020K2A9A2A06070856).

Data Availability Statement: The data presented in this study are available on request from the corresponding author.

Acknowledgments: E.-W.H. and his group members very much appreciate the financial support from the National Synchrotron Radiation Research Center (NSRRC) Neutron Program.

Conflicts of Interest: The authors declare no conflict of interest.

Nomenclature

Selective laser melting	SLM
Low-cycle fatigue	LCF
Additive manufacturing	AM
Spallation Neutron Source	SNS
Oak Ridge National Laboratory	ORNL
General structure analysis system II	GSAS-II
Body-centered cubic	BCC
Face-centered cubic	FCC
○	Measured tensile stress
□	Measured compressive stress
○	Calculated tensile stress
□	Calculated compressive stress
○	Reference calculated tensile stress
□	Reference calculated compressive stress
□	Calibrated compressive stress
E	Young's modulus
v	Poisson's ratio
σ_{33}	Stress along the axial loading direction
ε_{33}	Strain along the axial loading direction
σ_{11}	Stress components in the transverse direction
σ_{22}	Stress components in the transverse direction
ε_{11}	Strain components in the transverse direction
ε_{22}	Strain components in the transverse direction
α	BCC α-matrix
γ	FCC γ-phase
$\sigma_{ij}^{calculated}$	Calculated stress
f	Volume fraction of FCC γ-phase

σ_{ij}^{α}	Stress components of BCC α-matrix
σ_{ij}^{γ}	Stress components of FCC γ-phase
σ_Y	Yield strength of porous material
E_{eff}	Effective modulus of porous material
σ_{y0}	Yield strength of bulk material
E_0	Elastic modulus of bulk material
ρ_{rel}	Relative density of porous material
P(%)	Porosity
$\rho_{measured}$	Measured density
$\rho_{theoretical}$	Theoretical density

References

1. Thompson, S.M.; Bian, L.; Shamsaei, N.; Yadollahi, A. An overview of Direct Laser Deposition for additive manufacturing; Part I: Transport phenomena, modeling and diagnostics. *Addit. Manuf.* **2015**, *8*, 36–62. [CrossRef]
2. Shamsaei, N.; Yadollahi, A.; Bian, L.; Thompson, S.M. An overview of Direct Laser Deposition for additive manufacturing; Part II: Mechanical behavior, process parameter optimization and control. *Addit. Manuf.* **2015**, *8*, 12–35. [CrossRef]
3. Kok, Y.; Tan, X.P.; Wang, P.; Nai, M.L.S.; Loh, N.H.; Liu, E.; Tor, S.B. Anisotropy and heterogeneity of microstructure and mechanical properties in metal additive manufacturing: A critical review. *Mater. Des.* **2018**, *139*, 565–586. [CrossRef]
4. Huang, E.W.; Lee, S.Y.; Jain, J.; Tong, Y.; An, K.; Tsou, N.-T.; Lam, T.-N.; Yu, D.; Chae, H.; Chen, S.-W.; et al. Hardening steels by the generation of transient phase using additive manufacturing. *Intermetallics* **2019**, *109*, 60–67. [CrossRef]
5. Tseng, J.C.; Huang, W.C.; Chang, W.; Jeromin, A.; Keller, T.F.; Shen, J.; Chuang, A.C.; Wang, C.C.; Lin, B.H.; Amalia, L.; et al. Deformations of Ti-6Al-4V additive-manufacturing-induced isotropic and anisotropic columnar structures: Insitu measurements and underlying mechanisms. *Addit. Manuf.* **2020**, *35*, 101322. [CrossRef] [PubMed]
6. Starr, T.; Rafi, H.; Stucker, B.; Scherzer, C.M. *Controlling Phase Composition in Selective Laser Melted Stainless Steels*; International Solid Freeform Fabrication Symposium: Austin, TX, USA, 2012; pp. 439–446.
7. Yadollahi, A.; Shamsaei, N.; Thompson, S.M.; Elwany, A.; Bian, L. Effects of building orientation and heat treatment on fatigue behavior of selective laser melted 17-4 PH stainless steel. *Int. J. Fatigue* **2017**, *94*, 218–235. [CrossRef]
8. Song, B.; Zhao, X.; Li, S.; Han, C.; Wei, Q.; Wen, S.; Liu, J.; Shi, Y. Differences in microstructure and properties between selective laser melting and traditional manufacturing for fabrication of metal parts: A review. *Front. Mech. Eng.* **2015**, *10*, 111–125. [CrossRef]
9. Herzog, D.; Seyda, V.; Wycisk, E.; Emmelmann, C. Additive manufacturing of metals. *Acta Mater.* **2016**, *117*, 371–392. [CrossRef]
10. Kim, J.S.; LaGrange, T.; Reed, B.W.; Taheri, M.L.; Armstrong, M.R.; King, W.E.; Browning, N.D.; Campbell, G.H. Imaging of Transient Structures Using Nanosecond in Situ TEM. *Science* **2008**, *321*, 1472–1475. [CrossRef] [PubMed]
11. Tomkow, J.; Janeczek, A.; Rogalski, G.; Wolski, A. Underwater Local Cavity Welding of S460N Steel. *Material* **2020**, *13*, 5535. [CrossRef]
12. Garcias, J.F.; Martins, R.F.; Branco, R.; Marciniak, Z.; Macek, W.; Pereira, C.; Santos, C. Quasistatic and fatigue behavior of an AISI H13 steel obtained by additive manufacturing and conventional method. *Fatigue Fract. Eng. Mater. Struct.* **2021**, *44*, 3384–3398. [CrossRef]
13. Hunt, J.; Derguti, F.; Todd, I. Selection of steels suitable for additive layer manufacturing. *Ironmak. Steelmak.* **2014**, *41*, 254–256. [CrossRef]
14. Facchini, L.; Vicente, N., Jr.; Lonardelli, I.; Magalini, E.; Robotti, P.; Molinari, A. Metastable Austenite in 17–4 Precipitation-Hardening Stainless Steel Produced by Selective Laser Melting. *Adv. Eng. Mater.* **2010**, *12*, 184–188. [CrossRef]
15. Haghdadi, N.; Laleh, M.; Moyle, M.; Primig, S. Additive manufacturing of steels: A review of achievements and challenges. *J. Mater. Sci.* **2020**, *56*, 64–107. [CrossRef]
16. Lum, E.; Palazotto, A.N.; Dempsey, A. Analysis of the Effects of Additive Manufacturing on the Material Properties of 15-5PH Stainless Steel. In Proceedings of the 58th AIAA/ASCE/AHS/ASC Structures, Structural Dynamics, and Materials Conference, Grapevine, TX, USA, 9–13 January 2017. [CrossRef]
17. Lass, E.A.; Stoudt, M.R.; Williams, M.E. Additively Manufactured Nitrogen-Atomized 17-4 PH Stainless Steel with Mechanical Properties Comparable to Wrought. *Metall. Mater. Trans. A* **2019**, *50*, 1619–1624. [CrossRef]
18. LeBrun, T.; Nakamoto, T.; Horikawa, K.; Kobayashi, H. Effect of retained austenite on subsequent thermal processing and resultant mechanical properties of selective laser melted 17–4 PH stainless steel. *Mater. Des.* **2015**, *81*, 44–53. [CrossRef]
19. Mahmoudi, M.; Elwany, A.; Yadollahi, A.; Thompson, S.M.; Bian, L.; Shamsaei, N. Mechanical properties and microstructural characterization of selective laser melted 17-4 PH stainless steel. *Rapid Prototyp. J.* **2017**, *23*, 280–294. [CrossRef]
20. Rafi, H.K.; Starr, T.L.; Stucker, B.E. A comparison of the tensile, fatigue, and fracture behavior of Ti–6Al–4V and 15-5 PH stainless steel parts made by selective laser melting. *Int. J. Adv. Manuf. Technol.* **2013**, *69*, 1299–1309. [CrossRef]
21. AlMangour, B.; Yang, J.-M. Improving the surface quality and mechanical properties by shot-peening of 17-4 stainless steel fabricated by additive manufacturing. *Mater. Des.* **2016**, *110*, 914–924. [CrossRef]
22. Hsu, T.-H.; Chang, Y.-J.; Huang, C.-Y.; Yen, H.-W.; Chen, C.-P.; Jen, K.-K.; Yeh, A.-C. Microstructure and property of a selective laser melting process induced oxide dispersion strengthened 17-4 PH stainless steel. *J. Alloys Compd.* **2019**, *803*, 30–41. [CrossRef]

23. Nong, X.D.; Zhou, X.L.; Wang, Y.D.; Yu, L.; Li, J.H. Effects of geometry, location, and direction on microstructure and mechanical properties of 15–5PH stainless steel fabricated by directed energy deposition. *Mater. Sci. Eng. A* **2021**, *821*, 141587. [CrossRef]

24. Chae, H.; Huang, E.W.; Jain, J.; Wang, H.; Woo, W.; Chen, S.-W.; Harjo, S.; Kawasaki, T.; Lee, S.Y. Plastic anisotropy and deformation-induced phase transformation of additive manufactured stainless steel. *Mater. Sci. Eng. A* **2019**, *762*, 138065. [CrossRef]

25. Carneiro, L.; Jalalahmadi, B.; Ashtekar, A.; Jiang, Y. Cyclic deformation and fatigue behavior of additively manufactured 17–4 PH stainless steel. *Int. J. Fatigue* **2019**, *123*, 22–30. [CrossRef]

26. Chae, H.; Luo, M.Y.; Huang, E.W.; Shin, E.; Do, C.; Hong, S.-K.; Woo, W.; Lee, S.Y. Unearthing principal strengthening factors tuning the additive manufactured 15-5 PH stainless steel. *Mater. Charact.* **2022**, *184*, 111645. [CrossRef]

27. Benson, M.L.; Liaw, P.K.; Saleh, T.A.; Choo, H.; Brown, D.W.; Daymond, M.R.; Huang, E.W.; Wang, X.L.; Stoica, A.D.; Buchanan, R.A.; et al. Deformation-induced phase development in a cobalt-based superalloy during monotonic and cyclic deformation. *Phys. B Condens. Matter* **2006**, *385–386*, 523–525. [CrossRef]

28. Chae, H.; Huang, E.W.; Woo, W.; Kang, S.H.; Jain, J.; An, K.; Lee, S.Y. Unravelling thermal history during additive manufacturing of martensitic stainless steel. *J. Alloys Compd.* **2021**, *857*, 157555. [CrossRef] [PubMed]

29. Huang, E.W.; Chang, C.K.; Liaw, P.K.; Suei, T.R. Fatigue induced deformation and thermodynamics evolution in a nano particle strengthened nickel base superalloy. *Fatigue Fract. Eng. Mater. Struct.* **2016**, *39*, 675–685. [CrossRef]

30. Huang, E.W.; Barabash, R.I.; Ice, G.E.; Liu, W.; Liu, Y.-L.; Kai, J.-J.; Liaw, P.K. Cyclic-loading-induced accumulation of geometrically necessary dislocations near grain boundaries in an Ni-based superalloy. *JOM* **2009**, *61*, 53. [CrossRef]

31. An, K.; Chen, Y.; Stoica, A.D. VULCAN: A "hammer" for high-temperature materials research. *MRS Bull.* **2019**, *44*, 878–885. [CrossRef]

32. An, K.; Skorpenske, H.D.; Stoica, A.D.; Ma, D.; Wang, X.-L.; Cakmak, E. First In Situ Lattice Strains Measurements Under Load at VULCAN. *Metall. Mater. Trans. A* **2010**, *42*, 95–99. [CrossRef]

33. Huang, E.W.; Lee, S.Y.; Woo, W.; Lee, K.-W. Three-Orthogonal-Direction Stress Mapping around a Fatigue-Crack Tip Using Neutron Diffraction. *Metall. Mater. Trans. A* **2011**, *43*, 2785–2791. [CrossRef]

34. Lam, T.-N.; Wu, S.-C.; Chae, H.; Chen, S.-W.; Jain, J.; Lee, S.Y.; An, K.; Vogel, S.C.; Chiu, S.-M.; Yu, D.; et al. Phase Stress Partition in Gray Cast Iron Using In Situ Neutron Diffraction Measurements. *Metall. Mater. Trans. A* **2020**, *51*, 5029–5035. [CrossRef]

35. Toby, B.H.; Von Dreele, R.B. GSAS-II: The genesis of a modern open-source all purpose crystallography software package. *J. Appl. Crystallogr.* **2013**, *46*, 544–549. [CrossRef]

36. Morooka, S.; Tomota, Y.; Kamiyama, T. Heterogeneous Deformation Behavior Studied by in Situ Neutron Diffraction during Tensile Deformation for Ferrite, Martensite and Pearlite Steels. *ISIJ Int.* **2008**, *48*, 525–530. [CrossRef]

37. Cyr, E.; Lloyd, A.; Mohammadi, M. Tension-compression asymmetry of additively manufactured Maraging steel. *J. Manuf. Processes* **2018**, *35*, 289–294. [CrossRef]

38. Chen, W.; Voisin, T.; Zhang, Y.; Florien, J.B.; Spadaccini, C.M.; McDowell, D.L.; Zhu, T.; Wang, Y.M. Microscale residual stresses in additively manufactured stainless steel. *Nat. Commun.* **2019**, *10*, 4338. [CrossRef] [PubMed]

39. Wang, Y.; Tomota, Y.; Harjo, S.; Gong, W.; Ohmura, T. In-situ neutron diffraction during tension-compression cyclic deformation of a pearlite steel. *Mater. Sci. Eng. A* **2016**, *676*, 522–530. [CrossRef]

40. Young, M.L.; Almer, J.D.; Daymond, M.R.; Haeffner, D.R.; Dunand, D.C. Load partitioning between ferrite and cementite during elasto-plastic deformation of an ultrahigh-carbon steel. *Acta Mater.* **2007**, *55*, 1999–2011. [CrossRef]

41. Shen, C.J.L.H. 3D finite element analysis of particle-reinforced aluminum. *Mater. Sci. Eng. A* **2002**, *338*, 271–281. [CrossRef]

42. Dakhlaoui, R.; Baczmański, A.; Braham, C.; Wroński, S.; Wierzbanowski, K.; Oliver, E.C. Effect of residual stresses on individual phase mechanical properties of austeno-ferritic duplex stainless steel. *Acta Mater.* **2006**, *54*, 5027–5039. [CrossRef]

43. Jagle, E.A.; Sheng, Z.; Kurnsteiner, P.; Ocylok, S.; Weisheit, A.; Raabe, D. Comparison of Maraging Steel Micro- and Nanostructure Produced Conventionally and by Laser Additive Manufacturing. *Material* **2016**, *10*, 8. [CrossRef] [PubMed]

44. Asgari, H.; Mohammadi, M. Microstructure and mechanical properties of stainless steel CX manufactured by Direct Metal Laser Sintering. *Mater. Sci. Eng. A* **2018**, *709*, 82–89. [CrossRef]

45. Khairallah, S.A.; Anderson, A.T.; Rubenchik, A.; King, W.E. Laser powder-bed fusion additive manufacturing: Physics of complex melt flow and formation mechanisms of pores, spatter, and denudation zones. *Acta Mater.* **2016**, *108*, 36–45. [CrossRef]

46. Sallica-Leva, E.; Jardini, A.L.; Fogagnolo, J.B. Microstructure and mechanical behavior of porous Ti-6Al-4V parts obtained by selective laser melting. *J. Mech. Behav. Biomed. Mater.* **2013**, *26*, 98–108. [CrossRef]

47. de Terris, T.; Andreau, O.; Peyre, P.; Adamski, F.; Koutiri, I.; Gorny, C.; Dupuy, C. Optimization and comparison of porosity rate measurement methods of Selective Laser Melted metallic parts. *Addit. Manuf.* **2019**, *28*, 802–813. [CrossRef]

48. Gu, H.; Pal, D.; Rafi, K.; Starr, T.; Stucker, B. Influences of Energy Density on Porosity and Microstructure of Selective Laser Melted 17-4PH Stainless Steel. In Proceedings of the 24th Annual International Solid Freeform Fabrication Symposium, Austin, TX, USA, 12–14 August 2013.

49. Wits, W.W.; Carmignato, S.; Zanini, F.; Vaneker, T.H.J. Porosity testing methods for the quality assessment of selective laser melted parts. *CIRP Ann.* **2016**, *65*, 201–204. [CrossRef]

50. Xu, Y.; Zhang, S.H.; Cheng, M.; Song, H.W. In situ X-ray diffraction study of martensitic transformation in austenitic stainless steel during cyclic tensile loading and unloading. *Scr. Mater.* **2012**, *67*, 771–774. [CrossRef]

 *materials*

Article

Material Extrusion of Structural Polymer–Aluminum Joints—Examining Shear Strength, Wetting, Polymer Melt Rheology and Aging

Stephan Bechtel [1,*], Rouven Schweitzer [1], Maximilian Frey [2], Ralf Busch [2] and Hans-Georg Herrmann [1,3]

1   Chair for Lightweight Systems, Saarland University, Campus E3 1, 66123 Saarbrücken, Germany; rouven.schweitzer@gmx.de (R.S.); hans-georg.herrmann@izfp.fraunhofer.de (H.-G.H.)
2   Chair of Metallic Materials, Saarland University, Campus C6 3, 66123 Saarbrücken, Germany; maximilian.frey@uni-saarland.de (M.F.); r.busch@mx.uni-saarland.de (R.B.)
3   Fraunhofer Institute for Nondestructive Testing IZFP, Campus E3 1, 66123 Saarbrücken, Germany
*   Correspondence: stephan.bechtel@izfp-extern.fraunhofer.de

Abstract: Generating polymer–metal structures by means of additive manufacturing offers huge potential for customized, sustainable and lightweight solutions. However, challenges exist, primarily with regard to reliability and reproducibility of the additively generated joints. In this study, the polymers ABS, PETG and PLA, which are common in material extrusion, were joined to grit-blasted aluminum substrates. Temperature dependence of polymer melt rheology, wetting and tensile single-lap-shear strength were examined in order to obtain appropriate thermal processing conditions. Joints with high adhesive strength in the fresh state were aged for up to 100 days in two different moderate environments. For the given conditions, PETG was most suitable for generating structural joints. Contrary to PETG, ABS–aluminum joints in the fresh state as well as PLA–aluminum joints in the aged state did not meet the demands of a structural joint. For the considered polymers and processing conditions, this study implies that the suitability of a polymer and a thermal processing condition to form a polymer–aluminum joint by material extrusion can be evaluated based on the polymer's rheological properties. Moreover, wetting experiments improved estimation of the resulting tensile single-lap-shear strength.

Keywords: structural joints; aging; material extrusion; additive manufacturing; ABS; PETG; PLA; aluminum; polymer rheology; thermal joining

Citation: Bechtel, S.; Schweitzer, R.; Frey, M.; Busch, R.; Herrmann, H.-G. Material Extrusion of Structural Polymer–Aluminum Joints—Examining Shear Strength, Wetting, Polymer Melt Rheology and Aging. *Materials* **2022**, *15*, 3120. https://doi.org/10.3390/ma15093120

Academic Editor: Rubén Paz

Received: 23 March 2022
Accepted: 19 April 2022
Published: 26 April 2022

Publisher's Note: MDPI stays neutral with regard to jurisdictional claims in published maps and institutional affiliations.

1. Introduction

Additive manufacturing (AM) offers huge technical potential, especially with regard to sustainable and customized solutions [1,2]. One of the most common AM technologies is material extrusion (ME), where the extruded material (i.e., polymer) is dosed in a targeted way to generate a three-dimensional part layer-by-layer—cf. DIN EN ISO/ASTM 52900-2017. This process is also referred to as fused deposition modeling® (FDM®) or fused filament fabrication (FFF). Lightweight systems in particular often require joining multiple components or materials due to size limitations of the production process or locally varying demands concerning functionality and costs [3,4]. Recently, Frascio et al. [3] emphasized the need for suitable joining concepts with regard to additive manufactured adherends and adhesives. One approach, examined in this work, is to use the polymer processed by ME as (hotmelt) adhesive, resulting in a smooth transition between the joint and the subsequently generated polymer part. Challenges exist primarily with regard to reliability and reproducibility of the additively generated joints. For industrial applications, lack of knowledge about joint degradation and aging resistance represents the key issue regarding this technique.

From adhesive technology, wetting is known to be a crucial factor to buildup adhesive interactions between the substrate surface and the adhesive. Habenicht [5] distinguished between mechanical and specific adhesion. The former corresponds to a form fit on the microscale, while the latter results from chemical and physical adsorption of the macromolecule functional groups onto the adherent surface. Due to the absence of highly chemically reactive functional groups in the thermoplastics used for ME, only physical adsorption, e.g., Van der Waals forces, dipole–dipole-interactions or H-bonding, is relevant. Both mechanical and specific adhesion require the polymer melt to penetrate into the substrate's surface texture to establish a form fit or short-range interactions, respectively. Hence, wetting is an indispensable precondition for strong polymer–metal bonding [6]. In thermodynamic equilibrium, the wetting angle, φ, results from the interfacial tension between polymer and substrate, γ_{PS}, polymer and atmosphere, σ_{PA}, and substrate and atmosphere, σ_{SA}, according to Young's equation (cf. Equation (1)). The interfacial tension is defined as the mechanical work that needs to be applied to increase the interfacial area by 1 cm^2. Figure 1 shows a polymer drop on a metal surface where the interfacial tensions are in equilibrium in a vectorial manner. Adhesion is promoted by proper adherent surface preparation, which forms interlocking structures, increases the real surface and generates or exposes highly energetic surface functional groups.

$$\cos(\varphi) = \frac{\sigma_{SA} - \gamma_{PS}}{\sigma_{PA}} \tag{1}$$

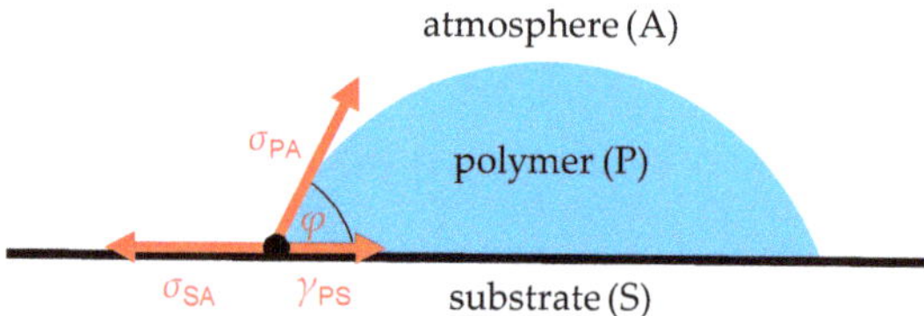

Figure 1. Polymer drop on substrate surface.

The work of adhesion, which is defined as the mechanical work required to separate two phases with a contact area of 1 cm^2, can be calculated according to the Young–Dupré equation (cf. Equation (2)). Although wetting is crucial to buildup the adhesive interactions, it is generally not possible to predict the strength of an adhesive joint solely based on the wetting conditions (Equation (2)) [5].

$$W_A = \sigma_{PA} + \sigma_{SA} - \gamma_{PS} = \sigma_{PA} \cdot [1 + \cos(\varphi)] \tag{2}$$

In general, clean metallic and oxidic surfaces possess a high surface tension, σ_{SA}, and are wetted properly by liquids with a low surface tension, such as polymers—σ_{PA} [7]. However, Equation (2) does not account for differences between the advancing and receding angle or for kinetically ruled processes. In thermal equilibrium, liquids with an elevated viscosity, such as polymer melts, wet the substrate gradually. The dynamics of wetting result in viscous friction in the drop and molecular adsorption and desorption processes at the contact line [8,9]. As a result, the contact angle decreases continuously until equilibrium is reached. Depending on the viscosity of the polymer melt, this process takes some 10 min (cf. e.g., [10]). Hence, wetting dynamics depend on polymer melt viscosity, which usually decreases with increasing shear rate and temperature [11]. Furthermore, the polymer melt surface tension, σ_{PA}, decreases with increasing temperature [12]. Hence, increasing the temperature promotes wetting in terms of thermodynamic equilibrium (cf. Equation (1)) and dynamics. Besides the polymer, the type of metal and its surface condition also affect wetting and joint strength due to differences in physical adsorption, interfacial tension (γ_{PS} and σ_{SA}), wetting dynamics and thermally induced internal stresses [5]. While the former two mainly depend on the substrate's surface functional groups, the latter two depend on the thermal conductivity and expansion coefficient, respectively.

Das et al. [13] emphasized the importance of polymer melt rheology in the context of ME. On the one hand, proper extrusion, wetting and interdiffusion between the adjacent traces is promoted by a low viscosity, η, and a high loss factor, $\tan(\delta)$. On the other hand, geometric accuracy and dimensional stability require high viscosity and a low loss factor. The loss factor, $\tan(\delta) = G''/G'$, equals the ratio of loss modulus, G'', and storage modulus, G', and specifies whether the polymer melt behaves dominant elastic ($\tan(\delta) < 1$) or dominant viscous ($\tan(\delta) > 1$). Based on the local thermal history and the viscoelastic polymer melt properties, some authors [14–16] derived a polymer healing degree or an effective weld time between adjacent layers in order to predict the interfacial strength. The underlying relaxation phenomena were correlated with, e.g., the reptation time t_d. This time corresponds to a long-range relaxation or diffusion, respectively, of a macromolecule and can be calculated according to Equation (3) based on the zero shear viscosity, η_0, and the plateau modulus, G_N^0 [17]. Another characteristic relaxation time is the rouse relaxation time, t_ro, which corresponds to short-range relaxations at the chain ends and equals the crossover time (cf. Equation (4)) [17].

$$t_\mathrm{d} = \frac{12 \cdot \eta_0}{\pi^2 \cdot G_\mathrm{N}^0} \tag{3}$$

$$t_\mathrm{ro} = t\big|_{G'=G'' \,\vee\, \tan(\delta)=1} \tag{4}$$

For hotmelts, adhesion interface performance is known to be highly sensitive to local temperature management during application. Due to their high thermal conductivity, Habenicht et al. [6] suggested preheating of metal substrates to the melting temperature, T_m, of the hotmelt prior to the joining process. Accordingly, Amancio-Filho and Falck [18] specified substrate temperatures in between the extrusion temperature, T_e, and the crystallization temperature, T_c, of the polymer processed by ME to optimize polymer–metal bonding. However, in their recent publications regarding the Addjoining$^\circledR$ process, Falck et al. [19,20] applied a primer onto the aluminum surface before the actual ME joining process was carried out at significantly lower substrate temperatures. Using carbon fiber reinforced polyamide 6 (CF-PA6) instead of acrylonitrile–butadiene–styrene copolymer (ABS) resulted in a higher tensile single-lap-shear strength. Chueh et al. [21] generated poly(ethylene terephthalate) (PET)–steel joints by ME of the polymer on a preheated (180 °C) structured metal substrate with undercuts, which was prepared by selective laser melting (SLM). After filling the surface structures with polymer, the ME process was paused to consolidate the polymer–metal interface by pressure and laser stimulation. By increasing substrate and extrusion temperature, Hertle et al. [22] increased the lap shear strength of ME-joined polypropylene (PP)–aluminum samples. The increased contact temperature resulted in improved filling of the microstructures of the electrochemically treated aluminum surface. Dröder et al. [23] joined ABS to surface structured aluminum substrates. Higher surface structures and substrate temperatures resulted in increased tensile single-lap-shear strength. In our previous work [24], we showed the ability of thermographic process monitoring to characterize the ME joining process in terms of wetting and joint strength, as exemplified by poly(lactic acid) (PLA)–aluminum joints. While Herlte et al. [22], Dröder et al. [23] and Bechtel et al. [24] observed a large influence due to substrate temperature, T_s, Falck et al. [19] reported only a minor effect. As Falck et al. applied a primer beforehand, the polymer was not deposited directly on the metal surface during the ME process. Due to the significantly lower thermal conductivity of the primer, substrate temperature dependence was reduced (cf., e.g., [5]). The presented studies all dealt with polymer–metal joints generated by ME; however, in terms of structural joining and industrial application of the joining concept, the following questions arise:

- *Which of the common thermoplastics for ME is most suitable to generate structural polymer–metal joints?* Direct comparison between the studies is not feasible, in particular due to different processing and substrate surface conditions.

- *Can structural polymer–metal joints be generated by ME on "simple" practical relevant metal surfaces (e.g., prepared by grid blasting)?* In the relevant studies, either a primer was applied beforehand or complex surface preparation methods were used. Moreover, none of the studies considers joint degradation.

This article focuses on the generation of structural polymer–aluminum joints by means of ME. Structural joints are characterized by high joint strength (shear strength greater than 7 MPa) and significant aging resistance [7]. Moreover, we examined how joint strength correlates to wetting and polymer (melt) properties for several thermoplastics in order to obtain appropriate thermal processing conditions. Hence, the originality of this work lies in the characterization and evaluation of ME-generated polymer–metal joints in terms of joint strength and aging resistance by taking into account wetting and polymer melt properties.

2. Materials and Methods

The hereafter described approach is a continuation of our previous work [24].

2.1. Aluminum Substrates

The aluminum substrates were received from the water-jet cutting company RS-Evolution (Saarwellingen, Germany). The deburred substrates were prepared to measure 25 mm by 115 mm from 2 mm thick sheet metal of EN AW-6082-T6. This medium-strength aluminum alloy has excellent corrosion resistance and is typically used for structural parts in, e.g., the transportation sector [25]. The substrates were grit blasted with corundum (Al_2O_3), size F150, which was received from Oberflächentechnik Seelmann (Dessau-Roßlau, Germany). Particle size was about 82 µm and laid within standardized range (45–106 µm)—cf. DIN EN 13887-2003. Sandblasting was performed with a ST 800-J (Auer Strahltechnik, Mannheim, Germany) at a pressure of 6 bar, a working distance of 10 cm and an angle of 90° to the surface.

2.2. Material Extrusion (ME)

The polymers were processed by a customized ME machine based on a desktop FFF platform (Ender 3, Creality 3D, Shenzhen, China). The extruder was equipped with a water-cooled heatsink, a "volcano"-type hotend and a brass nozzle with an inner diameter, w_{Po}, of 0.8 mm (all purchased from E3D Online, Oxfordshire, UK). Three thermoplastic polymers, common for ME, were considered:

- Acrylonitrile–butadiene–styrene copolymer (**ABS**), Extrafill™ (Fillamentum, 768 24 Hulín, Czech Republic), yellow-colored filament
- Poly(ethylene terephthalate) glycol comonomer (**PETG**), PolyLite™ (Polymaker, Shanghai, China), transparent filament
- Poly(lactic acid) (**PLA**), Ingeo™ 3D870 (Nature Works, Minnetonka, MN, USA), black-colored filament

ABS, PETG and PLA were extruded at a temperature, T_e, of 240 °C, 220 °C and 200 °C, respectively. Extruder temperature T_e was chosen based on polymer melt viscosity. During the deposition of the $d_{Po} = 0.3$ mm high layers, the extruder moved at a velocity, v_e, of 10 mm/s. The polymer-specific substrate temperatures, T_{s0}, of 100 °C, 80 °C, and 60 °C for ABS, PETG and PLA, respectively, were chosen based on the recommended build-plate temperatures in the datasheets [26–28]. In the joining and wetting experiments, the substrate temperature, T_s, was varied. If the substrate temperature at a certain layer i, $T_{s,i}$, differed from the polymer-specific substrate temperature, T_{s0}, it is specified in the corresponding section. Slicing was done with Ultimaker Cura 4.0 (Geldermalsen, The Netherlands).

2.3. Rheometry

Rheometry was performed with an Ar2000ex (TA Instruments, New Castle, DE, USA) rheometer using plate–plate configuration. A 1 mm thick polymer disc with a diameter of 25 mm was generated by ME and placed between the plates of the rheometer. After preheating to 200 °C, the disc was carefully compressed by the rheometer plates to prevent the polymer from squeezing out. Using deformation-controlled oscillation rheology, the deformation amplitude was set to 5%. After equalization for 2 min at the set measurement temperature, oscillation sweeps were carried out in the range from $\omega = 2\pi \cdot 10$ Hz down to $\omega = 2\pi \cdot 0.01$ Hz. In the first measurement cycle, the set temperature ranged from 200 °C to 130 °C for all polymers. In the second cycle, the maximum temperature was set to 250 °C, 240 °C and 220 °C for ABS, PETG and PLA, respectively, while the minimum temperature was kept unchanged at 130 °C. The third cycle was identical to the first cycle.

Rheometry was used to access the thermo–rheological properties of the polymer melts, which are crucial for wetting, cohesive properties [29–31] and adhesion interface performance [6]. In total, 5 specimens were tested for each polymer.

Providing that all relevant relaxation phenomena show the same temperature dependence, relaxation times, t_i, and frequency data, ω, were shifted with the horizontal shift factor a_T (cf. Equation (5)). Based on this time–temperature superposition [17], single master curves for modulus, $G^* = G' + iG''$, and viscosity, $\eta^* = \eta' + i\eta''$, at reference temperature T_0 were constructed according to Equation (6). The vertical shift factors, $b_T = T_0/T$, solely depend on measurement temperature, T, and reference temperature, T_0, while the horizontal shift factors, a_T, were fit to the Williams–Landel–Ferry equation with the WLF-constants C_1 and C_2 (cf. Equation (7)).

$$t_i(T_0) = \frac{t_i(T)}{a_T(T, T_0)}, \quad \omega(T_0) = a_T(T, T_0) \cdot \omega(T) \tag{5}$$

$$G^*(T_0, a_T\omega) = b_T(T, T_0) \cdot G^*(T, \omega), \quad \eta^*(T_0, a_T\omega) = \frac{b_T(T, T_0)}{a_T(T, T_0)} \cdot \eta^*(T, \omega) \tag{6}$$

$$\log[a_T(T, T_0)] = \frac{-C_1 \cdot (T - T_0)}{C_2 + (T - T_0)} \tag{7}$$

2.4. Wetting

To evaluate wetting behavior, an apparatus for contact angle measurement was added to the ME machine. This apparatus consists of illumination (7W LED spot with diffusor) and a camera (BFS-U3-04S2M (FLIR, Wilsonville, OR, USA), lens: LM50JC (Kowa, Nagoya, Japan)) triggered by the ME machine via specific commands in the G-Code (cf. Figure 2). After preheating the aluminum substrate to the specified first-layer substrate temperature, $T_{s,1}$, the primed extrusion nozzle was placed 1 mm above and 2 mm behind the front edge of the substrate and within the field-of-view of the camera. A polymer melt drop with a volume of about 5 µL (referring to the density at room temperature) was extruded before the extrusion nozzle was moved out of the camera's field-of-view (standby position). Dynamic wetting was recorded for up to 120 min at several substrate temperatures, $T_{s,1}$. The highest considered $T_{s,1}$ was equal to the polymer specific extrusion temperature T_e. $T_{s,1}$ was lowered in increments of 20 °C as long as proper deposition of the polymer melt drop could be achieved.

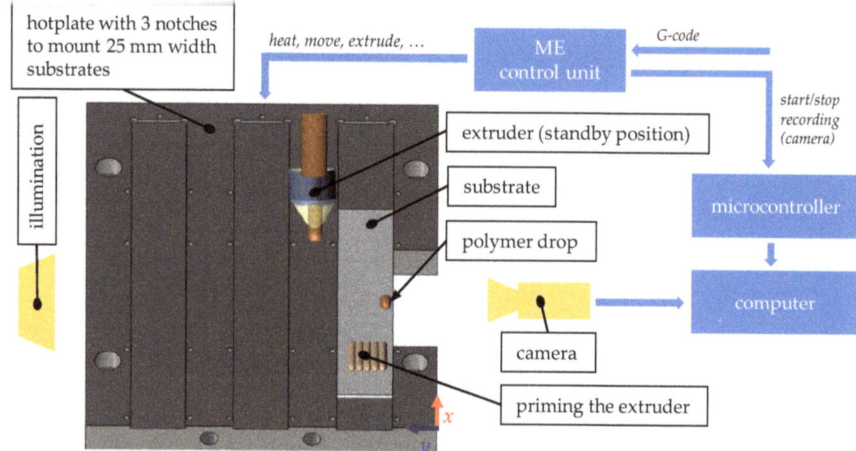

Figure 2. Description of the wetting experiment.

Contact angle analysis was done using MATLAB (MathWorks, Natick, MA, USA), as shown in Figure 3 based on the procedure presented by Andersen et al. [32,33]. After image segmentation, edge detection, drop separation, fitting (to the left and right of the apex, 4th-order polynomial) and baseline determination (non-wetted substrate surface on the left and right of the drop), the contact angle was determined at the triple point. This apparent macroscopic contact angle was determined by neglecting surface heterogeneity and roughness. Contact angle measurement was done 3 times for each combination of polymer and substrate temperature.

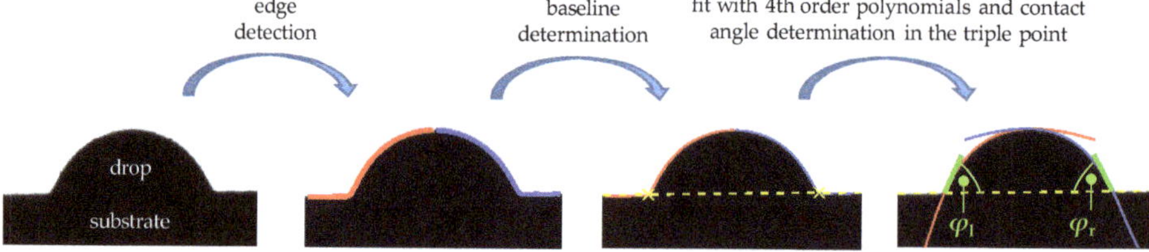

Figure 3. Determination of the contact angle, φ, as mean of the left, φ_l, and right, φ_r, contact angle.

2.5. Mechanical Performance

Tensile tests were carried out using a Kappa 100 DS (ZwickRoell, Ulm, Germany) equipped with a 100 kN loadcell at a controlled ambient temperature of 23 °C. In addition to adhesion interface performance, bulk properties of polymer samples prepared by ME were acquired. All tensile tests were driven-displacement controlled, with 1 mm/min crosshead speed.

Polymer bulk mechanical properties were obtained according to DIN EN ISO 527-2:2012 (type 1A). The extruded tracks were oriented parallel to the loading direction. The polymer was extruded on polyimide-taped aluminum substrates. Due to local excessive tension in the tapered regions [34], these regions were reinforced with an epoxy adhesive (Loctite EA3430, Henkel, Düsseldorf, Germany) to prevent failure outside the measurement area (parallel part). The strain measurement within the parallel part was not affected by the epoxy reinforcement. In total, 6 dog bone tensile test specimens were tested for each polymer.

The adhesion interface performance in polymer–aluminum assemblies was evaluated based on ISO 19095 (type B, without specimen retainer). Deviating from the standard, the joint width of the single-lap joints (SLJ) was increased from 10 mm to 20 mm to improve handling. The clamping length was set to 20 mm for the polymer part and 45 mm for the aluminum part to achieve comparable bending stiffness of both adherents. The dimensions and the ME buildup strategy of the polymer part are shown in Figure 4. The tensile single-lap-shear strength, $\tau_{SLJ} = F_B/A_J$, was calculated based on the breaking load, F_B, and the joint area, A_J. Failure patterns are classified as polymer part failure (PF, outside the joining area), cohesive failure (CF), adhesive failure (AF) and mixed adhesive and cohesive failure (ACF). The eccentric load path within the joint led to rotation during loading and caused, among other things, additional peel stresses (cf. e.g., [6]). These local stress concentrations caused early joint failure.

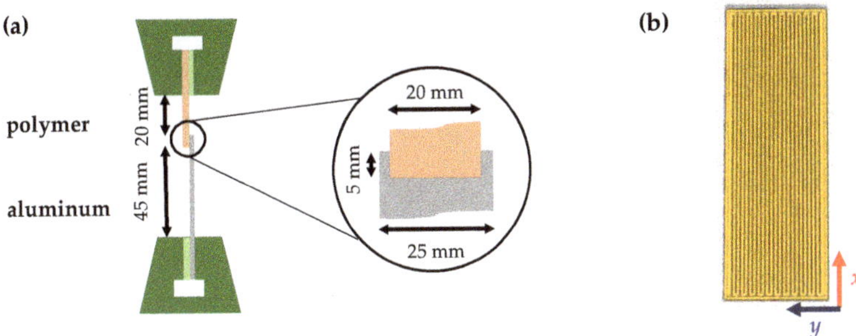

Figure 4. (a) Schematic test setup—tensile single lap shear strength; (b) ME buildup strategy of the polymer part.

The polymer–aluminum joints were prepared by fixing the aluminum substrates on an aluminum hot plate (cf. Figure 2) while extruding the polymer layer-wise on the substrate surface. The part of the polymer adherent beyond the overlap was supported by an aluminum spacer, partially taped with polyimide tape, which was removed after the ME process. The thermal joining process was evaluated by means of the first-layer substrate temperature, $T_{s,1}$. For higher layers, the substrate temperature was gradually decreased down to the polymer-specific substrate temperature, T_{s0}. Table 1 shows the substrate temperature increments for the explored temperature settings. As with wetting, the highest considered $T_{s,1}$ was equal to the polymer-specific extrusion temperature T_e. $T_{s,1}$ was lowered in increments of 20 °C until the adhesive strength of the joint was too low for handling.

Table 1. Substrate temperature at each layer i, $T_{s,i}$, during ME of the polymer part of the SLJ for the explored temperature settings.

Temperature Setting	$T_{s,1}/(°C)$	$T_{s,2}/(°C)$	$T_{s,3}/(°C)$	$T_{s,i}/(°C), i > 4$
1	240	150	100	$T_{s0}/(°C)$
2	220	150	100	$T_{s0}/(°C)$
3	200	150	100	$T_{s0}/(°C)$
4	180	150	100	$T_{s0}/(°C)$
5	160	150	100	$T_{s0}/(°C)$
6	140	100	$T_{s0}/(°C)$	
7	120	100	$T_{s0}/(°C)$	

Figure 5 shows the substrate and extruder temperature profiles during the production of an SLJ specimen in the case of a substrate temperature, $T_{s,1}$, of 200 °C and an extruder temperature, T_e, of 220 °C. Sample preparation was carried out on three identical substrates per sequence. In total, 6 specimens were tested for each set of parameters. The fresh SLJ were tested within 5 h after sample production to reduce aging effects.

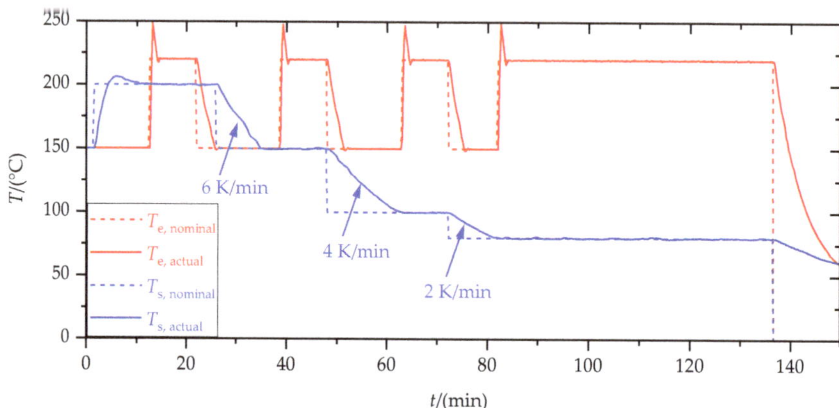

Figure 5. Temperature time profile of the nominal and actual temperature of the extruder, T_e, and the substrate, T_s, during production of an SLJ specimen. Cooling rates represent averaged values.

2.6. Aging

Two different moderate environments, which represent real operating conditions, were considered, and are referred to as aging and storing:

- Aging (moist–warm conditions): Darkness, 40 °C and humid air (75% r. h., setup above saturated NaCl solution [35])
- Storing (dry conditions): Darkness, 23 °C and dry air (<8% r. h., setup above silica gel desiccant)

Storing was considered as a reference, in order to separate the effect of internal thermal stresses from hygro–thermo–oxidative aging effects.

SLJs with the highest tensile single-lap-shear strength, τ_{SLJ}, in the fresh state were aged and stored for up to 100 days. Before mechanical testing, the specimens were acclimatized for at least 1 h at the test climate. Dog bone tensile test specimens were also aged for 100 days to achieve the mechanical polymer bulk properties in the aged state as a reference. For each aging time, 6 samples were tested.

3. Results and Discussion

3.1. Material Properties

Table 2 gives a summary of selected properties of the aluminum (Al) substrates and polymers used.

Table 2. Overview of selected thermal and mechanical properties of the materials used. The melting, crystallization and glass transition temperatures (onset) of the polymers were measured with a DSC3 (Mettler-Toledo, Columbus, OH, USA) at cooling and heating rates of 10 K/min. The thermal expansion coefficients α were not obtained for the polymers used; instead, typical ranges for ABS, PETG and PLA, respectively, are given. (*: for Al, the liquidus and solidus temperatures are given).

Property		Al	ABS	PETG	PLA
thermal expansion coefficient	$\alpha/(10^{-6}/K)$	23 [25]	108–234 [36]	120–123 [36]	126–145 [36]
density	$\rho/(g/cm^3)$	2.7 [25]	1.04 [26]	1.25 [27]	1.22 [28]
melting temperature	$T_m/(°C)$	650 * [25]	amorphous	amorphous	169 ± 0
crystallization temperature	$T_c/(°C)$	575 * [25]	amorphous	amorphous	119 ± 0
glass transition temperature	$T_g/(°C)$	–	99 ± 1	75.5 ± 0	58 ± 2

Table 2. *Cont.*

Property		Al	ABS	PETG	PLA
elastic modulus	$E/(\text{GPa})$	70 [25]	2.3 ± 0.1	2.1 ± 0.1	2.9 ± 0.1
yield strength	$\sigma_\text{y}/(\text{MPa})$	280 [25]	30 ± 1	47 ± 1	48 ± 1
elongation at yield	$\varepsilon_\text{y}/(\%)$		1.8 ± 0.2	4.1 ± 0.1	2.5 ± 0.2
surface roughness	$\text{Ra}/(\mu\text{m})$				
- blank		0.18 ± 0.02 [24]			
- sandblasted (FEPA 150)		1.9 ± 0.5 [24]			

Due to the viscoelastic nature of the polymers, their properties were a function of heating, strain rate and temperature. For the technical polymer components investigated, material properties depended, in addition to molecular weight, on additives, such as plasticizers (e.g., [37]). Moreover, the mechanical properties were particularly sensitive to process conditions and buildup strategy (e.g., [38]). While the investigated ABS and PETG were amorphous (no crystallization and melting event), PLA showed crystallization at a cooling rate of 10 K/min. At room temperature, all polymers were in the glassy state, while ABS ($T_\text{g} = 99\,^\circ\text{C}$) showed the highest and PLA ($T_\text{g} = 58\,^\circ\text{C}$) the lowest glass transition temperature, T_g. The thermal expansion coefficient, α, of the polymers was about an order of magnitude higher than that of aluminum. Hence, the thermal joining process resulted in thermally induced interfacial stresses. Internal stress buildup occurred below the glass transition temperature, T_g, in the amorphous phase and below the crystallization temperature, T_c, in the crystalline phase. Consequently, thermally induced interfacial stresses were particularly relevant for PLA and ABS due to crystallization and high glass transition temperature, T_g, respectively. With regard to the sensitivity to the polymer formulation, the process and buildup parameters, and the strain rate, the observed values for elastic modulus, E, yield strength, σ_y, and elongation at yield, ε_y, lay in the data range reported in the literature (e.g., [38–42]). The elongation at failure could not be determined reliably, as fracture behavior showed a high variation depending on failure position (parallel part or tapered regions). Moreover, if failure occurred in the tapered regions, strain measurement was inaccurate after the yield point, as failure proceeded outside the measurement marks of the video extensiometry.

Polymer Melt Rheology

Viscosity depended on shear rate, $\dot{\gamma}$, and temperature, T. In order to determine an appropriate and, in terms of processing viscosity, comparable extrusion temperature for all polymers, the shear rate, $\dot{\gamma}$, of the polymer melt during the extrusion process was obtained by Equation (8) as a function of extruder velocity, v_e, layer height, d_Po, and nozzle diameter, w_Po, as $\dot{\gamma} = 48\ 1/\text{s}$ [43].

$$\dot{\gamma} = \frac{32}{\pi} \cdot \frac{v_\text{e} \cdot d_\text{Po}}{w_\text{Po}^2} \tag{8}$$

Based on the Cox–Merz rule, $|\eta^*(\omega)| = \eta(\dot{\gamma} = \omega)$, which applies to the polymers used [15,44,45], the steady shear viscosity could be approximated based on the oscillatory measurements. For an extrusion temperature T_e of 240 °C, 220 °C and 200 °C, respectively, ABS, PETG and PLA showed comparable viscosity at $\omega = 48$ Hz (cf. Figure 6), which lies in a suitable range for the application of hotmelts and extrusion (i.e., 1×10^0 to 1×10^4 Pa s [5,46]).

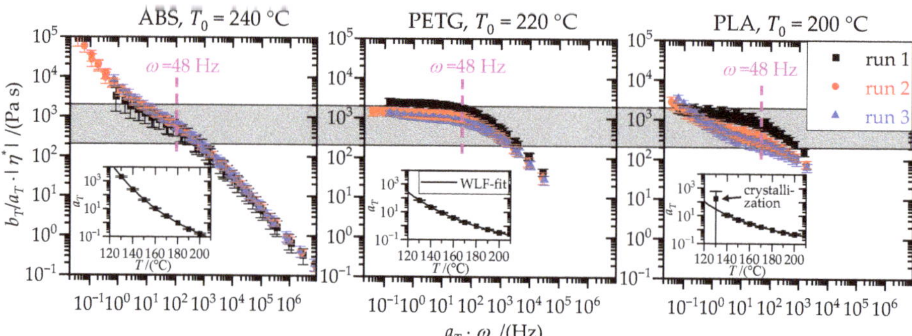

Figure 6. Frequency dependence of the complex viscosity, η^*, at a reference temperature T_0 of 240 °C (ABS), 220 °C (PETG) and 200 °C (PLA), respectively, for three subsequent measurement runs. The target viscosity range (gray area) and the shear rate (48 1/s) during the ME process are indicated. Horizontal shift factors, a_T, with corresponding WLF-fit are shown for the first run.

Comparing the three polymers in terms of modulus, G' and G'', or loss factor, $\tan(\delta)$, revealed pronounced differences, particularly in terms of the loss factor (cf. Figure 7). Especially for low frequencies, the loss factor, $\tan(\delta)$, and thus the viscous character of the polymer melt, was significantly higher for PETG than for PLA and ABS.

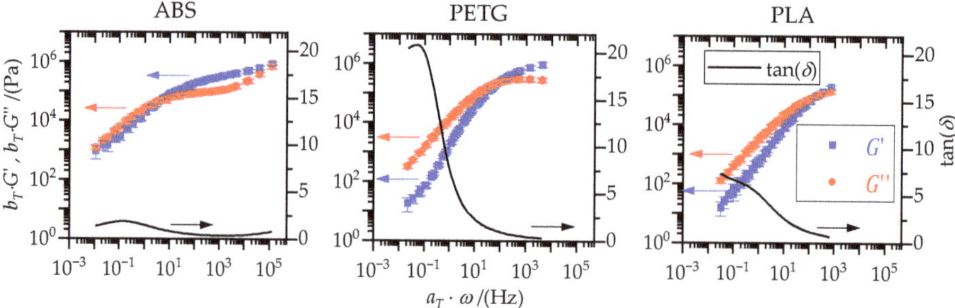

Figure 7. Frequency dependence of storage modulus, G', loss modulus, G'', and loss factor, $\tan(\delta)$, at a reference temperature T_0 of 180 °C for the first run. The loss factor was determined based on smoothed master curves of the storage and loss modulus.

For ABS as well as for PLA in the second and third run, no zero shear viscosity, η_0, could be determined, as there was a continuous increase in viscosity for small frequencies, ω (cf. Figure 6). The plateau modulus, G_N^0, could not be obtained for PLA because there was not a minimum in G' nor a maximum in G'' in the available frequency range (cf. Figure 7). Therefore, the reptation time, t_d, which is calculated based on zero shear viscosity, η_0, and plateau modulus, G_N^0, was not ascertainable for all polymers. Hence, only the rouse relaxation time, t_{ro}, which could be determined for all polymers, was considered. To account for the differences in viscous character of the polymer melts, the maximum of the loss factor, $\tan(\delta)\,|_{max}$, in the available frequency range, ω, was considered. The maximum of the loss factor, $\tan(\delta)\,|_{max}$ and the rouse relaxation time, t_{ro}, are listed in Table 3.

Table 3. Maximum of the loss factor, $\tan(\delta)\,|_{\max}$, and rouse relaxation time, t_{ro}, at a reference temperature T_0 of 180 °C for three subsequent measurement runs.

Property	ABS, Run 1	ABS, Run 2	ABS, Run 3	PETG, Run 1	PETG, Run 2	PETG, Run 3	PLA, Run 1	PLA, Run 2	PLA, Run 3
$\tan(\delta)\,|_{\max}$	1.9 ±0.1	1.3 ±0.0	1.3 ±0.0	23 ±5	25 ±6	14 ±3	8.0 ±1.8	3.3 ±0.3	2.8 ±0.2
$t_{\mathrm{ro}}\,(T_0)/(\mathrm{ms})$	1500 ±100	2020 ±180	2440 ±190	80 ±5	48 ±5	47 ±6	33 ±3	15 ±5	13 ±4

All polymers showed degradation phenomena, resulting in deviations between the subsequent measurement runs. PETG und PLA contain ester groups in their backbones that make them vulnerable to thermally activated, hydrolytic chain scission [47]. Shortening of the macromolecules leads to a reduced relaxation time, t_{ro}, and viscosity, η, (cf. Figure 6 and Table 3). The hydrolysis products are new chain ends with carboxyl-groups [47]. In ABS, the acrylnitril–styrol phase is more stable than the butadiene phase. Thermo–oxidative processes mainly take place at the butadiene double bonds [47]. However, the result is not a chain scission. Instead, subsequent reactions lead to functional groups and cross-linking, which explains the increase in relaxation time, t_{ro}, and viscosity, η, (cf. Figure 6 and Table 3).

Due to heat-induced polymer degradation, the first measurement run is most relevant. However, in the thermal joining and wetting process, the polymers were exposed to similar conditions by means of temperature and atmospheric contact. Hence, the observed degradation phenomena may also be decisive in these processes, in particular with regard to the following aspects:

- ABS and degraded PLA (second and third run) behaved like viscoplastic fluids. These types of fluids do not converge to a zero shear viscosity, η_0, for low shear rates. Instead, the viscosity continuously increases, as the fluids have a yield stress [46]. This is of great relevance for the thermal wetting and joining processes since there is no external force (except gravity) acting on the polymer (melt) after it leaves the extrusion nozzle.
- The carboxyl-(end-) groups resulting from hydrolysis of PETG and PLA can form strong physical bonds (H-bonding) to the metal substrate [5].

3.2. Wetting

The temporal change in contact angle, $\varphi(t)$, depended on substrate temperature, $T_{\mathrm{s},1}$, and polymer (cf. Figure 8). Equilibration took between a few minutes and some hours. According to Table 4, the equilibrium contact angles, φ_{eq}, lay in the range from 15 to 140°, representing optimal ($\varphi_{\mathrm{eq}} < 30°$) to insufficient ($\varphi_{\mathrm{eq}} > 90°$) wetting conditions [5].

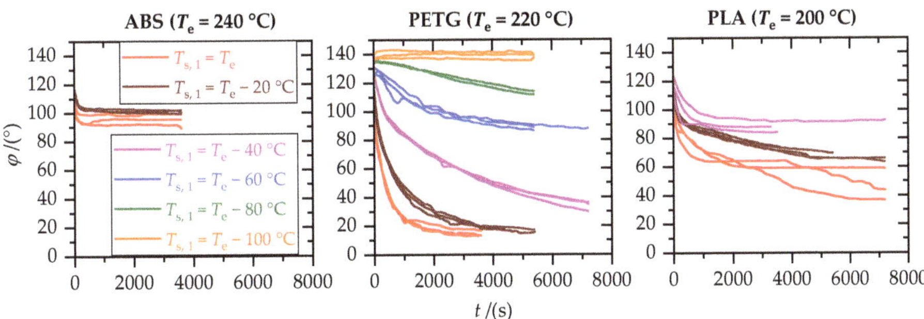

Figure 8. Contact angle, φ, between polymer drop and aluminum substrate as a function of wetting time, t, for several substrate temperatures, $T_{\mathrm{s},1}$.

Table 4. Equilibrium contact angles, φ_{eq}, for the explored temperature settings in terms of substrate temperature, $T_{s,1}$. (*: Equilibrium was not reached in the wetting experiment. Therefore, φ_{eq} was estimated as the last observed contact angle (cf. Figure 8)).

$T_{s,1}/(°C)$	$\varphi_{eq}/(°)$, ABS	$\varphi_{eq}/(°)$, PETG	$\varphi_{eq}/(°)$, PLA
240	95 ± 4	-	-
220	101 ± 1	14 ± 2	-
200	-	16 ± 1	46 ± 8
180	-	34 ± 3 *	67 ± 3
160	-	87 ± 1	88 ± 3
140	-	113 ± 1 *	
120	-	139 ± 1	

For all three polymers, wetting improved with increasing substrate temperature, $T_{s,1}$, which is consistent with temperature dependence of surface tension, σ_{PA} [12], and viscosity, η (cf. Figure 6). However, there were significant differences between the polymers. PETG showed good wetting, even for substrate temperatures, $T_{s,1}$, 40 °C below the extrusion temperature, T_e. Contrary, wetting through ABS was already poor for a substrate temperature, $T_{s,1}$, equal to the extrusion temperature, T_e. The polymer-specific wetting behavior is attributed to differences in surface tension, σ_{PA}, and polymer melt rheology. While the surface tension of the polymer melts and their temperature dependence was not in the scope of this work, the influence of polymer melt rheology on wetting can be discussed. The viscous character of the polymer melts, by means of $\tan(\delta)|_{max}$, was highest for PETG and lowest for ABS (cf. Table 3). Additionally, ABS and degraded PLA behaved like viscoplastic fluids (cf. Figure 6). Since there was no external force acting on the polymer (melt) drop, the yield stress required to start flow may not have been reached. Both the viscoplastic character and the low loss factor, $\tan(\delta)|_{max}$, were adverse for wetting.

3.3. Adhesion Interface Performance

3.3.1. Influence of Thermal Processing

Wetting of the aluminum substrates through the polymer melts depended significantly on wetting time and took up to several hours to reach equilibrium. Therefore, first, the influence of wetting time on tensile single-lap-shear strength, τ_{SLJ}, was investigated, exemplarily for PETG and $T_{s,1}$ = 200 °C (cf. Figure 9). Increasing the wetting time by 1 h decreased τ_{SLJ} by about 10 MPa. Hence, contrary to wetting, tensile single-lap-shear strength did not increase with wetting time. Considering the failure patterns reveals that ongoing degradation of the polymer weakened the mechanical bulk properties, leading to polymer part failure (PF). Hence, polymer–metal joints with no additional wetting time were considered.

Figure 10 shows the tensile single-lap-shear strength, τ_{SLJ}, as a function of substrate temperature, $T_{s,1}$. In conjunction with the improved wetting (cf. Table 4), τ_{SLJ} increased with increasing substrate temperature. Differences between the polymers were also consistent with wetting. ABS, which wetted the substrate insufficiently ($\varphi_{eq} > 90°$) at all substrate temperatures, showed the lowest tensile single-lap-shear strength. Contrary, the well-wetting PETG had the highest τ_{SLJ}. With increasing substrate temperature and tensile single-lap-shear strength, the failure pattern changed from adhesive (AF) over than mixed (ACF) and cohesive (CF) to polymer part failure (PF). For PETG, the shear strength reached a plateau for $T_{s,1}$ above 180 °C. This is in accordance with the equilibrium contact angle, φ_{eq}, which decreased only slightly between 180 and 220 °C (cf. Table 4). Moreover, for $T_{s,1}$ greater than 180 °C, cohesive failure (CF) dominated, which means the polymer–metal interface was no longer the weak point. For $T_{s,1}$ = 220 °C, there was even a small decrease in τ_{SLJ}, accompanied by an increase in polymer part failure (PF). Similar to the effect of the additional wetting time (cf. Figure 9), this decrease in τ_{SLJ} for the highest substrate temperatures is attributed to degradation of the mechanical polymer bulk properties.

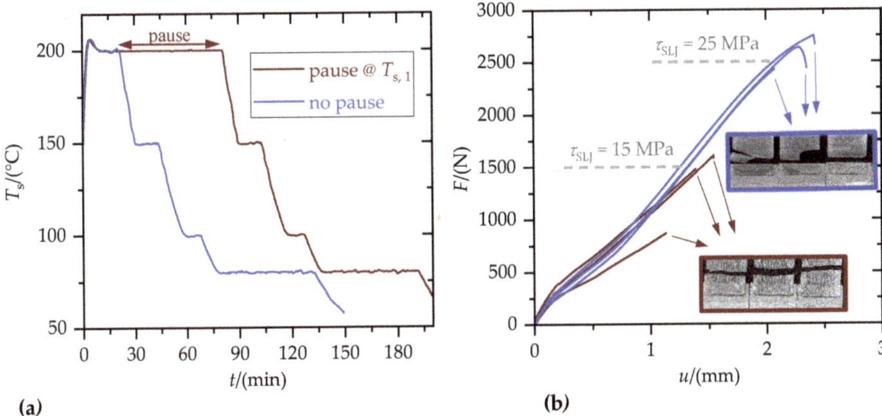

Figure 9. Influence of additional wetting time (1 h pause @ $T_{s,1}$) on tensile single-lap-shear strength, τ_{SLJ}. (**a**) Temperature–time profiles of the actual temperature of the substrate, T_s, during ME of the SLJ; (**b**) Corresponding force–displacement diagrams, $F(u)$, and SLJ fracture surfaces.

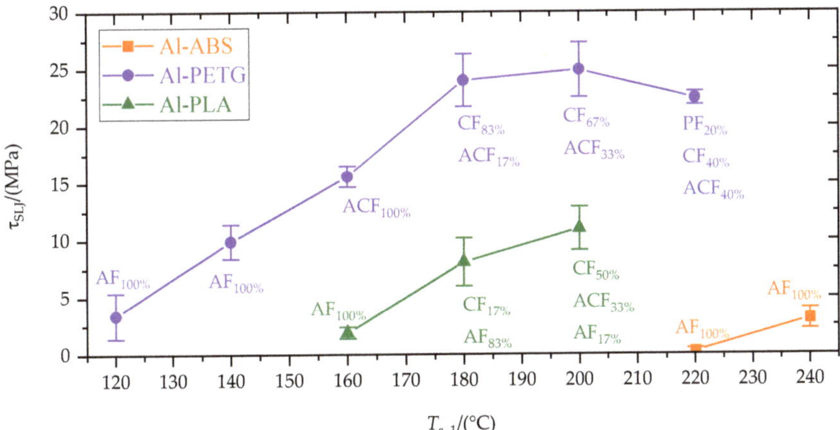

Figure 10. Tensile single-lap-shear strength, τ_{SLJ}, as a function of substrate temperature, $T_{s,1}$. Percentages of the failure patterns: polymer part failure (PF, outside the joining area), cohesive failure (CF), adhesive failure (AF) and mixed adhesive and cohesive failure (ACF) are given.

For PETG, tensile single-lap-shear strength, τ_{SLJ}, decreased for the highest substrate temperatures and long wetting times. Hence, in the thermal ME joining process, two opposing effects in terms of tensile single-lap-shear strength took place. On the one side, wetting improved with increasing substrate temperature and wetting time, but, on the other hand, degradation at high temperatures and long times weakened polymer bulk properties. Wetting time in the ME joining process was just a few minutes (cf. Figure 5) and, hence, significantly shorter than the time required to reach the equilibrium contact angle (cf. Figure 8). However, there was sufficient adhesion formation to reach high tensile single-lap-shear strength and cohesive failure (CF) in the case of PETG and high substrate temperatures ($T_{s,1} > 180\,°C$). Consequently, arrangement and orientation of the macromolecules in the wetted area, which is required to form adhesive interactions (i.e., mechanical adhesion and physical adsorption), proceeded much faster than the macroscopic wetting.

3.3.2. Aging Resistance

First, the effect of aging on the mechanical polymer bulk properties is presented. Figure 11 shows stress–strain diagrams of the polymers in the fresh and aged (100 days) state.

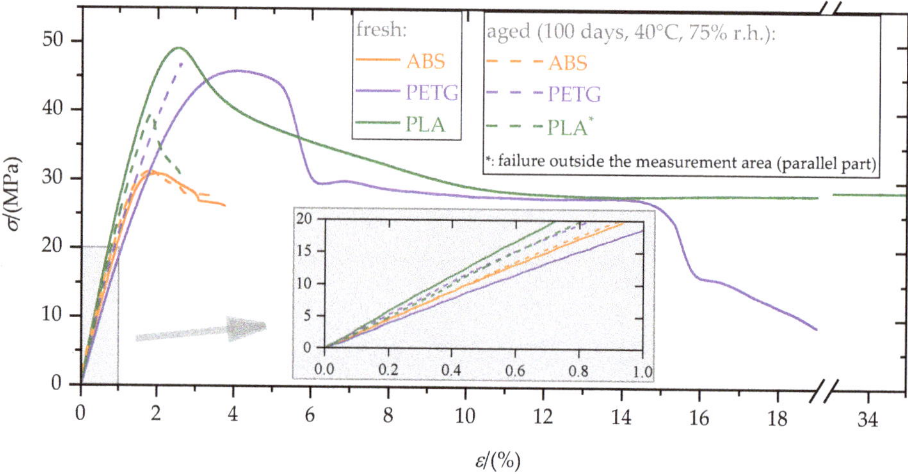

Figure 11. Exemplary stress–strain diagrams, $\sigma(\varepsilon)$, of the polymers in the fresh and aged state. Graphs in the enlarged section are shifted to (0/0).

Elastic modulus, E, yield strength, σ_y, and elongation at yield, ε_y, are given in Table 5. Aging effects were much more pronounced for PETG and PLA than for ABS. In particular, the yield strength of PLA as well as the elongation at yield of PLA and PETG decreased significantly due to aging. Moreover, while PLA and PETG showed predominantly ductile failure (elongation at failure, $\varepsilon_f > 10\%$) in the fresh state, they broke brittle ($\varepsilon_f < 10\%$) in the aged state. Hence, PLA and PETG became brittle due to aging, which made them more sensitive to notches and local stress concentrations. Reasons for the embrittlement could include the extraction of plasticizers, swelling and hydrolysis (ester groups) [47]. The observed elongation at failure for PLA in the fresh state was higher than usually reported in the literature (e.g., [38]). One explanation for this particularly high elongation at failure is strain crystallization [48]. Due to the low strain rate and material heating caused by plastic deformation, the mechanical glass transition, which depends on strain rate, could be reached, and hence, strain crystallization occurred.

Table 5. Elastic modulus, E, yield strength, σ_y, and elongation at yield, ε_y, of the polymers in the fresh and aged (100 days) state.

Property	ABS, Fresh	ABS, Aged	PETG, Fresh	PETG, Aged	PLA, Fresh	PLA, Aged
$E/(\mathrm{MPa})$	2260 ± 50	2320 ± 40	2100 ± 80	2210 ± 80	2880 ± 70	2640 ± 120
$\varepsilon_y/(\%)$	1.8 ± 0.2	1.7 ± 0.1	4.1 ± 0.1	2.5 ± 0.3	2.5 ± 0.2	1.8 ± 0.0
$\sigma_y/(\mathrm{MPa})$	29.6 ± 1.3	30.3 ± 1.0	46.9 ± 0.9	46.2 ± 3.4	48.1 ± 1.3	41.4 ± 2.0

Due to low τ_{SLJ} in the fresh state, Al–ABS-SLJ was not considered for aging and storing experiments. PETG- and PLA-SLJs, with the highest τ_{SLJ} in the fresh state, were aged and stored for up to 100 days (cf. Figure 12). The substrate temperature, $T_{s,1}$, was set to 200 °C for both polymers.

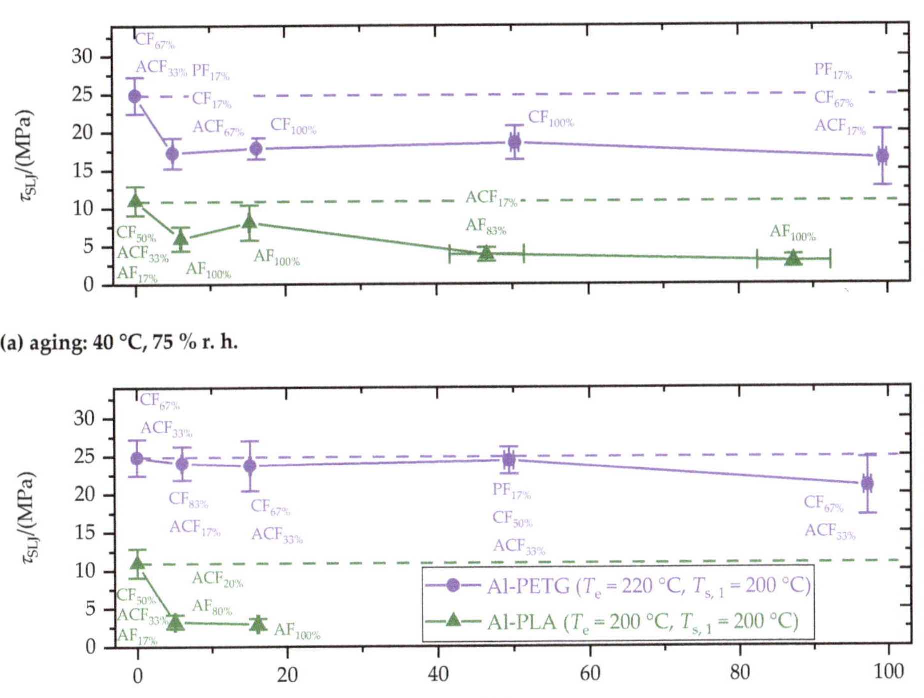

Figure 12. Tensile single-lap-shear strength, τ_{SLJ}, as a function of (**a**) aging and (**b**) storage time, t. Percentages of the failure patterns polymer: part failure (PF, outside the joining area), cohesive failure (CF), adhesive failure (AF) and mixed adhesive and cohesive failure (ACF) are given.

In the course of aging, τ_{SLJ} of Al–PETG-SLJ decreased within 5 days from 25 MPa to 17 MPa and then remained constant. Even in the aged state, no adhesive failure (AF) occurred. Hence, the decrease in τ_{SLJ} cannot be attributed to interfacial aging effects. Instead, embrittlement of the polymer component due to aging (cf. Figure 11) in combination with the stress concentrations at the joint edges led to the decrease in τ_{SLJ}. Storing had no significant effect on tensile single-lap-shear strength, τ_{SLJ}, of Al–PETG-SLJ.

Tensile single-lap-shear strength, τ_{SLJ}, of Al–PLA-SLJ decreased continuously from 11 to 3 MPa within 100 days of aging. Additionally, the failure pattern changed in favor of adhesive failure (AF), indicating a weakening of the polymer–metal interface. Contrary to PETG, storing caused an even faster decrease in τ_{SLJ}. Consequently, the decrease in τ_{SLJ} is attributed to thermally induced internal stresses. These stresses built up below the crystallization temperature, T_c, of PLA. Compared to storing, the mobility of the macromolecules increased during aging due to the higher temperature and the plasticizing effect of water. Moreover, swelling counteracted the thermally induced internal stresses. This explains the faster decrease in τ_{SLJ} during storing compared to aging.

3.4. Correlating Polymer Properties, Wetting and Adhesion Interface Performance

First, wetting was correlated with the polymer melt properties. Instead of the contact equilibrium angle, φ_{eq}, the term $\cos(\varphi_{\text{eq}}) + 1$ was considered, which is proportional to the work of adhesion (cf. Equation (2)). Figure 13a shows $\cos(\varphi_{\text{eq}}) + 1$ as a function of the rouse relaxation time, t_{ro}. Temperature dependence of the Rouse relaxation time is given by the horizontal shift factors, a_T. Taking the maximum loss factor, $\tan(\delta)|_{\text{max}}$, into account revealed a correlation between polymer melt rheology and wetting, which is almost independent of the polymer material (cf. Figure 13b). According to Figure 13b, wetting improved with decreasing Rouse relaxation time, t_{ro}, and increasing maximum loss factor, $\tan(\delta)\,|_{max}$. This is plausible, as t_{ro} describes the time dependence of the polymer dynamic processes and $\tan(\delta)\,|_{max}$ the viscous character of the polymer melt.

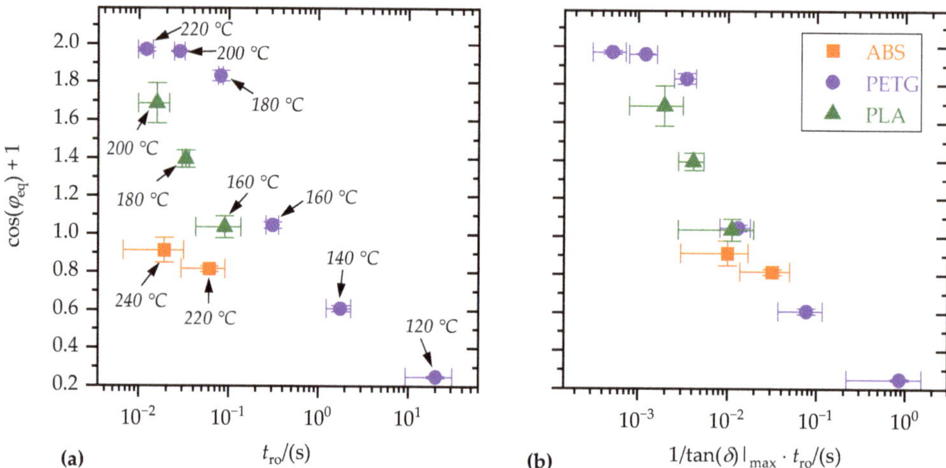

Figure 13. Equilibrium contact angle, φ_{eq}, as a function of (**a**) Rouse relaxation time, t_{ro}, and (**b**) Rouse relaxation time, t_{ro}, and maximum loss factor, $\tan(\delta)\,|_{max}$. Corresponding substrate temperatures, $T_{s,1}$, are indicated.

Figure 14a shows the tensile single-lap-shear strength, τ_{SLJ}, as a function of $\cos(\varphi_{eq}) + 1$. There were significant differences between the polymers. For a given equilibrium contact angle, φ_{eq}, single lap shear strength, τ_{SLJ}, was higher for PETG than for ABS and PLA. Again, by taking the maximum loss factor, $\tan(\delta)\,|_{max}$, into account, the differences between polymers were significantly reduced (cf. Figure 14b).

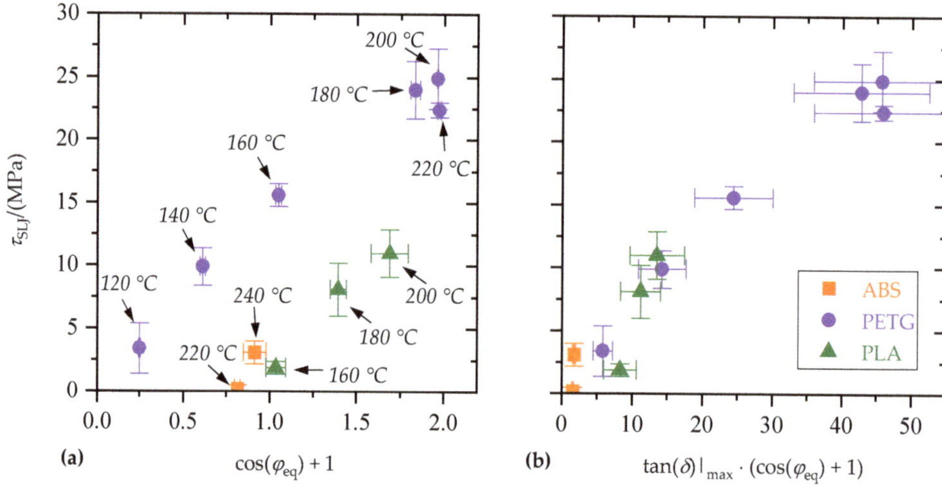

Figure 14. Tensile single-lap-shear strength, τ_{SLJ}, as a function of (**a**) equilibrium contact angle, φ_{eq}, and (**b**) equilibrium contact angle, φ_{eq}, and maximum loss factor, $\tan(\delta)\,|_{max}$. Corresponding substrate temperatures, $T_{s,1}$, are indicated.

By combining the correlations from Figures 13b and 14b, single-lap-shear strength, τ_{SLJ}, is shown as a function of the rheologically derived quantities of Rouse relaxation time, t_{ro}, and maximum loss factor, $\tan(\delta)|_{max}$, in Figure 15.

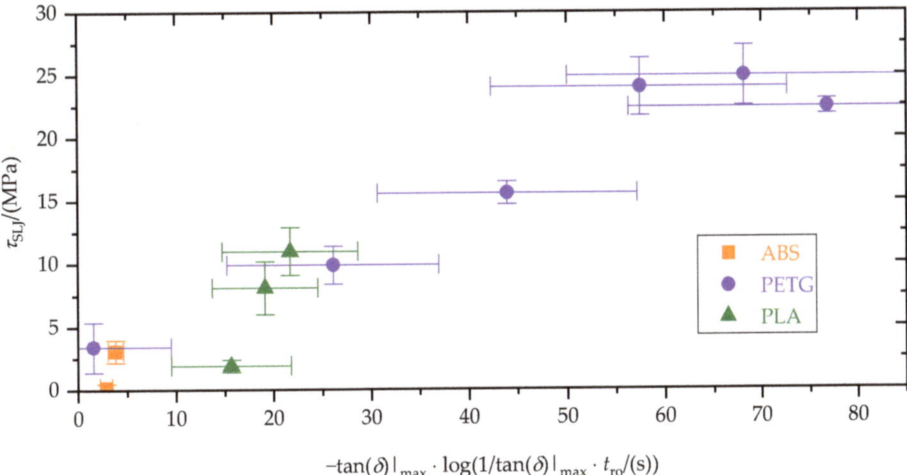

Figure 15. Tensile single-lap-shear strength, τ_{SLJ}, as a function of the rheologically derived quantities of Rouse relaxation time, t_{ro}, and maximum loss factor, $\tan(\delta)|_{max}$.

4. Conclusions

This study focused on the generation of structural polymer–aluminum joints by means of ME. Based on the relevant literature [3,19–24], the following questions arose:

- *Which of the common thermoplastics for ME is most suitable to generate structural polymer–metal joints?*
- *Can structural polymer–metal joints be generated by ME on "simple" practical relevant metal surfaces (e.g., prepared by grid blasting)?*

These questions were addressed by investigating ME-generated joints between grid-blasted aluminum substrates and the thermoplastics ABS, PETG and PLA as a function of thermal processing (substrate temperature) and aging. For all polymers, tensile single-lap-shear strength increased with increasing substrate temperature. However, there were significant differences between the polymers. For the given conditions and material combinations, PETG was the most suitable to generate structural polymer–metal joints. Appropriate thermal processing conditions for the joining were an extrusion temperature of 220 °C and a substrate temperature of 200 °C. For this case, cohesive failure dominated, and the demands of a structural joint in terms of joint strength and aging resistance were met. Increasing the substrate temperature beyond 200 °C or increasing the time PETG was exposed to the elevated temperatures led to pronounced polymer degradation and reduced joint strength. While storing (dry conditions) had no significant effect on PETG–aluminum joint strength, aging (moist-warm conditions) reduced the tensile single-lap-shear strength due to degradation of the mechanical polymer bulk properties. Hence, this decrease did not result in reduced adhesive strength. Contrary to PETG, ABS–aluminum joints in the fresh state as well as PLA–aluminum joints in the aged state did not meet the demands of a structural joint. PLA–aluminum joint strength decreased faster during storage than aging, which was attributed to internal stresses resulting from the thermal joining process. In particular, crystallization of PLA favored internal stress buildup.

Wetting is known to be crucial to buildup adhesive interactions between substrate surface and adhesive [5]. Polymer melt rheology is a key property in terms of wetting (substrate and adjacent polymer traces) and interdiffusion (between adjacent traces) [6,8,9,13].

Considering this, tensile single-lap-shear strength, τ_{SLJ}, equilibrium contact angle, φ_{eq}, and the rheologically derived quantities of Rouse relaxation time, t_{ro}, and maximum loss factor, $\tan(\delta)|_{\max}$, were correlated with each other. For the considered polymers and processing conditions, this study implied that the suitability of a polymer and a thermal processing condition to form a polymer–aluminum-joint by ME could be evaluated based on the polymer's rheological properties (cf. Figure 15). Moreover, taking into account wetting experiments allowed improved estimation of the resulting tensile single-lap-shear strength, τ_{SLJ}, (cf. Figure 14b). Remaining deviations between the polymers are attributed to differences in chemical structure and internal stresses. The former is decisive for the types of physical adsorption, and the latter mainly depends on crystallization tendency and glass transition temperature.

To reveal the effect of internal stresses, it would be interesting to vary the crystallization tendency of the polymer or adjust its thermal expansion coefficient by fillers or additives. Moreover, taking into account the temperature-dependent interfacial energies of the polymer melts could improve estimation of the adhesion interface performance based on polymer melt rheology and wetting. This study focused on joining the thermoplastics ABS, PETG and PLA to grit-blasted aluminum substrates. In order to reveal an optimal combination of metal substrate and polymer for the ME joining process, further polymers and metals should be tested. Finally, for a deeper understanding of the ongoing damage mechanism, combining numerical modeling (e.g., [49]) with a characterization of the deformation and failure process by nondestructive testing methods (e.g., [50]) is advisable.

Author Contributions: Conceptualization, S.B. and H.-G.H.; methodology, S.B. and R.S.; software, S.B. and R.S.; validation, S.B.; formal analysis, S.B.; investigation, S.B. and M.F.; data curation, S.B., M.F. and R.S.; writing—original draft preparation, S.B.; writing—review and editing, S.B.; visualization, S.B.; supervision, H.-G.H. and R.B.; project administration, S.B. and H.-G.H.; funding acquisition, H.-G.H. All authors have read and agreed to the published version of the manuscript.

Funding: This research was funded by the European Regional Development Fund (ERDF) grant number 14.2.1.4–2017/3UdS. We acknowledge support by the Deutsche Forschungsgemeinschaft (DFG, German Research Foundation) and Saarland University within the funding programme Open Access Publishing.

Institutional Review Board Statement: Not applicable.

Informed Consent Statement: Not applicable.

Data Availability Statement: The data presented in this study are available on request from the corresponding author.

Acknowledgments: Gratitude goes to Karen Lienkamp, chair of polymeric materials at Saarland University, for providing laboratory equipment (i.e., tensile tester and sandblaster).

Conflicts of Interest: The authors declare no conflict of interest.

References

1. Bromberger, J.; Kelly, R. Additive manufacturing: A long-term game changer for manufacturers. In *The Great Re-Make: Manufacturing for Modern Times*; McKinsey & Company: New York, NY, USA, 2017; pp. 59–66.
2. Ligon, S.C.; Liska, R.; Stampfl, J.; Gurr, M.; Mülhaupt, R. Polymers for 3D Printing and Customized Additive Manufacturing. *Chem. Rev.* **2017**, *117*, 10212–10290. [CrossRef] [PubMed]
3. Frascio, M.; Marques, E.A.D.S.; Carbas, R.J.C.; da Silva, L.F.M.; Monti, M.; Avalle, M. Review of Tailoring Methods for Joints with Additively Manufactured Adherends and Adhesives. *Materials* **2020**, *13*, 3949. [CrossRef] [PubMed]
4. Klein, B. Fügetechniken. In *Leichtbau-Konstruktion: Berechnungsgrundlagen und Gestaltung*; Klein, B., Ed.; Springer: Wiesbaden, Germany, 2013; pp. 284–317. ISBN 978-3-658-02271-6.
5. Habenicht, G. *Kleben: Grundlagen, Technologien, Anwendungen*; 5. Erweiterte und Aktualisierte Auflage; Springer: Berlin, Germany, 2006; ISBN 978-3-540-26273-2.
6. Habenicht, G.; Ahner, C. *Applied Adhesive Bonding: A practical Guide for Flawless Results*; Wiley-VCH: Weinheim, Germany; Chichester, UK, 2009; ISBN 978-3-527-32014-1.
7. Ebnesajjad, S. *Adhesives Technology Handbook*, 2nd ed.; William Andrew Pub: Norwich, NY, USA, 2008; ISBN 0815515332.

8. Ramiasa, M.; Ralston, J.; Fetzer, R.; Sedev, R. The influence of topography on dynamic wetting. *Adv. Colloid Interface Sci.* **2014**, *206*, 275–293. [CrossRef]

9. Lu, G.; Wang, X.-D.; Duan, Y.-Y. A Critical Review of Dynamic Wetting by Complex Fluids: From Newtonian Fluids to Non-Newtonian Fluids and Nanofluids. *Adv. Colloid Interface Sci.* **2016**, *236*, 43–62. [CrossRef]

10. Welygan, D.G.; Burns, C.M. Dynamic Contact Angles of Viscous Liquids. *J. Adhes.* **1980**, *11*, 41–55. [CrossRef]

11. Elias, H.-G. *Makromoleküle: Band 2: Physikalische Strukturen und Eigenschaften*; 5. Viskositat von Schmelzen; Wiley-VCH: Weinheim, Germany; New York, NY, USA, 2001; ISBN 9783527626502.

12. Adamson, A.W.; Gast, A.P. *Physical Chemistry of Surfaces*, 6th ed.; Wiley: New York, NY, USA, 1997; ISBN 0471148733.

13. Das, A.; Gilmer, E.L.; Biria, S.; Bortner, M.J. Importance of Polymer Rheology on Material Extrusion Additive Manufacturing: Correlating Process Physics to Print Properties. *ACS Appl. Polym. Mater.* **2021**, *3*, 1218–1249. [CrossRef]

14. Costa, S.F.; Duarte, F.M.; Covas, J.A. Estimation of filament temperature and adhesion development in fused deposition techniques. *J. Mater. Processing Technol.* **2017**, *245*, 167–179. [CrossRef]

15. Seppala, J.E.; Hoon Han, S.; Hillgartner, K.E.; Davis, C.S.; Migler, K.B. Weld formation during material extrusion additive manufacturing. *Soft Matter.* **2017**, *13*, 6761–6769. [CrossRef]

16. Bartolai, J.; Simpson, T.W.; Xie, R. Predicting strength of additively manufactured thermoplastic polymer parts produced using material extrusion. *Rapid Prototyp. J.* **2018**, *24*, 321–332. [CrossRef]

17. Dealy, J.M.; Read, D.J.; Larson, R.G. *Structure and Rheology of Molten Polymers: From Structure to Flow Behavior and Back Again*, 2nd ed.; Hanser Publications: Cincinnati, OH, USA, 2018; ISBN 9781569906118.

18. Amancio-Filho, S.T.; Falck, R. Verfahren zum Herstellen eines schichtförmigen Bauteils. Patent DE 10 2016 121 267 A1, 7 November 2016.

19. Falck, R.; Dos Santos, J.F.; Amancio-Filho, S.T. Microstructure and Mechanical Performance of Additively Manufactured Aluminum 2024-T3/Acrylonitrile Butadiene Styrene Hybrid Joints Using an AddJoining Technique. *Materials* **2019**, *12*, 864. [CrossRef]

20. Falck, R.; Goushegir, S.M.; dos Santos, J.F.; Amancio-Filho, S.T. AddJoining: A novel additive manufacturing approach for layered metal-polymer hybrid structures. *Mater. Lett.* **2018**, *217*, 211–214. [CrossRef]

21. Chueh, Y.-H.; Wei, C.; Zhang, X.; Li, L. Integrated laser-based powder bed fusion and fused filament fabrication for three-dimensional printing of hybrid metal/polymer objects. *Addit. Manuf.* **2020**, *31*, 100928. [CrossRef]

22. Hertle, S.; Kleffel, T.; Wörz, A.; Drummer, D. Production of polymer-metal hybrids using extrusion-based additive manufacturing and electrochemically treated aluminum. *Addit. Manuf.* **2020**, *33*, 101135. [CrossRef]

23. Dröder, K.; Reichler, A.-K.; Mahlfeld, G.; Droß, M.; Gerbers, R. Scalable Process Chain for Flexible Production of Metal-Plastic Lightweight Structures. *Procedia CIRP* **2019**, *85*, 195–200. [CrossRef]

24. Bechtel, S.; Meisberger, M.; Klein, S.; Heib, T.; Quirin, S.; Herrmann, H.-G. Estimation of the Adhesion Interface Performance in Aluminum-PLA Joints by Thermographic Monitoring of the Material Extrusion Process. *Materials* **2020**, *13*, 3371. [CrossRef]

25. Ostermann, F. *Anwendungstechnologie Aluminium*, 3rd ed.; Springer: Berlin/Heidelberg, Germany, 2014; ISBN 978-3-662-43806-0.

26. *ABS Extrafill—Technical Data Sheet*; Fillamentum: Hulin, Czech Republic, 2019.

27. *PolyLite (TM) PETG—Technical Data Sheet*; Polymaker: Shanghai, China, 2017.

28. *Ingeo Biopolymer 3D870 Technical Data Sheet*; NatureWorks: Minnetonka, MN, USA, 2017.

29. Coogan, T.J.; Kazmer, D.O. In-line rheological monitoring of fused deposition modeling. *J. Rheol.* **2019**, *63*, 141–155. [CrossRef]

30. Coogan, T.J.; Kazmer, D.O. Modeling of interlayer contact and contact pressure during fused filament fabrication. *J. Rheol.* **2019**, *63*, 655–672. [CrossRef]

31. Mackay, M.E. The importance of rheological behavior in the additive manufacturing technique material extrusion. *J. Rheol.* **2018**, *62*, 1549–1561. [CrossRef]

32. Andersen, N.K.; Taboryski, R. Drop shape analysis for determination of dynamic contact angles by double sided elliptical fitting method. *Meas. Sci. Technol.* **2017**, *28*, 47003. [CrossRef]

33. MATLAB Central File Exchange. Drop Shape Analysis—Fit Contact Angle by Double Ellipses or Polynomials. Available online: https://www.mathworks.com/matlabcentral/fileexchange/57919-drop-shape-analysis-fit-contact-angle-by-double-ellipses-or-polynomials (accessed on 24 August 2021).

34. Gordelier, T.J.; Thies, P.R.; Turner, L.; Johanning, L. Optimising the FDM additive manufacturing process to achieve maximum tensile strength: A state-of-the-art review. *RPJ* **2019**, *25*, 953–971. [CrossRef]

35. Wexler, A.; Hasegawa, S. Relative humidity-temperature relationships of some saturated salt solutions in the temperature range 0° to 50 °C. *J. Res. Natl. Bur. Stand.* **1954**, *53*, 19–26. [CrossRef]

36. *Ansys GRANTA EduPack*; ANSYS, Inc.: Canonsburg, PA, USA, 2021.

37. Farah, S.; Anderson, D.G.; Langer, R. Physical and mechanical properties of PLA, and their functions in widespread applications—A comprehensive review. *Adv. Drug Deliv. Rev.* **2016**, *107*, 367–392. [CrossRef] [PubMed]

38. Goh, G.D.; Yap, Y.L.; Tan, H.K.J.; Sing, S.L.; Goh, G.L.; Yeong, W.Y. Process–Structure–Properties in Polymer Additive Manufacturing via Material Extrusion: A Review. *Crit. Rev. Solid State Mater. Sci.* **2020**, *45*, 113–133. [CrossRef]

39. Vidakis, N.; Petousis, M.; Velidakis, E.; Liebscher, M.; Mechtcherine, V.; Tzounis, L. On the Strain Rate Sensitivity of Fused Filament Fabrication (FFF) Processed PLA, ABS, PETG, PA6, and PP Thermoplastic Polymers. *Polymers* **2020**, *12*, 2924. [CrossRef] [PubMed]

40. Algarni, M.; Ghazali, S. Comparative Study of the Sensitivity of PLA, ABS, PEEK, and PETG's Mechanical Properties to FDM Printing Process Parameters. *Crystals* **2021**, *11*, 995. [CrossRef]
41. Chacón, J.M.; Caminero, M.A.; García-Plaza, E.; Núñez, P.J. Additive manufacturing of PLA structures using fused deposition modelling: Effect of process parameters on mechanical properties and their optimal selection. *Mater. Des.* **2017**, *124*, 143–157. [CrossRef]
42. Vairis, A.; Petousis, M.; Vidakis, N.; Savvakis, K. On the Strain Rate Sensitivity of Abs and Abs Plus Fused Deposition Modeling Parts. *J. Mater. Eng. Perform.* **2016**, *25*, 3558–3565. [CrossRef]
43. Mackay, M.E.; Swain, Z.R.; Banbury, C.R.; Phan, D.D.; Edwards, D.A. The performance of the hot end in a plasticating 3D printer. *J. Rheol.* **2017**, *61*, 229–236. [CrossRef]
44. Quintans, J. Rheological Characterization of Unmodified and Chemically Modified Poly (Ethylene Terephthalate) Resins. Master's Thesis, New Jersey Institute of Technology, Newark, NJ, USA, 1998.
45. Bagheriasl, D.; Carreau, P.J.; Riedl, B.; Dubois, C.; Hamad, W.Y. Shear rheology of polylactide (PLA)–cellulose nanocrystal (CNC) nanocomposites. *Cellulose* **2016**, *23*, 1885–1897. [CrossRef]
46. Hepperle, J. Rheological properties of polymer melts. In *Co-Rotating Twin-Screw Extruder: Fundamentals, Technology, and Applications*; Kohlgrüber, K., Bierdel, M., Eds.; Hanser: Munich, Germany, 2008; ISBN 9781569904220.
47. Ehrenstein, G.W.; Pongratz, S. *Beständigkeit von Kunststoffen*; Hanser Verlag: München, Germany, 2007; ISBN 978-3-446-21851-2.
48. Zhang, X.; Schneider, K.; Liu, G.; Chen, J.; Brüning, K.; Wang, D.; Stamm, M. Structure variation of tensile-deformed amorphous poly(l-lactic acid): Effects of deformation rate and strain. *Polymer* **2011**, *52*, 4141–4149. [CrossRef]
49. Campilho, R.D.S.G.; Banea, M.D.; Neto, J.A.B.P.; Da Silva, L.F.M. Modelling of Single-Lap Joints Using Cohesive Zone Models: Effect of the Cohesive Parameters on the Output of the Simulations. *J. Adhes.* **2012**, *88*, 513–533. [CrossRef]
50. Summa, J.; Becker, M.; Grossmann, F.; Pohl, M.; Stommel, M.; Herrmann, H.G. Fracture analysis of a metal to CFRP hybrid with thermoplastic interlayers for interfacial stress relaxation using in situ thermography. *Compos. Struct.* **2018**, *193*, 19–28. [CrossRef]

 *materials*

Article

Requirements for Hybrid Technology Enabling the Production of High-Precision Thin-Wall Castings

Vladimír Krutiš [1,*], Pavel Novosad [2], Antonín Záděra [1] and Václav Kaňa [1]

[1] Institute of Manufacturing Technology, Brno University of Technology, 616 69 Brno, Czech Republic; zadera@fme.vutbr.cz (A.Z.); kana@fme.vutbr.cz (V.K.)

[2] Institute of Machine and Industrial Design, Brno University of Technology, 616 69 Brno, Czech Republic; 208950@vutbr.cz

* Correspondence: krutis@fme.vutbr.cz

Abstract: Prototypes and small series production of metal thin-walled components is a field for the use of a number of additive technologies. This method has certain limits related to the size and price of the parts, productivity, or the type of requested material. On the other hand, conventional production methods encounter the limits of shape, which are currently associated with the implementation of optimization methods such as topological optimization or generative design. An effective solution is employing hybrid technology, which combines the advantages of 3D model printing and conventional casting production methods. This paper describes the design of aluminum casting using topological optimization and technological co-design for the purpose of switching to new manufacturing technology. It characterizes the requirements of hybrid technology for the material and properties of the model in relation to the production operations of the investment casting technology. Optical roughness measurement compares the surface quality in a standard wax model and a model obtained by additive manufacturing (AM) of polymethyl methacrylate (PMMA) using the binder jetting method. The surface quality results of the 3D printed model evaluated by measuring the surface roughness are lower than for the standard wax model; however, they still meet the requirements of prototype production technology. The measurements proved that the PMMA model has half the thermal expansion in the measured interval compared to the wax model, which was confirmed by minimal shape deviations in the dimensional analysis.

Keywords: topological optimization; hybrid technology; additive manufacturing; investment casting

Citation: Krutiš, V.; Novosad, P.; Záděra, A.; Kaňa, V. Requirements for Hybrid Technology Enabling the Production of High-Precision Thin-Wall Castings. *Materials* **2022**, *15*, 3805. https://doi.org/10.3390/ma15113805

Academic Editor: Rubén Paz

Received: 15 April 2022
Accepted: 23 May 2022
Published: 26 May 2022

Publisher's Note: MDPI stays neutral with regard to jurisdictional claims in published maps and institutional affiliations.

1. Introduction

The requirements for the production of thin-wall castings (TWC) [1] are based on considerable pressure to decrease the weight of the parts produced, especially in the automobile and aviation industries. They are connected with the expression "light-weighting", which is not only a trend of replacing steel castings by aluminum alloys, but it also involves designing a part with the aim of finding a compromise between the design, manufacturability, properties, and the price of the part being produced [2]. This is why the requirements are increasing for the almost unlimited shape variability of the castings, thinner walls, and higher mechanical properties of the parts, which are connected with the internal and surface quality. Moreover, there is a growing demand for the fast supply of the first prototypes and verification series. Therefore, many foundries are forced to implement rapid prototyping technologies in order to comply with the new trends and technologies. Additive manufacturing (AM) development raises possibilities for foundry technology. Using these new methods requires the implementation of virtual engineering [3] and optimization methods in the initial development phases so that, if possible, the casting can be cast at the first attempt.

Recently, the shapes of castings have been designed to meet at least the basic requirements of the construction technology (DFM—design for manufacturability). It is the adaptation of the part construction to the production method to ensure efficient and quality production [4]. The construction technology was focused on the model partition with respect to the pre-selected production technology focused on the elimination of hot spots by an ideal connection and wall transition to the elimination of foundry defects. For these reasons, the shapes of the castings were rather conservative to meet the requirements for functionality and production technology. The development of new optimization methods in 3D design such as topological optimization or generative design and the development of 3D printing methods have shifted the design of part shapes to an area that previously only belonged to artistic castings.

The investment casting (IC) technology complies both with the specific requirements for casting thin-wall castings and the wide material variability and possibility to produce complex shapes. It is also a technology that is widely used for producing prototypes and can be well combined with the AM (in the sequel referred to as 3D printing) of models, ceramic cores or shells. This combination is called "hybrid technology" [5].

The main way of using hybrid technology in IC is the 3D printing of models, subsequent production of a ceramic shell, and its casting in a conventional way. An overview of the use of rapid prototyping (RP) for thin-walled castings, including individual methods, is published in the review [6]. The advantages of the hybrid technology of 3D printed model production lie mainly in low costs and time savings for prototype or small series production and the possibility of casting complex shapes, which are difficult to produce in a conventional way. Hybrid technology is in a sense a competitor to direct metal printing and the choice for this technology will depend mainly on the shape of the part and the quantity required [7]. The use of rapid prototyping in model printing was first mentioned in 1989 [8]. These are methods based on stereolithography (SLA). Mukhtarkhanov et al. (2020) enumerated the history, development, advantages, and disadvantages of this method for model production [9]. A far cheaper method exploited widely by engineers today is material extrusion AM-fused deposition modelling (FDM) [10] or fused filament fabrication (FFF) method, which uses plastics such as ABS or PLA to print models [11]. A disadvantage of the FDM and FFF methods is low surface quality of the models. Methods of solving the stair-stepping structure are dealt with, for example in [12]. Whether it is the SLA or FDM method, the main problem when using these methods and materials is the cracking of the ceramic shells in the melting and firing phase. As described in [10,13,14], this phenomenon is associated with the expansion of the material used or the construction and shape of the model infill. For this reason, materials are being developed that have low thermal expansion or that soften when certain temperatures are reached [15]. Polymethyl methacrylate (PMMA) in powder form seems to be an interesting material in terms of low thermal expansion, which is used in the binder jetting (BJ) technology [16] while maintaining surface quality, low ash content, and a high level of dimensional accuracy [17]. There is a lack of information available on this material and the BJ technology; therefore, this work partly deals with the measurement of material characteristics.

2. Materials and Methods

2.1. Topology Casting Optimization

The project is based on the requirements set out by a team of Formula Student designers from Brno University of Technology participating in a prestigious European competition among university teams, in which the aim is to build a single-seat racing car that must be easily controllable, powerful, reliable, and safe at the same time. The project focused on the possibility to change the production technology from a machined part to a casting, specifically for a new Formula concept designated as Dragon X. The main aim was to design a casting that would be lighter while maintaining the stiffness, would have smaller deformations, and its production would not be lengthy and costly. The part in question is an upright, which is one of the most important parts of the suspension system. The task of the uprights is to transfer the load from the driving forces to the vehicle suspension, to mount the brake caliper and, in the case of a steered axle, also the steering point. Considering the fact that it is unsprung mass, it is important to keep the weight of the upright low. However, it should not be implemented at the expense of decreasing the stiffness of the whole mounting, which is the most important factor in the design. If the bearing of the wheel is too flexible, it may distort the driving experience and the response of the whole car, which then becomes unpredictable and difficult to control for the driver [6]. When producing the design, topological optimization of the part was first performed by the ANSYS Workbench 18.1, which became an inspiration for the design and helped to improve the resulting parameters of the part. Subsequently, the modelling of the upright started in the SolidWorks 3D. Figure 1 shows an optimization chain procedure containing defined load conditions and a limit for material preservation being 20 % of the original volume.

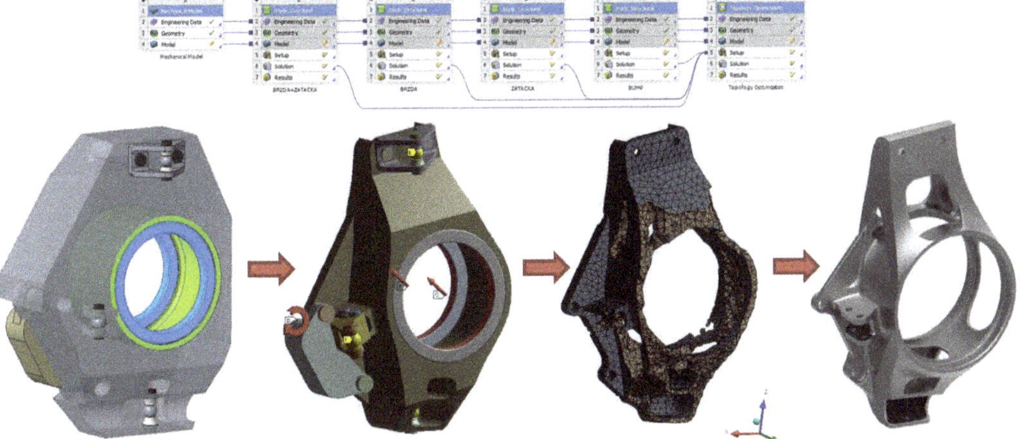

Figure 1. Topological optimization workflow.

The aluminum alloy EN AC-AlSi7Mg0.3 was chosen as the material for the future casting, considering the heat treatment T6 to increase the mechanical properties. FEM component analyses were performed during the topological optimization and for the final design. The upright design was simulated using the finite element method in the ANSYS Mechanical software and structural static calculations were performed.

The assessment criteria were strength and stiffness of the mounting and safety related to the ultimate limit of stress and strain, which was required to be at least 1.2, which corresponds to a maximum reduced stress of 200 MPa. An example of the FEM analysis for the stress conditions of turning and braking is shown in Figure 2. The casting design of an upright for the Dragon X vehicle had better maximum deformation results in all the stress conditions. Compared to the previous generation, the final weight of the part decreased by 12% while the stiffness increased by 25%. The final external dimensions of the upright are 150 × 80 × 40 mm. The optimized geometry was then explored with regard to manufacturability in a special module of the ProCAST software called Co-design. This approach enabled the verification of the manufacturability of the part with respect to the selected production technology (DFM) focusing on the wall thickness transitions, radii of the part, and occurrence of hot spots—Figure 3. Inappropriate selection of wall transitions, sharpness of angles, and size of the radii may lead to the formation of hot spots in the casting or formation of hot tears or cracks. Based on the analysis, the radii around the base were modified (increasing the radius from 1 mm to 2 mm) and in the center part of the hub the lightening holes were modified in order to make the subsequent drying of the ceramic shell easier. The last sections of the casting solidification were selected as sections for attaching the gating system so that, through these connections, metal could be fed from the gating system during the solidification. The castings were also subjected to a numerical simulation of the ceramic shell pre-heating, metal filling and solidification including a prediction of the occurrence of defects such as shrinkage cavities and porosities.

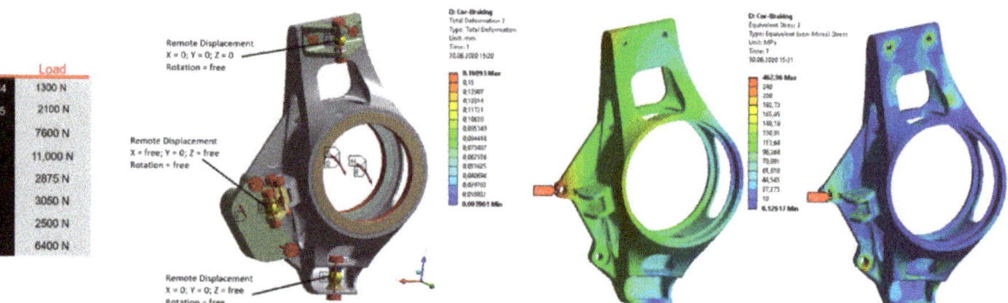

Figure 2. Assessment of the optimized casting shape for the stress conditions of turning and braking (distribution of deformation and equivalent stress).

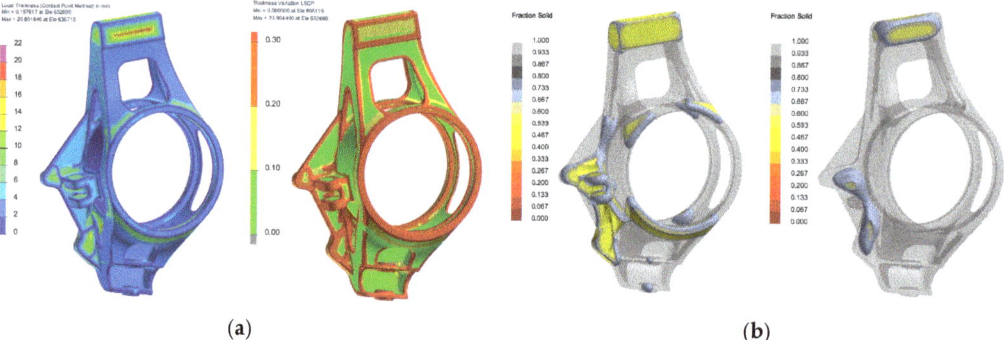

(**a**) (**b**)

Figure 3. Co-design results. (**a**) Wall thickness analysis; (**b**) Simulation of hot spot formation.

The casting model was finally modified with machining allowances, it was enlarged by 1.2% for the shrinkage of the aluminum alloy, the functional surfaces were supplied with machining allowances, and small holes for anchoring the part were sealed—Figure 4.

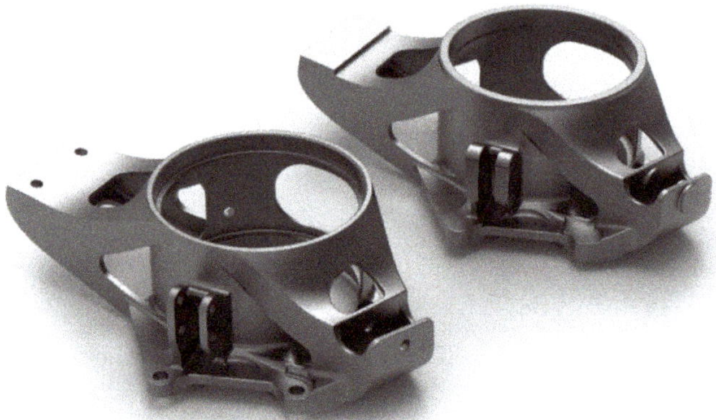

Figure 4. The final 3D data of the workpiece (**left**) and casting (**right**).

Figure 5 maps the development of the shape and production technology used for the upright, specifically from the Dragon 8 Formula to Dragon X. In recent years, the team experienced the production of uprights from aluminum alloy using additive technology for the Dragon 8 single-seater and aluminum milling for Dragon 9. Each technology has its pitfalls and benefits. Production using additive technology gives the designer a free hand in terms of the shape of the part. However, its disadvantage is a complex and expensive production technology, especially for larger parts and a large number of the parts required. During production using milling, the designer must bear in mind the limits of this technology (shape limits) and make sure that the design is machinable. An advantage is the availability of the production equipment. Production using aluminum casting should combine the advantages of both of the previous manufacturing processes and appears to be a promising possibility for future use [17].

	Dragon X	Dragon 9	Dragon 8	Dragon 8
Upright:	407 g	461 g	410 g	501 g
Assembly:	574 g	596 g	511 g	609 g
Method:	Cast	CNC	3D Print	CNC
Max Deformation:	0.36 mm	0.45 mm	0.56 mm	0.57 mm

Figure 5. Comparison of parameters of the different Formula Student versions.

2.2. Investment Casting Requirements

The investment casting technology is a "net shape" technology. It is regarded as one of the most precise foundry methods in terms of dimension tolerance (DCTG 4-8–DIN EN ISO 8062-3) and it is among the technologies that can produce a high-quality surface in the as-cast condition (Ra 1.6-12.5) [18]. Generally, it is suitable for smaller castings. It is highly effective for prototype parts and thin-wall castings. This method is commonly used to produce small and medium series production castings in the aviation and medical sectors, in transport and power engineering (power generation industries). Compared to conventional foundry methods, complex shapes can be produced that would be difficult to make due to a complex parting line. Introduction of 3D printing pushes this limit even further because, to make the model, there is no need for a metal mold (for wax models), which also has its design limits.

The following are the specifics of the investment casting technology related to the shape of the part [19].

- Min. radius on the outer and inner edges 1 mm;
- The smallest theoretical wall thickness 0.8 mm (depends on the casting size and shape, material, temperature, etc.) and the common casting thickness is 2–4 mm;
- No draft angle has to be factored;
- Gradual wall transition towards feeding (directed solidification towards the gate);
- Use transition ribs for thin walls (increasing stiffness, metal flow, decreasing deformation);

- If possible, do not close the internal casting cavities (with regard to the drying of the ceramic layers and subsequent removal of the shell);
- With the 3D printed model, model the gates right away (better connection of the part to the gating system).

The shape variability of 3D printing practically has no limits and fully supports modern trends using shape optimization tools such as topological optimization and generative design. However, certain requirements for the model and the 3D print material have to be met when using hybrid technology. The basic requirements for the model include:

- High surface quality of the printed model (connected with the 3D print method and the possible surface finish);
- High dimensional accuracy of the printed model (affects the final dimensions of the casting);
- Low thermal expansion of the material (higher expansion causes the ceramic shell to crack when being fired);
- Low content of ash matter after the model burns (high content of ash matter requires long firing. High ash content may have a negative effect on the resulting surface quality of the casting).

The casting assembly can be printed directly or a combination is used of wax down sprue and fusible connection with the printed model. In some cases, mechanical securing of the junction is necessary to prevent the models from breaking off the gating system. The subsequent dipping of the printed model in ceramic slurry is not different from dipping a wax model. In most cases, the wettability does not have to be modified by using special agents. Stucco coating of the ceramic material and the subsequent drying of the shell layers is the same as in the classic procedures. Unlike the technology that uses a wax model and subsequent de-waxing in a boiler-clave, 3D printed models are fired in annealing furnaces with an afterburning chamber. The afterburning temperature ranges from 600 to 800 °C based on the material of the model printed. What can be problematic is the de-waxing of combined gating systems if sufficient temperature shock is not ensured for the wax model. In that case, the shell may crack during the firing.

3. Results

3.1. Properties of a Model Made by 3D Printing

The hybrid casting production technology is based on the 3D printing of a model, which is then used for the production of a ceramic shell. In the last step, this shell is used as a mold for the production of a metal casting. As described above, for the hybrid technology for casting production using investment casting, it is necessary to ensure dimensional accuracy of the model, good surface quality and to use material with low thermal expansion to prevent the occurrence of cracks in the shell during the firing phase.

The upright model was 3D printed by the binder jetting (BJ) method using PMMA. Polymethyl methacrylate (PMMA) is an acrylic plastic with excellent burnout behavior. PolyPor B was used as a binder. The printing resolution used was 600 dpi, the thickness of the printed layer was 150 μm, and the final surface of the model was treated using wax infiltration. The printing was performed in cooperation with the Voxeljet company. The part printed is shown in Figure 6.

3.1.1. Assessment of Dimensional Changes in the Model

The investment casting method is directly intended for making "NET SHAPE" castings, when further shaping is not expected; therefore, the casting must already be made in the "as-cast condition" with narrow dimensional tolerances [19]. Knowledge of the behavior of the wax model during its production is, from the perspective of dimensional accuracy of the final casting, only one part of the necessary comprehensive knowledge of dimensional changes during the whole technological flow. This means that knowledge of dimensional changes during the production of the shell (coating, de-waxing, drying, and shell annealing) is also necessary as well as changes after the pouring of the liquid metal and during the solidification and cooling stages. What is not insignificant is the shape factor and constrained shrinkage [20].

Figure 6. The upright model printed by the BJ (binder jetting) method.

When making models using classic wax and a mold, shrinkage is always applied to the model (0.8–1.2%). For printed models, this shrinkage is not commonly applied. Only shrinkage of the cast material or a dimensional change in the shell are considered. Nevertheless, some materials and printing technologies are sensitive to temperature and with larger and less compact shape model deformations may occur. Therefore, it is also necessary to check dimensional deviations from the CAD data.

Optical measurement of the upright model was carried out by the Atos Core device using the best fit function for model positioning. Figure 7 shows an evaluation of dimensional changes compared with the CAD model.

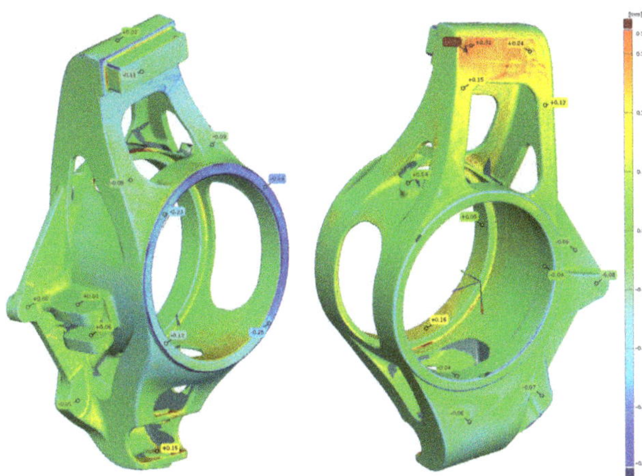

Figure 7. Dimensional changes of printed model (PMMA−BJ method).

By the ISO 8062-3 standard, for a nominal dimension of up to 160 mm of a casting made by investment casting, the permitted length dimensional tolerance (DCTG 5) is 0.62 mm. For geometric flatness tolerance (GCTG 5), for example, the maximum permissible tolerance (for dimensions of 100–300 mm) is 1.4 mm. The largest deviation measured on the model was the one in negative direction −0.29 mm, specifically on the upper surface of the central hub. However, in this part, there is a machining allowance of 1.1 mm, so this deviation is not a problem and is compensated for by the allowance. The maximum deviation from the CAD model was +0.37 mm in the recess part. Visually, this deviation looks like a thicker layer after a treatment of the PMMA model by hot wax infiltration. Again, this local spot does not constitute a production problem given the positive deviation and subsequent machining of this spot. The remaining shape deviations were in the range of ±0.2 mm. The dimensional accuracy of the model is thus satisfactory and the model can be used to make the final casting. The accuracy of the 3D printed model is sufficient and all the deviations in the critical parts of the functional surfaces were within the limits.

3.1.2. Assessment of Surface Roughness

The model roughness was measured on a test body in the shape of a wedge with an angle of 45°—Figure 8. The body was selected this simple so that it would be possible to compare the printed sample (Sample A) with the wax model injected into a metal mold (Sample B). All the roughness measurements were made using the Talysurf CCI Lite device. It is a contactless 3D profilometer based on the principle of coherence correlation interferometry. The device has an image sensor with a resolution of 1024 × 1024 pixels and three Mirau lenses with 10×, 20×, and 50× magnification. Assessed areas sized approximately 1.65 × 1.65 mm, 0.825 × 0.825 mm, and 0.33 × 0.33 mm correspond to these lenses. All the measurements presented below were made using a lens with 20×.

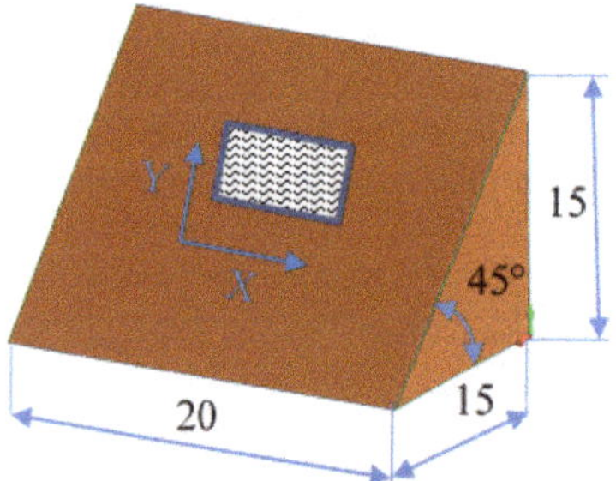

Figure 8. Sample geometry used for the roughness analysis (printing position).

The data obtained by the measurement were then processed in the TalyMap Gold software, which enables the creation of a 2D and 3D model of the analyzed surface. The program uses different ways of model surface treatment such as surface levelling or shape removal, interpolation of unmeasured points, etc., assessment of various surface structure parameters based on a number of standards, and data export in various formats for further processing. Table 1 summarizes the key measurement results. It contains a graphical representation of a 3D model of the sample structure including the Z axis, which is given in μm. It can be stated that the structure of the 3D printed sample, compared to the wax sample, is relatively heterogenous with uneven occurrence of recesses and projections. This is related to the technology of 3D print layers of 0.16 mm, which are then partly smoothed out by the final surface treatment—wax infiltration. The table also contains graphical evaluation of the average profile in the X axis and Y axis directions. The Y direction is transverse to the printing direction. From the entire scanned area, a total of 1024 basic profiles in the transverse direction were created, based on which the roughness height parameters were calculated in compliance with the ISO 4287 standard. Table 1 shows the roughness parameters calculated from all the extracted basic profiles.

Table 1. Surface roughness analysis.

Sample A	Sample B
Wax model – metal die	3D printed model – PMMA (binder jetting)
3D model of the sample structure	3D model of the sample structure

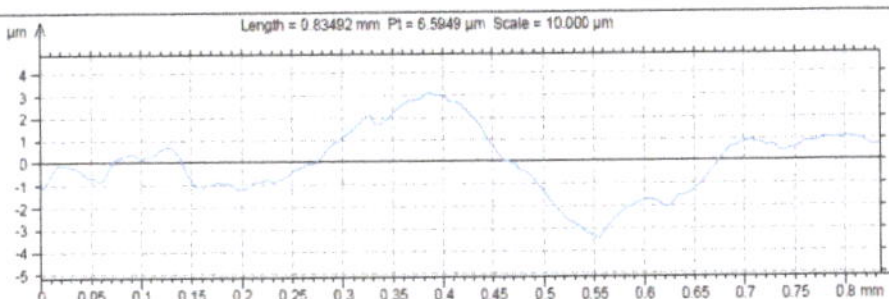

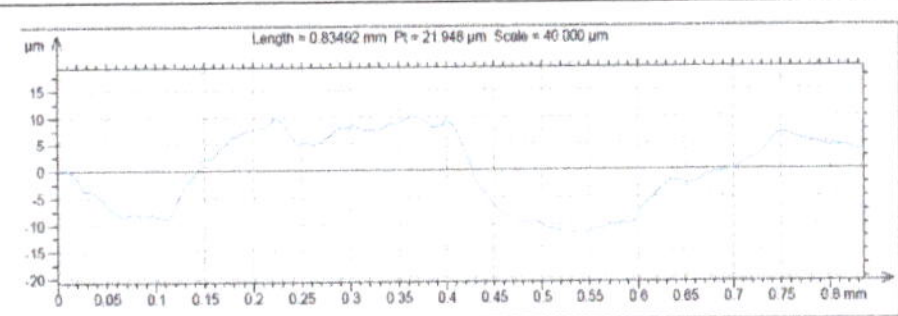

Basic profile along the X axis	Basic profile along the X axis

Roughness parameters along the X axis (Sample A)

ISO 4287		Context	Mean	Std dev
Amplitude parameters - Roughness profile				
Rp	µm	Gaussian filter, 0.8 mm, End-effects managed	6.4978	2.2816
Rv	µm	Gaussian filter, 0.8 mm, End-effects managed	8.9097	3.2849
Rz	µm	Gaussian filter, 0.8 mm, End-effects managed	15.407	4.6032
Rc	µm	Gaussian filter, 0.8 mm, End-effects managed	9.1498	2.6223
Rt	µm	Gaussian filter, 0.8 mm, End-effects managed	15.537	4.6126
Ra	µm	Gaussian filter, 0.8 mm, End-effects managed	2.6822	0.65408
Rq	µm	Gaussian filter, 0.8 mm, End-effects managed	3.3345	0.84969
Rsk		Gaussian filter, 0.8 mm, End-effects managed	-0.53273	0.53335
Rku		Gaussian filter, 0.8 mm, End-effects managed	3.2073	1.7899

Roughness parameters along the X axis (Sample B)

ISO 4287		Context	Mean	Std dev
Amplitude parameters - Roughness profile				
Rp	µm	Gaussian filter, 0.8 mm, End-effects managed	23.905	4.6995
Rv	µm	Gaussian filter, 0.8 mm, End-effects managed	37.363	8.4300
Rz	µm	Gaussian filter, 0.8 mm, End-effects managed	61.267	9.3446
Rc	µm	Gaussian filter, 0.8 mm, End-effects managed	36.299	6.8536
Rt	µm	Gaussian filter, 0.8 mm, End-effects managed	61.610	10.143
Ra	µm	Gaussian filter, 0.8 mm, End-effects managed	11.342	1.2883
Rq	µm	Gaussian filter, 0.8 mm, End-effects managed	14.252	1.7696
Rsk		Gaussian filter, 0.8 mm, End-effects managed	-0.63854	0.42180
Rku		Gaussian filter, 0.8 mm, End-effects managed	3.0898	1.0218

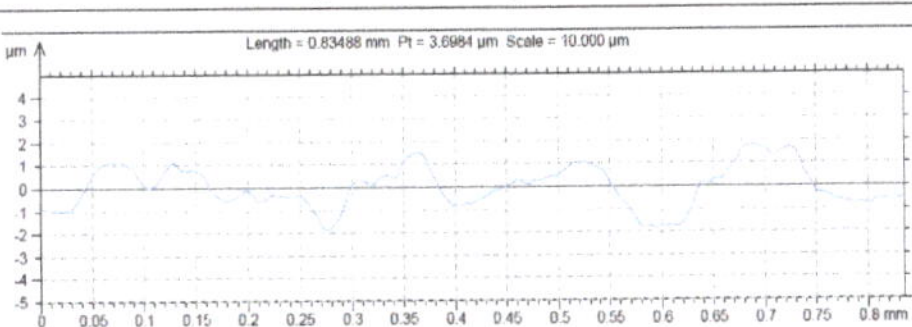

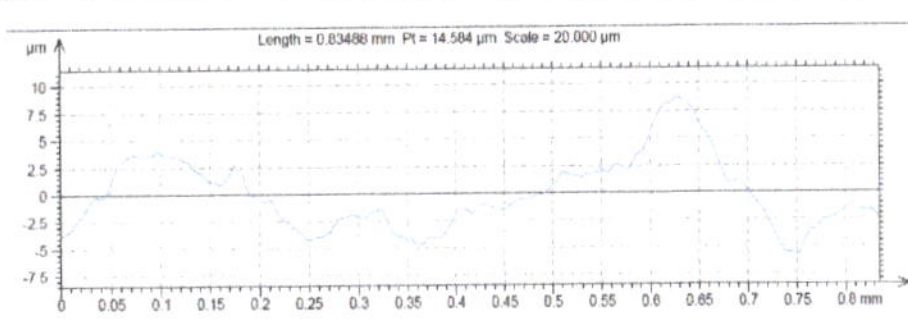

Basic profile along the Y axis	Basic profile along the Y axis

Roughness parameters along the Y axis (Sample A)

ISO 4287		Context	Mean	Std dev
Amplitude parameters - Roughness profile				
Rp	µm	Gaussian filter, 0.8 mm, End-effects managed	6.3911	1.8900
Rv	µm	Gaussian filter, 0.8 mm, End-effects managed	10.078	2.9873
Rz	µm	Gaussian filter, 0.8 mm, End-effects managed	16.469	3.9424
Rc	µm	Gaussian filter, 0.8 mm, End-effects managed	9.3980	2.6072
Rt	µm	Gaussian filter, 0.8 mm, End-effects managed	16.551	3.8900
Ra	µm	Gaussian filter, 0.8 mm, End-effects managed	2.6326	0.57472
Rq	µm	Gaussian filter, 0.8 mm, End-effects managed	3.3603	0.73810
Rsk		Gaussian filter, 0.8 mm, End-effects managed	-0.75398	0.57283
Rku		Gaussian filter, 0.8 mm, End-effects managed	3.7576	1.5124

Roughness parameters along the Y axis (Sample B)

ISO 4287		Context	Mean	Std dev
Amplitude parameters - Roughness profile				
Rp	µm	Gaussian filter, 0.8 mm, End-effects managed	20.493	5.1674
Rv	µm	Gaussian filter, 0.8 mm, End-effects managed	28.793	10.055
Rz	µm	Gaussian filter, 0.8 mm, End-effects managed	49.287	12.287
Rc	µm	Gaussian filter, 0.8 mm, End-effects managed	30.589	11.422
Rt	µm	Gaussian filter, 0.8 mm, End-effects managed	49.462	12.170
Ra	µm	Gaussian filter, 0.8 mm, End-effects managed	9.0969	2.5517
Rq	µm	Gaussian filter, 0.8 mm, End-effects managed	11.198	2.8912
Rsk		Gaussian filter, 0.8 mm, End-effects managed	-0.52679	0.49586
Rku		Gaussian filter, 0.8 mm, End-effects managed	3.0258	0.95908

Sample A represents a casting model made by the conventional method of injecting wax into a metal mold. The Ra roughness measured on the model along the X axis was 2.68 µm and 2.63 µm along the Y axis. The results suggest that the roughness expressed by the Ra parameter both in the transverse and longitudinal direction is comparable. Additionally, the Rz parameter, expressing the average size of the projections and recesses on the surface, is completely comparable in the transverse and longitudinal directions. The average size of the projections on the surface (Rp) is comparable to the average size of the recesses (Rv). Such even surface roughness of the wax model is related to the roughness of the metal mold made by the conventional method of metal cutting. In wax models with similar surface roughness, it is possible to achieve the final Ra roughness of up to 1.6 µm, which can be compared to a conventionally turned or milled surface.

It is clear from the basic profile of Sample B (Table 1—column on the right) made by the 3D print technology that the projections on the sample surface (red color) are not distributed evenly on the surface but in lines corresponding to the motion trajectory of the printhead. More significant differences between the roughness of the conventional wax model and the 3D printed model can be observed on the Rz parameter. The values in Table 1 indicate that the heights of the recesses and projections in the 3D printed model are approx. 4 times greater, and therefore the average value of the Rz parameter is 4 times higher. The Ra roughness parameter for the 3D printed model in the longitudinal X direction is 11.34 µm, and in the transverse Y direction the Ra is 9.09, which is approximately 3 times higher than for the wax model molded in the metal mold. Such high values of the Ra or Rz roughness parameters indicate significant differences between the roughness of the castings. In terms of the casting surface quality, these are limit values. With such surface roughness, it is possible to detect by mere sensitive assessment (touch) considerable differences in the surface quality of the castings made by the wax injection technology and polymer 3D printing.

The roughness of the 3D printed models is significantly affected primarily by the printhead step size, which is, however, often limited by the total printing time of the part and therefore also the production costs. With some materials, it is possible to reduce the model roughness by the smoothing technology using, for example, immersing or steaming in isopropyl alcohol or by penetration of the model with synthetic wax.3.1.3 assessment of thermal expansion of model material.

Thermal expansion of the models is directly connected to the formation of cracks in the ceramic molds. The thermal expansion coefficient is required to be low and the contact pressure of the model on the shell should be lower than the shell strength—MOR (modulus of rapture). The printed models require such infill and structure that allow the model to collapse inwards and thus reduce the contact stress [21].

The measurement was made using the Setsys Evolution TMA vertical dilatometer. The test samples are cylinders with a diameter of Ø6 mm and a height of 6 mm (fully filled samples). A temperature gradient of 3 °C/min was used in the measurement and the absolute elongation of the sample was monitored. The measurement was only made for temperatures of up to 120 °C because at higher temperatures the sample stuck to the measuring probe. The measurement was made for the PMMA material (taken directly from the printed BJ model infiltrated with wax) and the A7-FR/1200 model wax used for the injection in a mold technology. A comparison was also made for the PolyCAST™ material, which is a specially developed material (filament-(Polyvinyl butyral)) for FDM model printing and also for the investment casting technology. The elongation values depending on temperature are shown in Figure 9. The curves are evaluated in one figure but, with regard to the order differences in the dilatations measured, the scales are shown separately for each material.

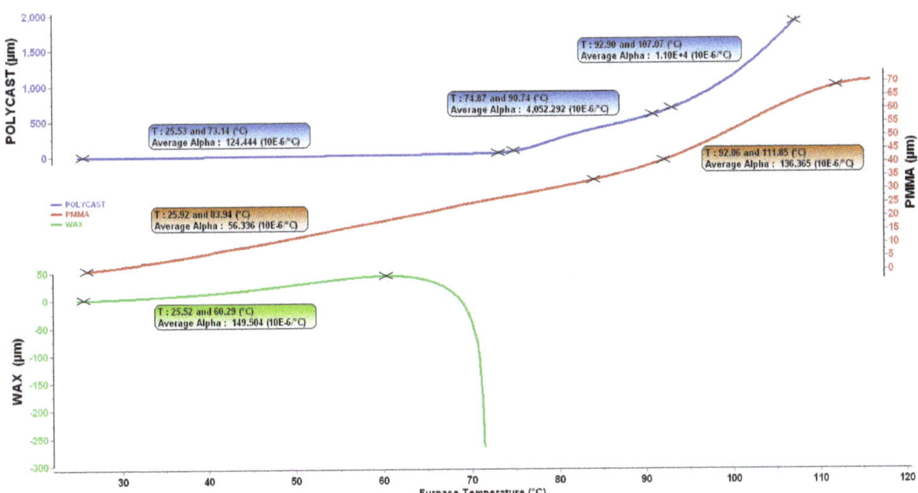

Figure 9. Elongation of samples depending on temperature.

From the values obtained, we can determine the thermal expansion coefficient for each material. The dilatation curves are not linear dependencies in the whole temperature interval and it is therefore necessary to consider the coefficient for a specific temperature interval. Amorphous polymers have a random molecular structure and can gradually soften with increasing temperature. The break temperature is called the glass transition temperature (Tg) [22] and, on the temperature dependent curve, it is the area where a change in the slope (rise) of the curve occurs. Tg is an important characteristic of polymer materials as it represents a point at which a change in the properties of the polymer occurs. More precisely, it is an area, not a specific temperature. Nevertheless, it is usually given as one specific number. Table 2 summarizes the coefficients of thermal expansion and the Tg temperature for each material measured and for specific temperature intervals.

Table 2. Comparison of thermal expansion for selected materials.

Material	Temperature Interval [°C]	Thermal Expansion Coeff. [10^{-6}/°C]	Tg [°C]
PMMA (BJ)	25.92–83.94	56.34	92
	92.06–111.85	136.37	
WAX	25.52–60.29	149.51	-
Polycast ™ (FDM filament)	25.53–73.14 74.87–90.74 92.9–107.07	124.44 4052 11000	75

The measurements show that the greatest thermal expansion occurs in the interval up to 60 °C for the model wax. On the contrary, the smallest coefficient in this temperature range belongs to the PMMA material, whose expansion is 2.5 times lower than that of wax. Melting of the model waxes occurs in the temperature range of 60–70 °C, which is apparent as a significant break on the curve around 68 °C. The thermal expansion coefficients of the PMMA and PolyCast™ materials increase at higher temperatures. For PMMA, this increase is not sudden as it is for PolyCast™. The softening temperature for this material given by the producer [23] is 68 °C. From this temperature, the collapse of this material can be assumed. Whether the expansion of the model will cause a rupture in the shell when using the FDM/FFF method depends on the structure and percentage of the filling, air content and, last but not least, shape of the part.

4. Discussion

The paper presents a combination of topological optimization and so-called co-design when designing and optimizing an upright for investment casting technology. It was verified that after topological optimization it is necessary to use other virtual tools that will help us solve the manufacturability of the part with the help of hybrid technology. This chaining will not only allow the production of castings on the first try, but in particular significantly shorten the product development cycle. The co-design helped with the adjustment of the wall transitions and pointed out the insufficiently large radii in certain parts of the castings. In reality, this could result in porosity in the casting, or with insufficient radii, it could cause cracks or fissures.

For the hybrid production technology, there is a detailed description of the model production phase and the requirements imposed on the model and its properties. The 3D print technology proved the suitability of the BJ method for making the model. An assessment is made of the model properties primarily with respect to dimensional accuracy, surface quality, and dilatations in the model firing phase. Dimensional changes in the model after printing do not exceed ± 0.35 mm, which is compensated for by allowances on the treated surfaces. The Ra roughness values of the PMMA model with an infiltration wax layer do not exceed 12.5 μm. In comparison with the sample injected in a mold (Ra 2.3 μm), this roughness is considerably worse. When measuring the roughness of the infiltrated sample, no significant difference was found in the printing direction or in the transverse printing direction. This is due to the wax's ability to fill surface irregularities. The ceramic slurry is able to copy any surface irregularities, and to improve the surface roughness of the 3D printed models, it is therefore necessary to pay close attention to pattern post-processing. The issue with wax infiltration is the viscosity of the synthetic wax, the thickness of the layer, and the dripping of the wax so that it does not accumulate in certain parts. The difference found in the roughness of the model produced by wax injection and the printed model is not a problem for prototype parts; however, for small series production in the automotive or aerospace industry, it would be necessary to pay more attention to surface treatment or print settings. The method of coating and its effect on the resulting surface quality will be the subject of further research.

The PMMA material used for the BJ technology shows lower expansion than wax injected in a mold. This property has a positive effect on decreasing the risk of shell cracking during the firing of the model. The better dimensional stability of the PMMA model at room temperatures also allows more optimal transport and storage of the models. The requirements for the model quality were therefore evaluated as suitable. The PolyCAST™ material was included in the measurement only subsequently and it will be necessary to perform additional measurements of thermal expansion with regard to the adhesion of the sample to the measuring probe. This could affect the measurement at the softening temperatures of this material and thus distort the measurement result.

5. Conclusions

The work presented the first stage of the hybrid investment casting technology, which is the model production. The work is focused on the design, optimization, and production of the upright model from the PMMA (acrylic plastic) material, which will be used in the next stage for the production of ceramic shells. Based on the investigations carried out in this work, the following conclusions can be drawn:

- Topological optimization of the casting design must be complemented by technological co-design to ensure that the design is technologically feasible.
- Due to its dimensional stability, the PMMA material is more accurate than standard wax models, which makes it possible to eliminate the use of levelling tools for complex shapes.

- The surface quality of the printed model is significantly lower in comparison to wax models produced by injection into metal molds. On the other hand, it still meets the needs of prototype castings. To improve the surface roughness, it would be necessary to increase the print resolution, which would require a change in the 3D printer or the printing technology. Another way is to adjust the infiltration process of the surface layer, either by repeated dipping into a wax suspension, or by adjusting its viscosity.

The production of a ceramic shell and an aluminum prototype casting of the upright is another stage of the work, which will focus on the final comparison of the entire process of hybrid technology. Further studies will be targeted on the evaluation of dimensional accuracy, surface quality, and internal defects in the manufactured metal casting.

Author Contributions: Conceptualization, V.K. (Vladimír Krutiš) and A.Z.; methodology, V.K. (Vaclav Kana) and A.Z.; software, V.K. (Vladimír Krutiš) and P.N.; investigation, V.K. (Vladimír Krutiš), P.N. and V.K. (Vaclav Kana); data curation, V.K. (Vladimír Krutiš) and V.K. (Vaclav Kana); writing—original draft preparation, V.K. (Vladimír Krutiš); writing—A.Z. and P.N., visualization, V.K. (Vladimír Krutiš) and P.N.; All authors have read and agreed to the published version of the manuscript.

Funding: This work was supported by the specific research project of BUT FME Brno, no. FSI-S-22-8015 research on melting and metallurgical processing of high-temperature alloys.

Conflicts of Interest: The authors declare no conflict of interest.

References

1. Campbell, J. Thin wall castings. *Mater. Sci. Technol.* **1988**, *4*, 194–204. [CrossRef]
2. Halonen, A. Trends in Lightweighting with Metal Castings. Casting Source, May–June, 2020, 32. Available online: https://lightweighting.co/wp-content/uploads/2020/06/Trends-in-Lightweighting-with-Metal-Castings.pdf (accessed on 20 March 2022).
3. Krutiš, V.; Šprta, P.; Kaňa, V.; Zadera, A.; Cileček, J. Integration of Virtual Engineering and Additive Manufacturing for Rapid Prototyping of Precision Castings. *Arch. Foundry Eng.* **2021**, *21*, 51–55. [CrossRef]
4. Bralla, J.G. *Design for Manufacturability Handbook*, 2nd ed.; McGraweHill Handbooks: New York, NY, USA, 1999.
5. Kumar, P.; Ahuja, I.P.S.; Singh, R. A Framework for Developing a Hybrid Investment Casting Process. *Asian Rev. Mech. Eng.* **2013**, *2*, 49–55.
6. Pattnaik, S.; Karunakar, D.B.; Jha, P.K. Developments in investment casting process, A review. *J. Mater. Process. Technol.* **2012**, *212*, 2332–2348. [CrossRef]
7. Mueller, T.J. What impact will 3D metal printing have on investment casting. In *Transactions of the American Foundry Society, One Hundred Twenty-Second Annual Metalcasting Congress*; American Foundry Society: Schaumburg, IL, USA, 2018; pp. 305–310. ISBN 978-0-87433-464-7.
8. Greenbaum, P.Y.; Khan, S. Direct investment casting of rapid prototype parts: Practical commercial experience. In Proceedings of the 2nd European Conference on Rapid Prototyping, Nottingham, UK, 15–16 July 1993; pp. 77–93.
9. Mukhtarkhanov, M.; Perveen, A.; Talamona, D. Application of Stereolithography Based 3D Printing Technology in Investment Casting. *Micromachines* **2020**, *11*, 946. [CrossRef] [PubMed]
10. Kumar, P.; Ahuja, I.P.S.; Singh, R. Application of fusion deposition modelling for rapid investment casting—A review. *Int. J. Mater. Eng. Innov.* **2012**, *3*, 204–227. [CrossRef]
11. Carneiro, V.H.; Rawson, S.D.; Puga, H.; Meireles, J.; Withers, P.J. Additive Manufacturing Assisted Investment Casting: A Low-Cost Method to Fabricate Periodic Metallic Cellular Lattices. *Addit. Manuf.* **2020**, *33*, 101085. [CrossRef]
12. Chohan, J.S.; Singh, R.; Boparai, K.S. Vapor smoothing process for surface finishing of FDM replicas. *Mater. Today Proc.* **2020**, *26*, 173–179. [CrossRef]
13. Cheah, C.M.; Chua, C.K.; Lee, C.W.; Feng, C.; Totong, K. Rapid prototyping and tooling techniques: A review of applications for rapid investment casting. *Int. J. Adv. Manuf. Technol.* **2005**, *25*, 308–320. [CrossRef]
14. Chen, X.; Li, D.; Wu, H.; Tang, Y.; Zhao, L. Analysis of ceramic shell cracking in stereolithography-based rapid casting of turbine blade. *Int. J. Adv. Manuf. Technol.* **2011**, *55*, 447–455. [CrossRef]
15. Note, Investment Casting with PolyCast TM. Available online: https://polymaker.com/Downloads/Application_Note/PolyCast_Application_Note_V1.pdf (accessed on 8 March 2022).
16. Why PMMA 3D Printing for Investment Casting is the Future. Available online: https://www.voxeljet.com/3d-printing-solution/investment-casting/#anchor-2 (accessed on 7 March 2022).
17. Doležal, T. Těhlice Formulového Vozu z Hliníkové Slitiny. Bachelor's Thesis, Brno University of Technology, Faculty of Mechanical Engineering, Brno, Czech Republic, 2020.

18. Pooley, R.R.; Smart, R.F. *Investment Casting*, 1st ed., The Institute of Materials, The University Press Cambridge: Cambridge, UK, 1995; p. 486.
19. Horáček, M. Accuracy of castings manufactured by the lost wax process. *Foundry Trade J.* **1997**, *171*, 424–429.
20. Voight, R.C. Factors Influencing the Dimensional Variability of Investment Castings. In Proceedings of the 45th Technical Meeting of the Investment Casting Institute, Dallas, TX, USA; 1997; pp. 42–60.
21. Jacobs, P. QuickCast 1.1 & Rapid Tooling. In Proceedings of the NASUG 1995 Annual Meeting and Conference, Tampa, FL, USA, 4–7 June 1995.
22. Moylan, J. Considerations for Measuring Glass Transition Temperature. Available online: https://www.element.com/nucleus/2017/08/15/18/45/considerations-for-measuring-glass-transition-temperature (accessed on 7 March 2022).
23. Polymaker. PolyCast Data Sheet. Available online: https://filament2print.com/gb/index.php?controller=attachment&id_attachment=313 (accessed on 3 March 2022).

MDPI
St. Alban-Anlage 66
4052 Basel
Switzerland
Tel. +41 61 683 77 34
Fax +41 61 302 89 18
www.mdpi.com

Materials Editorial Office
E-mail: materials@mdpi.com
www.mdpi.com/journal/materials